POSTHARVEST DECAY

Control Strategies

POSTHARVEST DECAY

Control Strategies

Edited by

SILVIA BAUTISTA-BAÑOS

Instituto Politécnico Nacional. Centro de Desarrollo de Productos Bióticos, Yautepec Morelos, Mexico

ELSEVIER

Amsterdam • Boston • Heidelberg • London
New York • Oxford • Paris • San Diego
San Francisco • Singapore • Sydney • Tokyo
Academic Press is an imprint of Elsevier

Academic Press is an imprint of Elsevier
32 Jamestown Road, London NW1 7BY, UK
225 Wyman Street, Waltham, MA 02451, USA
525 B Street, Suite 1800, San Diego, CA 92101-4495, USA

British Library Cataloguing-in-Publication Data
A catalogue record for this book is available from the British Library

Library of Congress Cataloging-in-Publication Data
A catalog record for this book is available from the Library of Congress

ISBN : 978-0-12-411552-1

For information on all Academic Press publications
visit our website at elsevierdirect.com

Typeset by TNQ Books and Journals Pvt Ltd.
www.tnq.co.in

CONTENTS

PREFACE

Fruits and vegetables are living organisms that after harvest obtain their energy through the respiration process. This postharvest metabolism causes the commodity to ripen leading eventually to senescence. The positive characteristics that make fruit suitable for consumption also make them susceptible to disease. During this ripening period, fruit is prone to develop rots caused by microorganisms that hasten commodity ripening, damage their internal and/or external appearance, cause off-odors, produce mycotoxins and contaminate adjacent commodities. Although the economic losses due to fungal infection in fruits and vegetables during the postharvest chain are variable and not well documented, they usually reach anywhere from 30 to 50% and, on some occasions, rots can lead to total loss of the produce. Both fungi and bacteria cause rots; however, in general, fungal infections are reported to have a greater ability to infect a broader range of hosts throughout the whole postharvest chain. Several fungal species affect commercially important fruits and vegetables, such as *Alternaria alternata*, which is present in tomatoes, peppers, mangoes and other crops. The genus *Penicillium*, including species such as *P. italicum*, *P. digitatum* and *P. expansum*, has a wide range of hosts such as citrus fruits and apples. *Aspergillus* is also considered a serious postharvest fungus of temperate and tropical fruits and vegetables. Probably the most important aspect of the above-mentioned fungi is that they are also considered allergenic. *Monilia fructicola*, *M. laxa* and *Botrytis cinerea* are important pathogens of various economically important temperate fruits such as apples, grapes, kiwifruit and peaches, among others. Other important fungi are *Colletotrichum gloeosporioides*, *Botryiodiplodia theobromae* and various species of the genus *Thielaviopsis*, which can be highly infectious for fruits grown under tropical and subtropical conditions, such as mangoes, papayas, citrus and pineapples. Crown rot disease, caused by a complex fungal association, is also a serious disease that affects bananas and certain vegetables while *Rhizopus stolonifer* is a saprophytic fungus that can totally invade fruits and vegetables in less than three days.

Infection by these fungi can occur during crop blossom, at harvest or during handling operations. Control of these fungi has always included the use of fungicides. However, in spite of developing new chemicals with apparently less side effects, producers are under heavy pressure to reduce or eliminate their application. To control these fungi, other commercially viable alternatives or possible future options that work integrated with synthetics or alone have been put forward. They include physical means including heat, irradiation, and low temperatures, hypobaric, modified and controlled atmospheres, biological control with antagonistic microorganisms and GRAS (generally regarded as safe) compounds such as sanitizers, salts, volatiles, and so forth. To date, postharvest pathology research is also focusing on enhancing the individual potential of the host to respond to the attack of the pathogen – resistance inducers. Products such as jasmonates, chitosan, salycilic acid and harpin among others are under study. Depending on the particular pathogen, control may vary; however, all the contributors to this book agree that integration of one or more different strategies to achieve higher levels of disease control should be proposed.

The information in this book covers a wide range of topics related to each selected fungus, such as economic importance, morphology and taxonomy, biology of the infection process, the influence of preharvest agronomic practices and the environment, and the effect of handling operations for the infection and disease development to occur and the strategic controls for each host–pathogen, including traditional and non-traditional alternatives.

This book is aimed at a wide audience including researchers involved in the study of postharvest handling of agricultural commodities and undergraduate and graduate students whose area of research deals with postharvest topics. Growers, managers and operators working at packinghouses and storage, retail and wholesale facilities can also benefit from this book.

The collaborators of this book are experienced researchers in the field of postharvest pathology of fruits and vegetables. Contributors are high-level scientists with many years of research experience related to each selected fungus. The authors belong to scientific associations, participate and collaborate as authors of articles published in international journals and serve as advisors on scientific committees in the postharvest pathology area. I am deeply grateful to all of them for being willing to spare time in their busy agendas as lecturers and researchers to attend to my invitation. I am sure that the information stated in this book will serve present and future generations involved in the postharvest pathology field.

I am also grateful to the editors Patricia Osborn, Carrie Bolger and the Elsevier Production Department for their guidance from beginning to the end of this project.

Silvia Bautista-Baños

LIST OF CONTRIBUTORS

Asgar Ali
Centre of Excellence for Postharvest Biotechnology (CEPB), School of Biosciences, The University of Nottingham Malaysia Campus, Semenyih, Selangor D. E., Malaysia

Laura L. Barrera-Necha
Centro de Desarrollo de Productos Bióticos, Instituto Politécnico Nacional Carretera Yautepec-Jojutla km. 8.5, Col. San Isidro, Yautepec Morelos, Mexico

Silvia Bautista-Baños
Centro de Desarrollo de Productos Bióticos, Instituto Politécnico Nacional Carretera Yautepec-Jojutla km. 8.5, Col. San Isidro, Yautepec Morelos, Mexico

Elsa Bosquez-Molina
Universidad Autónoma Metropolitana-Iztapalapa, Col. Vicentina, Mexico

Luc de Lapeyre de Bellaire
CIRAD, Research Unit Banana, Plantain and Pineapple Cropping Systems, Montpellier, France

Deena Errampalli
Agriculture and Agri-Food Canada, SCPFRC, Vineland Station, ON, Canada

Erica Feliziani
Marche Polytechnic University, Department of Agricultural, Food, and Environmental Sciences, Ancona, Italy

I.G.N. Hewajulige
Senior Deputy Director/Principal Research Scientist, Food Technology Section, Industrial Technology Institute, Colombo, Sri Lanka

Ludivine Lassois
Forest, Nature and Lanscape Department, Liège University, Gembloux Agro-Bio Tech, Gembloux, Belgium

Marta Mari
CRIOF, DipSA, University of Bologna, Cadriano, Bologna

Camilla Martini
CRIOF, DipSA, University of Bologna, Cadriano, Bologna

Lluís Palou
Laboratori de Patologia, Centre de Tecnologia Postcollita (CTP), Institut Valencià d'Investigacions Agràries (IVIA), Montcada, València, Spain

Maribel Plascencia-Jatomea
Departamento de Investigación y Posgrado en Alimentos, Universidad de Sonora, Hermosillo, Sonora, México

Gianfranco Romanazzi
Marche Polytechnic University, Department of Agricultural, Food, and Environmental Sciences, Ancona, Italy

Yasmeen Siddiqui
Laboratory of Food Crops, Institute of Tropical Agriculture, Universiti Putra Malaysia, Serdang Selangor D. E., Malaysia

Martín Ernesto Tiznado-Hernández
Coordinación de Tecnología de Alimentos de Origen Vegetal, Centro de Investigación en Alimentación y Desarrollo, Hermosillo, Sonora, México

Rosalba Troncoso-Rojas
Coordinación de Tecnología de Alimentos de Origen Vegetal, Centro de Investigación en Alimentación y Desarrollo, Hermosillo, Sonora, México

John Martin Velez-Haro
Departamento de Investigación y Posgrado en Alimentos, Universidad de Sonora, Hermosillo, Sonora, México

R.L.C. Wijesundera
Senior Professor of Plant Pathology, Department of Plant Sciences, University of Colombo, Colombo, Sri Lanka

María Susana Yépiz-Gómez
Departamento de Investigación y Posgrado en Alimentos, Universidad de Sonora, Hermosillo, Sonora, México

Jiuxu Zhang
Florida Department of Citrus, Lake Alfred, Florida, USA

CHAPTER 1

Rhizopus stolonifer (Soft Rot)

Silvia Bautista-Baños[1], Elsa Bosquez-Molina[2], Laura L. Barrera-Necha[1]

[1]Centro de Desarrollo de Productos Bióticos, Instituto Politécnico Nacional Carretera Yautepec-Jojutla km. 8.5, Col. San Isidro, Yautepec Morelos, Mexico; [2]Universidad Autónoma Metropolitana-Iztapalapa, Col. Vicentina, Mexico

Contents

OVERALL IMPORTANCE OF THE FUNGUS

Rhizopus stolonifer is one of the most common and fastest-growing species in the *Zygomycota* phylum. Disease caused by this fungus is known as soft rot, black mold and Rhizopus rot. It often grows within a few days in moist and humid conditions because its spores are quite common in the air. *Rhizopus stolonifer* is an important postharvest pathogen that contributes to considerable spoilage of a great variety of fruits and vegetables (Table 1.1). Genera susceptible to this fungus include *Allium*, *Ananas*, *Brassica*, *Cucumis*, *Cucurbita*, *Fragaria*, *Lycopersicon*, *Phaseolus*, *Pisum*, *Solanum*, and

Postharvest Decay
http://dx.doi.org/10.1016/B978-0-12-411552-1.00001-6

Table 1.1 Summary of some Plant Species and Organs Affected by *Rhizopus stolonifer* and Overall Expressed Symptomatology*

Plant Species	Plant Organ	Symptoms	Reference
Aonla	Fruit	Water-soaked lesion and rapid decay	Rathod, 2010
Berry fruits (strawberry, blackberry, raspberry wineberry etc.)	Fruit	Discoloration, rapid softening, collapse and juice leaking	Mass, 1998
Carrots and parsnips	Roots	Brownish tissue, water-soaked and soft, liquid exudation	Snowdon, 1991
Cucurbits (melon, squash, zucchini, pumpkin etc.)	Fruit	Water-soaked appearance, tissue softening and collapse	Zitter et al., 1998
Chilli	Fruit	-	Meon and Nik, 1988; Adikaram, 1993; Thind and Jhooty, 1985
Date	Fruit	-	Al-Hindi et al., 2011;
Durian	Fruit	-	Adikaram, 1993; Teng-fei et al., 2009
Eggplant	Fruit	Water-soaked areas, brownish liquid, characteristic odor	Zacharia and Philip, 2010
Grapes	Fruit	Soft pedicels, watery rots and complete decay	Latorre et al., 2002
Guava	Fruit	Oily lesion and water soaked, quick lesion expansion	Ooka, 1980
Jackfruit	Flowers, fruit	Watery brown spots, black rotten, shrunken and mummification	Nelson, 2005; Hossain et al., 2010
Mango	Fruit	Water-soaked lesion and rapid decay	Teng-fei et al., 2009; Rathod, 2010
Papaya	Fruit	Soft and watery surface, intact fruit interior	Alvarez and Nishijima, 1987; Adikaram, 1993; Akinmusire, 2011; Singh et al., 2012

Peas and beans	Fruit	Decay tissue typically watery, fluid leakage	Snowdon, 1991
Periwinkle	Fruit	-	Saini et al., 1996
Red-mombin	Fruit	-	Bautista-Baños et al., 2000
Sarpaghanda	Inflores-cence, fruit	-	Shukla et al., 2006
Stone fruits (peach, pears, plum, sweetcherry, apricots, nectarines etc.)	Fruit	Soft and watery tissue	Ogawa et al., 2000; Kwon and Lee, 2006
Soursop	Fruit	-	Amusa et al., 2003
Sugar beet	Foliage, roots	Wilting foliage, crown collapse. Brow internal decay, spongy cavities and vinegar odor	Jacobsen, 2006
Sunflower	Flower, seeds	Dark spots on the back of heads, watery tissue. Dry, shrivelled and shred heads	Yildrim et al., 2010; Harveson, 2012
Sweet potato	Roots	Dark brown and soft tissue	Snowdon, 1991; Edmunds and Holmes, 2009
Taros and tannias	Roots	Broken skin, soft pale tissue, cheese watery consistency yeasty odor	Snowdon, 1991
Tomato	Fruit	Watery soaked	Kwon et al., 2001; Mahovic et al., 2006; Akinmusire, 2011
White yam tubers	Roots	Rapid collapse of cell walls turning brown, soft and wet	Aboagye-Nuamah et al., 2005; Nahunnaro, 2008

-; Not reported.
*Symptoms for all the above-mentioned fruits and vegetables also include host covered by coarse, gray and hairy white mycelia that usually form a mass of visible black sporangia at their tips.

several more. Other *Rhizopus* species such as *R. oryzae* and *R. artocarpi* are also reported to cause serious rots on fruits such as papaya, banana, mulberry roots and jackfruit, among others (McMillan, 1986; Yoshisda et al., 2003; Kwon et al., 2012). Due to its fast growth, it is considered one of the most devastating fungi during the storage period of various horticultural commodities.

So far, there are limited data on the exact losses attributed to *R. stolonifer*. In this respect, Ceponis and Butterfield (1979) reported that 80% of the total losses in the New York City tomato market was due to *R. stolonifer* and *Alternaria alternata*. In Nigeria, Fajola (1979) conducted a survey in five states of postharvest fruit rot diseases of tomato, finding that 85% of the total loss was due to soft rot diseases with *R. stolonifer* among the major pathogens in this country. Shehu and Muhammad (2011) and Bayewu et al. (2007) identified *R. stolonifer* among the most pathogenic in the mycoflora associated with storage disease of onion bulbs and papaya fruit, respectively. Results of a 3-year survey carried out in the Central province of Sri Lanka showed total postharvest losses from 5 to 50% due to microorganism infection. Among the identified fungi, *R. stolonifer* was considered as the one to cause fast and extensive rotting of a wide range of vegetable and fruit hosts (Adikaram, 1993). In markets in Asiut, Egypt, Abdel-Mallek et al. (1995) and Ziedan and Farrag (2008) reported that *R. stolonifer* from diseased tomatoes and peaches was isolated at a frequency of 35.9% and 56.5%, respectively. This last value is only exceeded by *Aspergillus niger* and *A. alternata*. This fungus is also reported to infect severely roots such as white yam (*Dioscorea rotundata*) in Ghana. In this study, out of the 10 microorganisms isolated from this tuber, *R. stolonifer* incidence was in a range of 20 to 55% depending on the market sampled (Aboagye-Nuamah et al., 2005). In other studies, Bautista-Baños et al. (2008) reported that in tomatoes collected from 15 different local markets in the state of Morelos, Mexico, *R. stolonifer* was the most predominant fungus, isolated from almost 50% of the samples. Out of the 15 fungal genera identified on imported fruits in Shangai, China markets (Teng-fei et al., 2009), *Rhizopus* was identified in durian and mango; however, it was not the most predominant. In guava fruits harvested in Aksum, Ethiopia, *Pestalotia psidii* had a frequency of 32.1% followed by *R. stolonifer* with 16% and *Aspergillus niger* 10.4% (Mathew, 2010). Out of 18 fungal species isolated from banana, grape, litchi and mango collected at a market in Dhaka, Bangladesh, *R. stolonifer* was the second most common fungus with a frequency of 37% (Bashar et al., 2012). A monthly market survey in

Gorakhpur, India, found the presence of *R. stolonifer* in papaya fruit throughout the year (Singh et al., 2012).

TAXONOMY AND MORPHOLOGY

Rhizopus stolonifer belongs to the class Zygomycetes, order Mucorales, family Mucoraceae, genus *Rhizopus* and species *R. stolonifer*. Based on the morphological classification carried out by Shipper (1984), the common features of the *R. stolonifer* group are: complex and well-developed rhizoids, sporangiophores, sporangium and sporangia. The shape of the main columnella (mycelium) is conical-cylindrical; it is gray or brownish, up to 140 μm in height and branches until a tangled mycelium is formed (Fig. 1.1A). This mycelium sends up from various points aerial erect hyphae forming whitish growth in cultures turning eventually into black spots, named sporangiophores, about 1–3 mm long and up to 20–25 μm in diameter (Fig. 1.1B). Each sporangiophore bears a single spherical sporangium of up to about 250 μm in diameter that contains numerous sporangiospores of up to 13 μm in length and diameter in a range of 100–275 μm. Rhizopus spores are of different shapes: angular, subglobose and ellipsoidal, with well- marked ridges on the surface (Fig. 1.1C) depending on maturity (Swingle, 1903). However, in this respect, Hernández-Lauzardo et al. (2005) confirmed the globose type as the principal shape of *R. stolonifer*, which varies in size, diameter and area according to isolate. Further taxa characterization of various species of the genus *Rhizopus* (Bengston Frye and Reinhardt, 1993), based on DNA differences, gave *R. stolonifer* a mol % G + C value of 37.8.

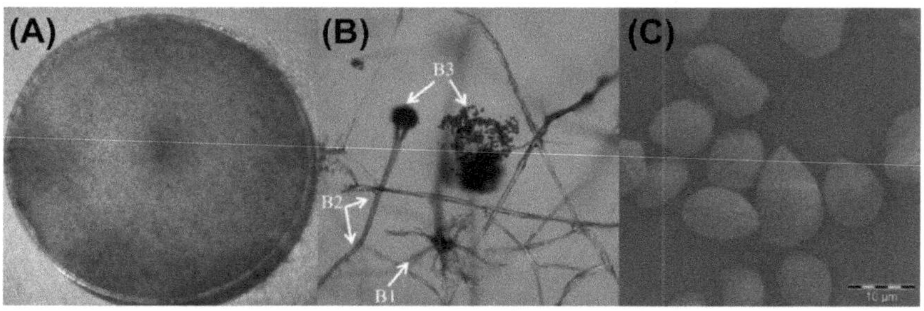

Figure 1.1 *Rhizopus stolonifer* morphology: (A) in Petri plate, (B) optical photomicrography showing (B1) branched rhizoid, (B2) columnella, (B3) sporangiophores and sporangiospores, and (C) scanning electron micrography showing shape diversity of sporangiospores.

BIOLOGY OF THE *R. STOLONIFER* INFECTION PROCESS

For infection to occur, *R. stolonifer* commonly requires, among others, injuries caused by hail, cracking and mechanical wounds (Vicente et al., 2005). In this respect, Nguyen-The et al. (1989) demonstrated that microcrackings on the surface of nectarine fruits are the main entrance for *R. stolonifer* spores; however, they were unable to germinate unless a nutrient solution was available. On sweet potatoes, rhizopus rot development was higher in the bruise wound-type, with infection susceptibility also being associated with cultivar and storage period (Holmes and Stange, 2002). According to Clark and Hoy (1994), susceptibility to *R. stolonifer* infection was clearly associated with sweet potato genotypes. For most fruits and vegetables, symptoms of infection caused by this fungus may be dramatically seen in matter of 3–6 days (Fig. 1.2A–C). Overall, rhizopus rot appears water soaked and exudes a clear leachate. Mena-Nevarez et al. (2012) reported that on infected mango fruit, leachate production was four times higher than on non-inoculated fruit after 35 days of storage, while in orange fruits it was ten times higher. During the initial day of infection, the surface may be covered with thin, fluffy, cotton-like fungal structures. Once *R. stolonifer* has gained entry into wounded tissue, its mycelia spread around the site of infection. Eventually, dark sporulation forms a mass of black sporangia at their tips, covering the entire surface of the commodity. Lisker et al. (1996) reported that germinating *Rhizopus* spores in inoculated grapes were observed after 12–15 h; one h later, germ tubes started to develop and, after 24 h, fruit was covered by hyphae. For tomatoes, no germination took place by 5 h (Fig. 1.3A) postinoculation but, after this time, mycelia grew inter- and intracellularly (Fig. 1.3B); after 24 h, mycelia spread all over the fruit cell walls with a concomitant disappearance of the spore (Fig. 1.3C).

Infection by *R. stolonifer* is closely associated with enzymatic activities, which play an important role in its colonization of fruit. Polygalacturonase (PG) is considered the primary hydrolase involved in the softening process of tomatoes, netted muskmelons and dates during Rhizopus infection (Bruton et al., 1998; Stevens et al., 2004; Al-Hindi et al., 2011) (Table 1.2). In addition, a variety of other macerating enzymes have been assigned roles in fruit cell wall degradation in infected dates such as xylanase, cellulose and amylase (Al-Hindi et al., 2011). As reported by Villanueva (2004) and Vicente (2004), other factors associated with the enzymatic activity are the fungus structure and spore load. In those studies, the activity of PG on infected tomato markedly increased after 48 h in tomatoes inoculated with mycelia (8.8 μL galacturonic acid/min/mg of protein) compared with those inoculated with spore

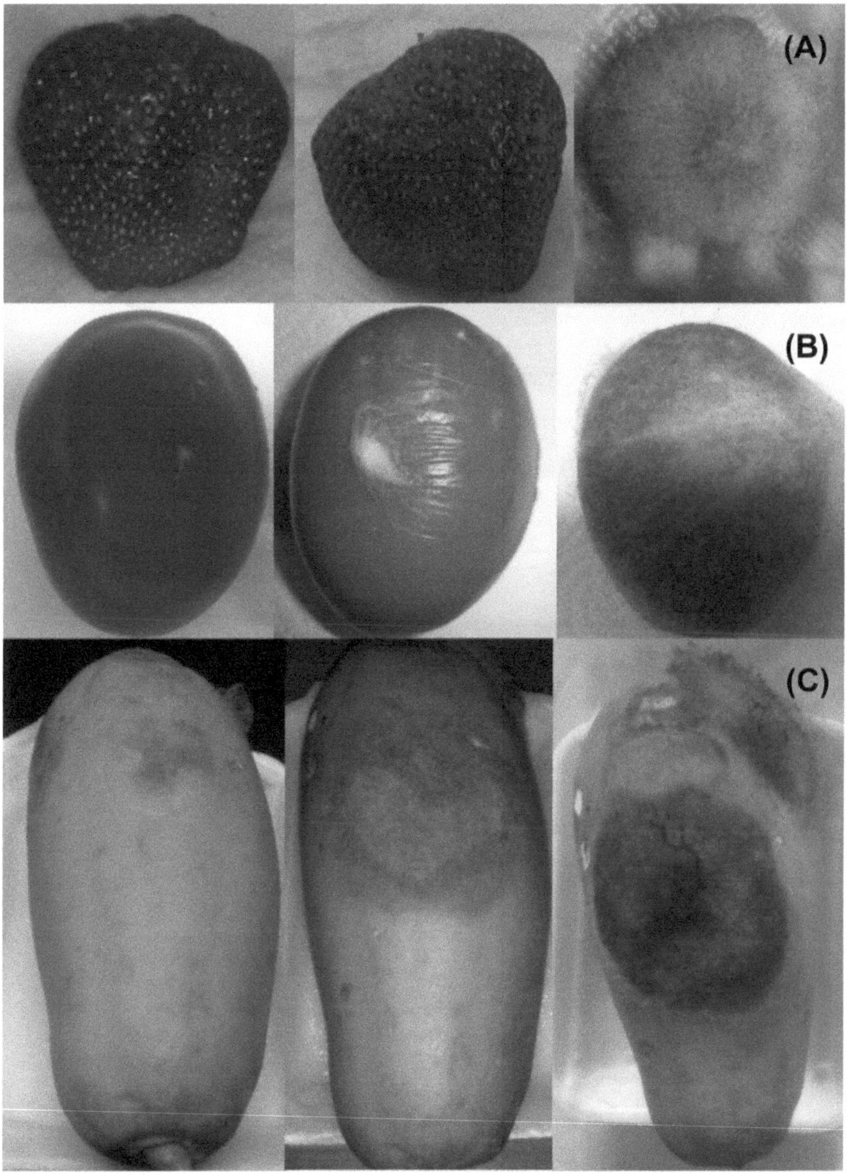

Figure 1.2 *Rhizopus* rot development on (A) strawberry (1, 2 and 3 days), (B) tomato (1, 2 and 4 days), and (C) papaya (1, 3 and 6 days) fruit during ambient storage (20–25°C).

suspension (6.7 μL of galacturonic acid/min/mg of protein). Likewise, tomatoes inoculated with 10^7 spores mL^{-1} concentration showed the highest pectin methylesterase activity after 24 h (1.8 μL/acid/min/mg of protein) compared to tomatoes inoculated at 10^5 spores mL^{-1} (1.4 μLl/acid/min/mg of protein).

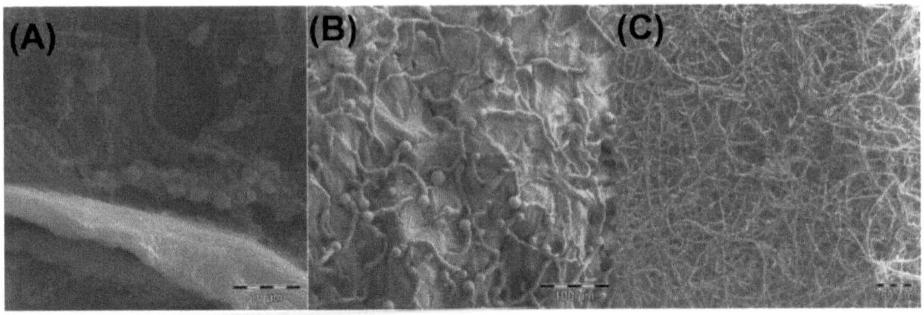

Figure 1.3 Scanning electron micrographs showing the infection process of wounded tomato inoculated by *Rhizopus stolonifer*: (A) non-germinated spores after 0–4 h post-inoculation, (B) germinating spores after 8–12 h, and (C) tomato surface completely covered by mycelia after 24 h.

Table 1.2 *Rhizopus stolonifer* Cell Wall Degrading Enzyme Production on Different Horticultural Commodities

Commodity	Enzyme	Quantity	Reference
Date	Xylanase	172 units/100 mL	Al-Hindi et al., 2011
	Polygalacturonase	189 units/100 mL	
	Cellulase	67units/100 mL	
	Amylase	43 units/100 mL	
Melon	Polygalacturonase	0.05 units/mg protein	Bruton et al., 1998
Tomatoes	Polygalacturonase	Mycelia: 8.8, spores: 6.7 (μL galacturonic acid/min/mg of protein)	Vicente, 2004
Tomatoes vars. 'Better boy', 'Floradade'	Polygalacturonase	var. 'Better boy': 8.79 var. 'Floradade': 6.7 (× 10^6 μL mol/g dry wt)	Stevens et al., 2004

ENVIRONMENTAL AND AGRONOMIC FACTORS FAVORING *R. STOLONIFER* INFECTION

Overall, *Rhizopus* species including *R. stolonifer* are naturally found in soil, debris and air. Fungal dispersal mechanisms of *R. stolonifer* are wind, air currents and some invertebrates such as mites and insects, among others. In this respect, in a study about detection of *R. stolonifer* in various vineyards, *Rhizopus* was the main fungus identified in soils collected between rows or beneath plants, and a high incidence of airborne spores coincided with the high incidence of fruit infection in the vineyard and the proximity of nearby stone-fruit orchards (Lisker et al., 1996). A notable incidence of *R. stolonifer* has also been reported in flowers. Harris and Dennis (1980) reported 30% isolation of this fungus in strawberry flowers. Few studies exist about the preharvest factors and their influence on the incidence of *Rhizopus* rot. For the reports published about agronomic practices, irrigation may increase or reduce the susceptibility of fruits to be infected by this fungus (Teviotdale et al., 1995; 2001). Studies carried out on the effect of this practice on hull rot disease of almond, reported a marked reduction when irrigation of the tree crops was limited before harvest. During two consecutive years, hull rot disease was almost undetected when preharvest irrigation terminated 52 days before harvest. In this study, an increase in *R. stolonifer* incidence was clearly observed as the number of days between last irrigation and harvest was reduced. Additional studies also demonstrated that deficit irrigations, manner and magnitude of water reduction before harvest, also had a direct effect on hull rot disease.

POSTHARVEST FACTORS FAVORING *R. STOLONIFER* INFECTION

Postharvest handling operations are the main reason *R. stolonifer* succeeds in entering and infecting most horticultural commodities. It is continuously reported that overall *R. stolonifer* infection gains better access through wounds, injuries and abrasions. A study carried out by Holmes and Stange (2002) reported that disease incidence on sweet potatoes may be associated with the type of injury. The bruise-wound type injury followed by the puncture-wound type caused the highest infection rate over the 2-year assessment period. With respect to the optimum conditions for infection to occur during postharvest handling, temperature is a very important factor. Overall temperature above 15°C accelerates infection occurring in a short period of time, but in lower temperatures, infection is only delayed. Pierson (1966) reported that in peaches the maximum growth rate of *R. stolonifer* was achieved at 25°C, while lower temperatures of 0, 4.4 and 18°C only delayed the appearance of symptoms by a few days. Spores already attached to fruits are a source of future inoculum for future infections; therefore, a source of contamination at packing plants may be the water systems used for all commodities requiring washing as a usual postharvest operation.

CONTROL

Conventional Alternatives

Fungicides

The use of chemicals is a common method to reduce the incidence of *Rhizopus stolonifer* on various horticultural commodities (Table 1.3). Its chemical

Table 1.3 Summary of Fungicides Applied during the pre- and Postharvest Handling of Various Horticultural Commodities and their Level of Control on *Rhizopus stolonifer*

Horticultural Commodity	Chemical Applied/Rate	Time of Application	Average Efficiency	Reference
Beetroot		Postharvest		Miles et al. (1977)
	TBZ DCNA (2-6 Dichloro-4-nitroniline) (BayMeb 6447®) SOPP (sodium-o-phenyl-phenate) 1000ppm		Total at 5°C storage. At higher temperatures limited control	
Nectarine		Postharvest		Qing and Shiping (2000)
cv. 'Nectarine'	Iprodione 100 and 1000 μg mL^{-1}		25 and 100%, respectively	
Nectarine		Postharvest		Adaskaveg et al. (2005)
cv. 'Red Diamond'	Fludioxonil 1.8 g L^{-1} Iprodione 7.2 g L^{-1}	(curative)	97% 93%	
		(preventive)		
	Fludioxonil 1.8 g L^{-1} Iprodione 7.2 g L^{-1}		30% 20%	
Papaya		Postharvest:		
cvs. 'Intenzza', ' Siluet', 'Sensation', 'Maradol', 'Red Lady'	Imazalil (Fruitgard IS-7,5®) 300 cc/hl Textar 60T® 75 cc/hL Micele-L® 450 cc/hL Copper oxychloride 70% 150cc/hl	Transport storage	Copper oxychloride 35 – 78% 30 – 18%	Perera et al. (2010)
Peach		Postharvest		Wells (1972)
	Dicloran (0-900 μg mL^{-1}) Benomyl (0-100 μg mL^{-1})		Diameter lesion proportionally reduced with concentration	

Peach		Postharvest		Northover and Zhou (2002)
cvs. 'Redhaven', 'Harrow Beauty', 'Babygold 7'	Azoxystrobin 259 g L^{-1} (SC Quadris®, Abound®)		24%	
	Benomyl 50% (Benlate)		0%	
	Captan 75% (Maestro®)		9%	
	Cyprodyl 75% (Vangard®)		33%	
	Dicloran 75% (Botran®)		94%	
	Febuconazole 75% (Indar®)		10%	
	Fenhexamid 50% (Elevate®)		0%	
	Fluodioxonil 0-50% (Medallion®)		0-100%	
	Fluodioxonil 25% + cyprodnil 37.5%		98%	
	Iprodione 50% (Rovral®)		81%	
	Myclobutanyl 40% (Nova®)		0%	
	Propiconazole 250 g l^{-1} (Topas®, Orbit®)		52%	
	Sulfur 80% (Kumulus®)		0%	
	Tebuconazole 45% (Elite®)		91%	
Peach		Postharvest		Adaskaveg et al. (2005); Förster et al. (2007)
cv. 'Elegant Lady'	Fludioxonil 1.8 g L^{-1}	(curative)	86%	
	Iprodione 7.2 g L^{-1}		36%	
	Tebuconazole 1.6 g L^{-1}		82%	
		(Preventive)		
	Fludioxonil 1.8 g L^{-1}		29%	
	Iprodione 7.2 g L^{-1}		52%	
	Tebuconazole 1.6 g L^{-1}		48%	
Peach		Postharvest		
cv. 'Andross'	Clawbor®		35-61%	
	Great®		32-82%	
	3, 4.5 and 6 mg L^{-1}		(as concentration increased lesion development decreased)	

Continued

Table 1.3 Summary of Fungicides Applied during the Pre- and Postharvest Handling of Various Horticultural Commodities and their Level of Control on *Rhizopus stolonifer*—cont'd

Horticultural commodity	Chemical applied/rate	Time of application	Average efficiency	Reference
Peach		Postharvest		De Carvalho et al. (2009)
cv. 'Diamond'	Dicloran		85%	
	Iprodione		70%	
Plum		Postharvest		Adaskaveg et al. (2005)
cv. 'Casselman'	Fludioxonil 1.8 g L^{-1}		37%	
	Fludioxonil 300 mg L^{-1}		100%	
Pomme fruit		Postharvest		Syngenta (2004)
	Fludioxonil (Scholar 50WP®) 300 ppm		100%	
Strawberry		Preharvest		Washington et al. (1992)
cv. 'Tioga'	Fluazinam, SC08558®, Dichlofluanil, Benomyl, Fosetyl-aluminium, Iprodione, Procymidone, Thiram (ca 3000 L/ha $^{-1}$)		(not consistent control)	
Strawberry			Preharvest:	Sallato et al. (2007)
cv. 'Camarosa'	Boscalid (500, 600, 700 g ha^{-1}), Boscalid + pyraclostrobin (102 + 77, 207 + 102, 252 + 128 g ha^{-1}) cyprodynil + fluodixonil (375 + 250 g ha^{-1}), fenhexamid (750 g ha^{-1}) iprodione (750 g ha $^{-1}$)	Preharvest (during flowering, fruit set and harvest). postharvest (fruit immersion)	Boscalid 700 g ha^{-1} (55%) Iprodione (45%) postharvest: Boscalid 700 g ha^{-1} (80%) Boscalid + pyraclostrobin 102 + 77 (75%), 207 + 102 (80%), 252 + 128 (95%)	

Sweet potato		Postharvest		Welch et al. (1996)
	SOPP 0.5%, 1% solutions		90–100%	
	Dicloran (Botran®) 1.5 lbs/100 gal of water		98–100%	
Table grapes		Postharvest		Lisker et al. (1996)
cvs. 'Perlette', 'Thompson seedless'	Orthophenyl phenate (OPP) 0–150 mg L^{-1}		90–100%	
Table grapes		Postharvest		Serey et al. (2007)
cv. 'Shami'	BAS 600 KBF® + metrafenone 100 + 150 mg L^{1}		54.6%	
	BAS 600 KBF® + boscalid 200 + 300 mg L^{-1}		55.4%	
	BAS 600 KBF® + pyraclostrobin 100 + 100 mg l^{-1}		73.1%	
	Boscalid 600 mg L^{-1}		63.9%	
	Boscalid + pyraclostrobin 200 + 100 mg L^{-1}		75.6%	
	Boscalid + kresoxim methyl 100 + 200 mg L^{-1}		72.8%	
	Cyprodinil + fluodioxonil 60 + 40 mg L^{-1}		92.2%	
	Captan 400 mg L^{-1}		35.3%	
Tomato		Postharvest		Abdel-Mallek et al. (1995)
'V_2 Type'	Benlate® Rovral® Sumisclex® 0–1000 µg mL^{-1}		Rovral ® (64.7%) Sumisclex® (43.7 %) Benlate (39.1%) (lesion development decreased as concentration increased)	

control has depended on a series of fungicides that belong to the benzimidazole (thiabendazole and benomyl) and azole (imazalil, propiconazole and thiocarbamates) chemical groups. Ortho-phenylphenol (OPP) and its sodium salt, sodium ortho-phenylphenate (SOPP), are also fungicides used as postharvest treatments to control *R. stolonifer* on fruits and vegetables; however, their action does not control new injures that may be present after their application.The fungicide iprodione of the chemical group dicarboximide is said to reduce *Rhizopus* rot in various horticultural commodities; however, in some countries, this fungicide is banned or never obtained registration. Dichloran, an aromatic hydrocarbon, is a broad-spectrum fungicide used to reduce *Rhizopus* rot; at this time, the EPA (Environmental Protection Agency) has revoked tolerances for its use. Other registered fungicides such as captan of the chemical group thiophthalamides are rarely used since visible residues are left on the surface of the treated commodity. New fungicides such as fluodioxonyl (phenylpirrol), tebuconazole (triazole), fenhexamide (hydroxyanilide) and azoxystrobin (strobulins) are mentioned to achieve notable control of this fungus in various horticultural commodities such as pomme and stone fruits postharvest.

The use of synthetic fungicides should be part of an integrated control. Treatments with dichloran, benomyl and paraffin-based wax mixture were approximately 95% more effective than the fungicides or wax alone in peaches and nectarines (Wells, 1972). Similar results were achieved in pear fruits with the mixture of the fungicide iprodione and wax oil (Adaskaveg et al., 1992). In other studies, a significant improvement in the control of *Rhizopus* rot on sweet potatoes of about 90% was achieved with the mixture of boscalid, pyraclostrobin and *Pseudomonas syringae* (BioSave 10LP) (Edmunds and Holmes, 2009).

Sanitizers

Surface and water disinfectants have also been considered for controlling *Rhizopus stolonifer.* Of the various sanitizers, chlorine (HOCl) is the most widely applied treatment in packing line wash water. It is used as a general disinfectant and it could be added in water solutions (dip, spray, or drench) and as a volatile (gas). Other forms of chlorine also used are sodium hypochlorite (NaHClO) (bleach) and calcium hypochlorite $Ca(ClO)_2$. According to the literature review, the efficacy of these treatments is proportional to the type and form of application of the sanitizer, concentration, exposure time and cultivar (Table 1.4). For example, the efficacy of chlorine was approximately 50% greater on tomatoes than on nectarines, peaches and

Table 1.4 Summary of Sanitizers Applied during the Pre- and Postharvest Handling of Various Horticultural Commodities and their Level of Control on *Rhizopus stolonifer*

Horticultural Commodity	Sanitizer/Concentration	Time of Application	Average Efficiency	Reference
Apricot cv. 'Tyrinthos'	Peracetic acid 250 mg L^{-1} (1, 2, 4, 8 min)	Postharvest	1 min = 95% 2 min = 100% 4 min = 100% 8 min = 100%	Mari et al., 2004
Nectarine cv. 'Star Red Gold'	Peracetic acid 250 mg L^{-1} (1, 2, 4, 8 min)	Postharvest	1 min = 70% 2 min = 100% 4 min = 100% 8 min = 100%	Mari et al., 2004
Nectarine cv. 'Red Diamond'	Chlorine (100 ppm) Perasan® (80ppm)	Postharvest	50% 60%	Adaskaveg, 2009
Peach cv. 'Rome Star'	Peracetic acid 250 mg L^{-1} (1, 2, 4, 8 min)	Postharvest	1 min = 90% 2 min = 90% 4 min = 100% 8 min = 100%	Mari et al., 2004
Peach cvs 'Aurora', 'Dourado', 'Flor da Prince'	Ozone (0.1 µl L^{-1}) 4–7 days	Postharvest	0%	Marchi et al., 2008
Peach cv. 'Elegant lady'	Chlorine (100 ppm) Perasan® (80 ppm)	Postharvest	20% 40%	Adaskaveg, 2009
Strawberry cvs 'Chandler', 'Oso Grande1', 'Oso Grande 2', 'Sweet Charlie'	Chlorine 120 mg L^{-1} (inoculated, non-inoculated, sanitized and non-sanitized)	Postharvest	Sanitized + non-inoculated = 85–95% Non sanitized + inoculated = 0% Sanitized + inoculated = 80–100%	Ferreira et al., 1996

Continued

Table 1.4 Summary of Sanitizers Applied during the Pre- and Postharvest Handling of Various Horticultural Commodities and their Level of Control on *Rhizopus stolonifer*—cont'd

Horticultural Commodity	Sanitizer/Concentration	Time of Application	Average Efficiency	Reference
Strawberry cv. 'Seascape'	Ozone/ionization (0.05, 0.1, 0.5, 1 ppm/10^6 cm^3)	Postharvest	0.5, 1 ppm = 90% (control proportionally reduced with concentration)	Nicoué et al., 2004
Strawberry cvs 'Seascape' and mixture of six cultivars	Chlorine (gas) (10, 20 mg L^{-1} moist air)	Postharvest	cv. 'Seascape' (10 mg L^{-1}) = 57%, (20 mg L^{-1}) = 74%; mixture of cvs (10 mg L^{-1}) = 66%, (20 mg L^{-1}) =79%	Avis et al., 2006
Strawberry	Peracetic acid (100 µL L^{-1}) Harvest seasons (Jan., Feb., March), time of spraying (3, 2, 1 day before harvest) storage time (24, 17, 19 days)	Preharvest fruit development	Storage infection (control proportionally increased with storage time)	Narciso et al., 2007
Table grapes cvs 'Alphonse Lavalle', Thompson Seedless', 'Zeiny'	Ozone alone. Concentration according to ozone exposure: 0.02–0.3 mg g^{-1} (0.1, 5, 10, 20, 30, 40, 60, 80 min) Ethanol 70% + ozone	Postharvest	Ozone alone (control according to exposure time); ethanol + ozone = 100% for all cultivars	Sarig et al., 1996

Table grapes cv. 'Shami'	Calcium hypochlorite powder (1, 5, 10 g 2 kg of fruit)	Postharvest	1 g 2 kg fruit = 90% 5 g 2 kg fruit = 100% 10 g 2 kg fruit = 100%	Lisker et al., 1996
Tomato cv. 'V_2 type'	Sodium hypochlorite (2, 5%)	Postharvest	Percentage lesion inhibition 2% = 47% 5% = 44%	Abdell-Mallek et al., 1995
Tomato cv. 'Agriset', 'FTE12'	Chlorine (50, 100, 150, 200 ppm) Two methods of chlorinated water: flume and shower Two positions of stem scar; up and down	Postharvest	Flume 95–100% Shower 95–100% Up 80–100% Down 100%	Vigneault et al., 2000
Tomato	Chlorine (0, 30, 40, 50, 90, 100, 160, 180 mg L^{-1})	Postharvest	Inconsistent control	Bartz et al., 2001

strawberries (Vigneault 2000; Mari et al., 2004; Avis et al., 2006; Adaskaveg, 2009). For the control of this fungus, calcium and sodium hypochlorite have also been tested on table grapes cv. 'Shami' and tomatoes 'V_2 type', respectively (Abdell-Mallek et al. 1995; Lisker et al., 1996), resulting in an average efficiency of 100% on table grapes and 47% on tomatoes. Scientific research has proved that ozone (O_3) is another method for controlling decay caused by *R. stolonifer*. Although this disinfectant method is generally used for controlling food-borne pathogens, experimentally, it has given a good level of control on strawberries cv. 'Seascape' (Nicoué et al., 2004) and, when combined with ethanol at 70%, complete control of *R. stolonifer* on table grapes cv. 'Alphonse Lavalle', 'Thompson Seedless' and 'Zeiny' was achieved (Sarig et al., 1996). Another sanitizer reportedly under experimentation is peracetic acid ($C_2H_4O_3$) (PAA). Overall, this sanitizer when used as a cleanser and disinfectant, controls various pathogenic microorganisms. Experimentation on this matter has shown that postharvest application of PAA at a concentration of $250\,mg\,L^{-1}$ gave a total inhibition of *R. stolonifer* after 2 min of exposure on apricots, nectarines and peaches (Mari et al., 2004). Additionally, it was reported that preharvest application of commercial PAA (Oxidate®) during flowering and fruit set generally reduced postharvest rots of strawberries (Narciso et al., 2007).

Alternatives Under Experimentation

Antagonists

Overall, the biological control of *R. stolonifer* involves the use of antagonistic microorganisms such as bacteria, yeast and one fungus (Table 1.5). Bacteria such as *Pantoe agglomerans*, *Enterobacter cloacae*, *Pseudomonas syringae* and *Bacillus* spp. have shown an important reduction of *R. stolonifer* on apricot, nectarine, peach, pear, and sweet potato (Wilson et al., 1987; Nunes et al., 2001a, 2002; Northover and Zhou, 2002; Bonaterra et al., 2003; Edmund and Holmes 2009; Wang 2013a, b), while various yeasts of the genera *Pichia*, *Kloeckera*, *Candida*, *Cryptoccoccus*, *Metschnikowia* and *Acremonium* also exert significant control of this fungus on various species of stone fruit, strawberry, sweet cherry, table grape and tomato (McLaughlin et al., 1992; Qing and Shiping, 2000; Shena et al., 2000; Qin et al., 2004; Zhang et al., 2007b; Zhao et al., 2008) and only one fungus classified as *Trichoderma harziarum* has been reported to control *Rhizopus* rot on apple, peach, and tomato during storage (Batta, 2007; El-Katatny and Emam, 2012). As reported, the efficacy of the antagonist depends on various factors including antagonistic concentration, type of strain and culture conditions. Overall, literature

Table 1.5 Summary of Antagonists Applied during the Postharvest Handling of Various Horticultural Commodities and their Level of Control on *Rhizopus stolonifer*

Horticultural Commodity	Antagonist (Concentration CFU mL^{-1})	Average Efficiency	Reference
Apple cv. 'Golden Delicious'	*Candida sake* ($7.5 \times 10^5, 1.6 \times 10^6, 2.6 \times 10^6$)	80–100% (as concentration of antagonist increased control increased)	Viñas et al., 1998
Apple cv. 'Golden Delicious'	*Pantoea agglomerans* ($2 \times 10^7, 8 \times 10^7, 10^8$)	50–100% (as concentration of antagonist increased control increased)	Nunes et al., 2002
Apple cv. 'Golden Delicious'	*Trichoderma harziarum* (4.6×10^8)	32–86% (lesion diameter reduction)	Batta, 2007
Apricot	29 endophytes (1.8×10^8–9.8×10^7)	Isolates: 6CE = 92%, JCAA = 51% effectiveness index	Pratella et al., 1993
Apricot cv. 'Reale d'Imola '	*Pantoea agglomerans* (10^8)	90–100%	Bonaterra et al., 2003
Nectarine cv. 'Nectarina'	*Pichia membranefaciens* ($5 \times 10^3, 5 \times 10^6, 5 \times 10^7, 5 \times 10^8$) and/or Ca 2%	Antagonist alone at the highest concentration, and combined with Ca =100%	Qing and Shiping, 2000
	Pichia membranefaciens (5×10^6) and/or Iprodione (100 μg mL^{-1})	80%	
Nectarine cvs 'Independence', 'Venus'	*Pantoea agglomerans* (10^8)	80–100%	Bonaterra et al., 2003
Peach	*Enterobacter cloacae* (1.5×10^{12})	70%	Wilson et al., 1987
Peach cv. 'Loring	*Kloeckera apiculata, Candida guilliermondi* (10^8)	10–80% (control according to strain) 75%	McLaughlin et al., 1992

Continued

Table 1.5 Summary of Antagonists Applied during the Postharvest Handling of Various Horticultural Commodities and their Level of Control on *Rhizopus stolonifer*—cont'd

Horticultural Commodity	Antagonist (Concentration CFU mL^{-1})	Average Efficiency	Reference
Peach	29 endophytes (1.8×10^8–9.8×10^7)	Isolates: 2CC = 60%, 3CB = 71%, 6CE = 58% effectiveness index	Pratella et al., 1993
Peach cv. 'Redhaven', 'Harrow Beauty'	*Pseudomona syringae* (9×10^{10})	85–97%	Northover and Zhou, 2002
Peach cv. 'Mohasan'	*Trichoderma harziarum* (4.6×10^8)	22–29% (lesion diameter reduction)	Batta, 2007
Peach cv. 'Baihua'	*Cryptoccoccus laurentii* (10^6–10^9)	25–80% (control according to antagonist concentration)	Zhang et al., 2007a
Peach cv. 'Baihua'	Heat treatment (37°C, 48h) + *Cryptoccoccus laurentii* (10^8)	100%	Zhang et al., 2007b
Peach	*Bacillus subtilis* (10^7)	Lower lesion diameter	Zhou et al., 2011
Peach cv. 'Baifeng'	*Bacillus subtilis* (10^8)	Storage days: 1d = 55% 2d = 30% 3d = 0%	Wang et al., 2013a
Peach cv. 'Baifeng'	*Bacillus cereus* (10^8)	25–45% (after 1 and 2 days storage)	Wang et al., 2013b
Pear cv. 'Blanquilla'	*Pantoea agglomerans* (2×10^7, 8×10^7, 10^8)	100%	Nunes et al., 2001
Pear cv. 'Spadona'	*Trichoderma harziarum* (4.6×10^8)	No significant reduction	Batta, 2007

Strawberry cv. 'chunxing'	*Cryptoccocus laurentii* (10^6, 10^7, 10^8, 10^9)	10–65% (control according to antagonist concentration)	Zhang et al., 2007c
	Hot water 55°C, 30 sec + *C. laurentii* 10^8	50%	
Strawberry cv. 'fengxiang'	*Rhodotorula glutinis* (10^8)	*R. glutinis* 20°C = 40%	Zhang et al., 2010
	R. glutinis (10^8) + *salycilic acid* (100 µg mL^{-1})	Combination 60%	
		R. glutinis 4°C = 85%	
	Both at 20°C and 4°C	Combination 95%	
Sweet cherries	*Trichosporon pullulans*	80%	Qin et al., 2004
	Cryptoccocus laurentii	50%	
	Rhodotorula glutinis	70%	
	Pichia membranefaciens (10^8)	75%	
Sweetpotato cv. 'Hernandez'	*Pseudomona syringae* (Bios-Save) (22 and 70.5 oz per100 gal)	20–70% (control according to strain and antagonist concentration)	Edmunds and Holmes, 2009
Table grapes cv. 'Perlette' 'Thompson seedless'	*Pichia guillermondi* *Hanseniaspora uvarum*	0–80% (control according to inoculation type)	Ben-Aire et al., 1991
Table grapes cv. 'Thompson seedless'	*Kloeckera apiculata* *Candida guilliermondi* (5×10^8)	100% in naturally infected fruits 35–90% (control according to yeast preparation and strain)	McLaughlin et al., 1992
Table grapes cv. 'Thompson seedless'	*Candida oleophila* *Metschnikowia* spp (various isolates) (10^8, 10^7, 10^6)	At the highest concentration significant control and according to strain	Schena et al., 2000
Table grapes cv. 'Thompson seedless'	*Candida guilliermondi* *Acremonium cephalosporium* (10^8, 10^7, 10^6)	60–80% (control according to antagonist concentration)	Zahavi et al., 2000

Continued

Table 1.5 Summary of Antagonists Applied during the Postharvest Handling of Various Horticultural Commodities and their Level of Control on *Rhizopus stolonifer*—cont'd

Horticultural Commodity	Antagonist (Concentration CFU mL^{-1})	Average Efficiency	Reference
Tomato cv. F4, F1-121	*Pichia guilliermondii* (10^9)	90–95%	Chalutz et al., 1991
Tomato	*Candida oleophila* *Metschnikowia* spp (various isolates) (10^8, 10^7, 10^6)	At the highest concentration significant control and according to strain	Schena et al., 2000
Tomato cv. 'Ziyu	*Pichia guilliermondii* (10^8, 10^7, 10^6, 10^5)	55–90% (control according to antagonist concentration)	Zhao et al., 2008
Tomato cv. 'Kassel Rock'	*Saccharomyces cerevisiae* (3×10^8)	85–90%	Abd-Alla et al., 2009
Tomato cv. 'Cerasiforme'	*Pichia guilliermondii* (10^8)	55%	Zhao et al., 2010
Tomato	*Trichoderma harzianum* (10^8, 10^7, 10^6)	50–90% (highest control at the highest concentration)	El-Katatny & Emam, 2012

reports that higher concentrations of the antagonist provided a lower incidence of *Rhizopus* disease and lesion diameter in most of the tested cultivars. The efficacy of the antagonist was also associated with specific strains as reported with the yeast *Kloeckera apiculata* and the bacterium *Enterobacter cloacae* (Wilson et al., 1987; McLaughlin et al., 1992), *Pseudomonas syringae* (Edmunds and Holmes, 2009) and various unidentified yeast isolates (Shena et al., 2000). According to McLaughlin et al. (1992), the efficacy of *K. apiculata* for controlling *R. stolonifer* of the cv. 'Thompson Seedless' in grapes was 40% greater when cultures were not filtrated. Another study by Zhao et al. (2008) about the performance of the antagonist indicated that its efficacy is associated with fruit maturity. In this study, the maturity stage of tomatoes was directly associated with the level of efficacy against *R. stolonifer*, with the mature green tomatoes being where the antagonist *P. guilliermondii* produced 100% inhibition compared with the 80% achieved in red and pink tomatoes. In other studies, it was stated that the combination of the antagonist with short hot water dips, calcium chloride or essential oils may enhance the efficacy of the treatments by 90–100%. This was reported in tomatoes (Abd-Alla et al., 2009), peaches and strawberries (Zhang et al., 2007a,c). So far, few commercial biofungicides that control *R. stolonifer* based on antagonists are available in the market; however, Biosave® 10LP and BioSave 11LP, composed of different strains of the bacterium *P. syringae,* and Shemer®, containing the yeast *Metshnikowia fructicola* as the active ingredient, have proven to be efficient in controlling this fungus in a wide range of commodities including sweet potatoes, peaches, apricot, grapes and strawberries, among others (Shena et al., 2000; Edmunds and Holmes 2009). It is known that the use of antagonists often needs to be supplemented with salts and additives to enhance the activity of the biocontrol agents. In this respect, studies by Tian et al. (2002) showed that the combination of calcium with the antagonist *C. gulliermondii* and *C. membranefasciens* resulted in significant reduction in *R. stolonifer* infection and lesion diameter in peaches and nectarine fruits. In other studies, it has been reported that, among several additives assayed, only 2% sodium bicarbonate enhanced the biocontrol performance of Aspire (yeast *Candida oleophila*) against *R. stolonifer* infection of peach (Droby et al., 2003). Resistance inducers such as salicylic acid have been shown to enhance the potential activity of the antagonist *Rhodotorula glutinis* against *R. stolonifer* on strawberries. Zhang et al. (2010) demonstrated that this combination increased the activity of the enzyme peroxidase and β,1-3- glucanase, thereby reducing infection levels significantly. Overall, microorganisms used for biocontrol produced secondary metabolites that are effective in the control of pathogens. In this respect, pyrrolnitrin isolated from the yeast *P. cepacia* was effective in controlling *Rhizopus* rot infection in strawberries. According to Takeda (1990), the onset of infection in treated strawberries was delayed for various days according to storage temperature.

Plant Derivatives

Extracts and Essential Oils

For controlling *R. stolonifer* on a variety of horticultural commodities, a diversity of plant species belonging to different botanical families including, among others, Amaryllidaceae, Annonaceae, Arecaceae, Bromeliaceae, Caricaceae, Lamiaceae, Lauraceae, Leguminosae, Meliaceae, Myrtaceae, Pinaceae, Sapotaceae, Solanaceae and Urticaceae have been tested as extracts, essential oils and/or volatiles. Generally, for plant extracts used as fungicides, it has been reported that *Rhizopus* rot development is dependent on plant species, plant organ and/or concentration (Table 1.6). Bautista-Baños et al. (2000) reported that among 19 different plant species tested, the fungicidal effect of aqueous leaf extracts was 80% stronger than that of aqueous stem extracts on red-mombin (*Spondias purpurea*). In this study, the plant species *C. edulis*, *C. papaya* and *P. guajava* also showed the highest control of *R. stolonifer*. Similarly, Amienyo and Ataga (2007) reported different levels of control of this fungus among five plant species tested, in which the aqueous extract of rhizomes of *Z. officinale* gave 71% fungal control. Okigbo and Ogbonnaya (2006) reported that among five plant species the aqueous extracts of leaves of *O. gratissimum* and *A. melegueta* had the best fungicidal effect. In other studies, it was also demonstrated that powders of leaves of *P. dulce* and rhizomes of *Z. officinale* significantly reduced this fungus on strawberries and tomato fruits, respectively (Bautista-Baños et al., 2003; Taiga 2009; Chuku et al., 2010). Overall, in the reported studies, water has been the main solvent used to extract the active ingredient (Feliziani et al., 2013; Romanazzi et al., 2013); however, ethanol and crude extracts have also exhibited significant control of this fungus in tests (Stange et al., 2001; Nahunnnaro, 2008).

In general, for essential oils, the plant species *Thymus vulgaris* has frequently obtained notable results against this fungus (Table 1.7). In this case, experiments have demonstrated an effective control of *R. stolonifer* of 40–80% although, as previously mentioned, its efficacy is also dependent on various factors such as the type of horticultural commodity, cultivar, concentration, and mode of application, as seen on nectarines, papaya, and strawberry fruits (Reddy et al., 1998; Perera et al., 2010; Nabigol and

Table 1.6 Summary of Plant Species extracts and Powders Applied during the Postharvest Handling of Various Horticultural Commodities and their Level of Control on *Rhizopus stolonifer*

Horticultural Commodity	Plant Species and organ	Solvent and Concentration	Average Efficiency	Reference
Red-mombin types: red, yellow, orange	Cherimoya (*Anonna cherimola*) Bromelia (*Bromelia hemisphaerica*) Papaya (*Carica papaya*) White sapote (*Casimiroa edulis*) Hawthorn (*Crataegus mexicana*) black sapote (*Dyospiros ebenaster*) Sapote mamey (*Pouteria sapota*) Guava (*Psidium guajava*) Manzanita (*Arctostaphylos polifolia*) Naseberry (*Achras sapota*) Capuli (*Prunis capuli*) cainito (*Chrysophyllum cainito*) Avocado (*Persea americana*) (leaves and stems)	Water (1:5 w/v)	Type red: *C. edulis* (leaves) 100%, *C. limon* and *C. mexicana* (leaves and stems) 80–90%, *D. ebenaster* (leaves) 70%, *P. dulce* (stems) 80%, *P. sapota* (leaves) 80% Type yellow: *A. cherimola*, *B. hemisphaerica*, *C. papaya*, (leaves) 100%; *A. sapota*, *C. edulis*, *P. sapota*, *P. guajava* (leaves) 90%; *A. polyfolia*, *P. capuli*, *P. americana*, *C. cainito*, *C mexicana*, *S. purpurea* (leaves) 80% Type orange: 0%	Bautista-Baños et al., 2000
Strawberry	Huamuchil (*Pithecelllobium dulce*) (leaves and seeds)	Water (2:10 w/v) Powder 16.6 g kg^{-1} fruit	60–80% (control according to organ and harvest date)	Bautista-Baños et al., 2003
Strawberry cv. 'Camarosa'	Nettle extract (*Urtica dioica*) Siberian fir (*Abies sibirica*)	(1% w/v) Water (10% w/v)	*U. dioica* 46% (control index) *A. sibirica* 63% (control index)	Romanazzi et al., 2013
Sweet cherry cv. 'Sweet Heart'	Nettle extract (*Urtica dioica*) Siberian fir (*Abies sibirica*)	Water (10 g L^{-1})	5–14% (control index)	Feliziani et al., 2013

Continued

Table 1.6 Summary of Plant Species extracts and Powders Applied during the Postharvest Handling of Various Horticultural Commodities and their Level of Control on *Rhizopus stolonifer*—cont'd

Horticultural Commodity	Plant Species and organ	Solvent and Concentration	Average Efficiency	Reference
Sweetpotato cvs. 'Beauregard', 'Hernandez', 'Jewel', 'White Delight'	Peel, fresh flesh and cured tissue of sweet potatoes	Aqueous-acetone (2:1) g^{-1}	(Control according to cv. and amount of extract)	Stange et al., 2001
Sweetpotato	Ginger (*Zingiber officinale*) rhizome Soursop (*Annona muricata*) leaves Bitter kola (*Garcinia cola*) fruit Christmas tree (*Alchornea cordifolia*) bark Garlic (*Allium sativum*) bulb	Water (10%)	71% 38% 47% 43% 45%	Amienyo and Ataga, 2007
Tomato	Ginger (*Zingiber officinale*) rhizome Garlic (*Allium sativum*) bulb	Powder 1, 2, 3 g	*Z. officinalis* (1 g =less than 20%, 2 and 3 g = 100%) *A. sativum* (0%)	Chuku et al., 2010
Yam	Grains of paradise (*Aframomum melegueta*) leaves Clove basil (*Ocimum gratissimum*) leaves	Water and ethanol 70% (10, 25, 50, 100%)	*O. gratissimum* (water extracts 35–65%; ethanol extract 69–99%) *A. melegueta* (water extracts 30–60%; ethanol extract 67–96%) Control according to dipping time and injury	Okigbo and Ogbonnaya, 2006
Yam	Plantain (*Musa paradisiaca*) peels Bitter leaf (*Vernonia amygdalia*) leaves	Ash Leaf juice	Number of spots 4–7 Number of spots 17–27 (control according to storage days)	Nahunnaro, 2008
Yam	Neem (*Azadirachta indica*) leaves Tobacco (*Nicotiana tabacum*) leaves Aloe (*Aloe barbadensis*) leaves	Cold and hot water 25, 50, 75, 100%	3–4 mm rot depth (control according to concentration)	Taiga, 2009

Table 1.7 Summary of the Plant Essential Oils Applied during the Postharvest Handling of Various Horticultural Commodities and their Level of Control on *Rhizopus stolonifer*

Horticultural Commodity	Essential Oil/ Concentration	Average Efficiency	Reference
Nectarine cvs. 'Flavela',' Flanola'	*Aloe vera* alone *Aloe vera* + thymol (3 l + 1 mL L^{-1})	'Flavela' 4500, 'Flanola' 4800 infection volume (mm^3) 'Flavela' 4000, 'Flanola' 3500 infection volume (mm^3)	Navarro et al., 2011
Orange cv. 'Washington Navel'	Oregano (*Origanum vulgare*) Thyme (*Thymus vulgaris*) Vervain (*Verbena officinalis*) 150, 200, 1000, 2000 ppm	Thyme and oregano 100 and 64 % inhibition at 2000 ppm, respectively	Camele et al., 2010
Papaya cv. 'Maradol'	Thyme (*Thymus vulgaris*) Mexican lime (*Citrus aurentifolia*) 0.1, 0.12, 0.15% (w/w)	Thyme and Mexican lime 20 and 40%, respectively (Control according to concentration)	Bosquez-Molina et al., 2010
Papaya cvs. 'Intenza', 'Siluet', 'Sensation', 'Maradol', 'Red Lady'	Red thyme (*Thymus zigis*) 300–500 cc h L^{-1} Cinnamon (*Cinnamomum zeylanicum*) 20–150 cc hL^{-1}	During transportation. *T. zigis*, 5.4 cm radial growth – 8.6% inhibition. *C. zeylanicum*, 5.4 cm radial growth – 11.2% inhibition During storage. *T. zigis*, 6.9 cm – 9.6% *C. zeylanicum*, 6.8 cm – 11.3%	Perera et al., 2010
Peach cv. 'Early grand'	Peppermint (*Agonis flexuosa*) Sweet basil (*Ocimum basilicum*) (Both 1,2, 3 4 mL/ box)	*A. flexuosa*: 1 = 70%, 2 = 94%, 3 and 4 mL/box = 100% *O. basilicum*: 1 = 60%, 2 = 90%, 3 = 98% 4 mL/box = 100%	Ziedan and Farrag, 2008

Continued

Table 1.7 Summary of the Plant Essential Oils Applied during the Postharvest Handling of Various Horticultural Commodities and their Level of Control on *Rhizopus stolonifer*—cont'd

Horticultural Commodity	Essential Oil/ Concentration	Average Efficiency	Reference
Strawberry Two clonal stocks 'Laval -1', 'Laval-2'	Thyme (*Thymus vulgaris*) 50, 100, 200 ppm. Two storage periods	200 ppm 7 days – 90% and 14 days – 80% (higher control at the highest concentration and at 7 days storage for both clones)	Reddy et al., 1998
Strawberry cv. 'Selva'	Thyme (*Thymus daenensis*), thyme (*Thymus carmanicus*) Three application methods: spray, dipping and fumigation	*T. daenensis*. spray and dipping 80%, fumigation 40% *T. carmanicus*. spray and dipping 60%. fumigation 35% (Control according to application method)	Nabigol and Morshedi, 2011
Strawberry cv. 'Hongyan'	Tea tree oil (*Melaleuca alternifolia*) 0-9 g L^{-1} air	20%	Shao et al., 2013
Tomato cv 'Celebrity'	White thyme (*Thymus vulgaris*), oregano (*Origanum vulgare*), lemongrass (*Cymbogon* sp) Emulsion: 0, 100, 500, 1000, 5000, 10 000 mg L^{-1} Fumigation: 50 mg L^{-1}	0% inhibition	Plotto et al., 2003
Tomato cv. 'Kassel Rock'	Peppermint (*Mentha × piperita*) 1% Combinations: *Saccharomyces cerevisiae* + carnauba wax + peppermint oil 1% *S. cerevisiae* + peppermint oil 1% *S. cerevisiae* + paraffin wax + peppermint oil 1%	90% 100% 100% 100%	Abd-Alla et al., 2009
Tomato	Citronella oil 1, 5, 10, 15, 20 μL mL^{-1}	40% at 10 μL mL^{-1}, 80 % at 15 μL mL^{-1} and 100% at 20 μL mL^{-1}	Matan, 2010

Morshedi, 2011; Navarro et al., 2011). Other essential oils tested such as oregano, cinnamon and peppermint have shown good control of *R. stolonifer* on orange fruits and tomatoes (Abd-Alla et al., 2009; Camele et al., 2010). A synergistic effect with the combination of peppermint essential oil, paraffin wax and the antagonist *S. cerevisae* was also observed on tomatoes (Abd-Alla et al., 2009).

Volatiles

Alternative treatments explored to control postharvest diseases of horticultural products are those of exposure to volatile compound solutions. Margosan et al. (1994, 1997) demonstrated a significant control of postharvest rots on strawberries ('Chandler' and 'Swede') when heated dilute ethanol solutions were used for controlling postharvest decay caused by *R. stolonifer*. In other studies, the effect of benzaldehyde, a volatile produced during ripening of some stone fruits, and other aroma compounds such as hexanal, methyl salicylate and benzyl acetate were tested on *R. stolonifer* of nectarines and plums, resulting in a limited effect for only 24 h storage since after this period disease infection was observed (Caccioni et al., 1995). Another volatile used in the food industry to inhibit microbial growth is acetic acid (AA) applied as a vapor of glacial acetic acid or as vinegar, in this study, the application of AA vapor on 'Harbrite' peaches, prevented *Rhizopus* rot only with the concentration of 2.7 mg L^{-1} (Sholberg and Gaunce 1996). In that research, higher concentrations also caused injuries to the fruit skin. In further studies, fumigation of eight different cultivars of cherries with AA (1.9 $\mu L\ L^{-1}$), formic acid (1.3 $\mu L\ L^{-1}$) and propionic acid (2.5 $\mu L\ L^{-1}$) significantly reduced *Rhizopus* rot by 55, 90 and 60%, respectively (Sholberg, 1998, 2009).

Irradiation

Rhizopus soft rot was effectively reduced when tomatoes were inoculated following UV-C irradiation; however, the positive effect on tomatoes decreased the later the treatments were performed in the ripening process (Table 1.8). Additional studies have shown that a combined treatment with a biocontrol agent like the yeast *Debaryomyces hansenii* significantly reduced the incidence of *Rhizopus* soft rot of strawberry, tomato and sweet potato (Stevens et al., 1997; Pan et al., 2004). On the other hand, there is evidence that the application of low-dose UV-C induces resistance in the host against pathogens, e.g. the retarded degradation of tomatine (a compound which is toxic to some fungi). Likewise, it has been demonstrated that UV-C treated

Table 1.8 Summary of the Postharvest Irradiation Treatments Applied during the Postharvest Handling of Various Horticultural Commodities and their Level of Control on *Rhizopus stolonifer*

Horticultural Commodity	UV-C Dose (kJm^{-2})	Average Efficiency	Reference
Peach cv. 'Baihua'	Microwave power 2450 MHz (1, 2, 3 min alone and combined with *Cryptoccoccus laurenti*	76.3%	Zhang et al., 2004
Strawberry cv. 'Seascape'	UVC-irradiation ($4.1\ kJ\ m^{-2}$) Irradiation + heat (air oven; 45°C, 3 h in air)	UVC-Irradiation alone = 55% Irradiation + heat = 80%	Pan et al., 2004
Tomato cvs. 'Tuskegee 80-130', 'Floradade'	UVC-irradiation $3.6\ kJm^{-2}$, 6.7 min combination: Irradiation + $CaCl_2$ 2% Irradiation + *Debaryomyces hansenii* 10^8 cell/mL + $CaCl_2$ 2%	UVC-irradiation = 50–80% Irradiation + $CaCl_2$ 2% = 60–90% Irradiation + *Debaryomyces hansenii* 10^8 cell/mL + $CaCl_2$ 2% = 80–97% (Control according to cultivar)	Stevens et al., 1997
Tomato cvs. 'Better Boy', 'Floradade'	$3.6\ kJm^{-2}$ + dark $3.6\ kJm^{-2}$ + light (Both 4.8 min)	$3.6\ kJm^{-2}$ + dark = 80% $3.6\ kJm^{-2}$ + light = 0%	Stevens et al., 1998
Tomato cv. 'Better Boy'	UVC-irradiation $3.6\ kJm^{-2}$, 5 min	PG secreted by *R. stolonifer* showed a 50% reduction in activity	Stevens et al., 2004

tomatoes are resistant to polygalacturonase activity secreted by *R. stolonifer* (Stevens et al., 1997, 1998). Another alternative that has been explored is the application of microwave energy, which provides rapid heating and is already widely used in the food industry. A study carried out by Zhang et al. (2004) demonstrated that microwave power alone or in combination with a yeast antagonist like *C. laurentii* reduced the percentage of decayed fruit by 42% and 75%, respectively, compared to untreated fruit.

Chitosan

Several studies have proven that chitosan reduces postharvest fruit decay caused by *R. stolonifer* in for example, strawberries, tomato, raspberries, peach and papaya (El Gaouth et al., 1992; Zhang and Quantick, 1998; Bautista-Baños and Bravo Luna, 2004; Park et al., 2005; Hernández-Lauzardo et al., 2010) (Table 1.9). As reported for the above mentioned fruits, the average efficacy of chitosan applied alone ranged from 50 to 90%. It is also known that chitosan is a potent elicitor of several plant defense processes against phytopathogens, with a film-forming property and excellent compatibility with different substances, making it one of the most versatile materials. In this latter context, chitosan has been combined with other active substances, such as potassium sorbate (PS) to enhance its antimicrobial efficacy in order to control the growth of *R. stolonifer* on fresh strawberries (Park et al., 2005). It was recently reported that chitosan has an interesting inhibitory effect on *R. stolonifer* of table grapes when combined with oregano essential oil (OE) (Dos Santos et al., 2012). In this study, the minimum inhibitory concentration (MIC) of chitosan and OE were first determined and then, based on this, different mixtures were tested, finding that the application of chitosan and OE at sub-inhibitory concentrations (chitosan ½ MIC + OE ¼ MIC; chitosan ½ MIC + OE ½ MIC) inhibited mycelial growth and spore germination, causing morphological changes in spores and mycelia of *R. stolonifer*, as well as the natural mycoflora inhibition of grapes, in addition to preserving the quality of grapes stored at both room and cold temperatures. Over the last few years, there has also been increasing interest in researching and using resistance inducers as alternatives to synthetic fungicides, and chitosan has again proved to be the most promising compound according to Romanazzi et al. (2013). These authors compared the efficacy of a water-soluble commercial chitosan formulation and practical grade chitosan dissolved in dilute acids and some resistance inducers in the control of gray mold and *Rhizopus* rot of strawberries. The highest postharvest disease control was 91% obtained with the practical grade chitosan solutions and the commercial chitosan formulation followed by benzothiadazole, calcium and organic acids. These treatments are of particular interest for their application in IPM.

Hypobaric Atmosphere

Among the new technologies that are being investigated to reduce disease development and infection caused by *R. stolonifer*, pressure treatments (hypobaric or hyperbaric) have also been explored. Hypobaric treatments

Table 1.9 Summary of Chitosan Treatments Applied during the Postharvest Handling of Various Horticultural Commodities and their Level of Control on *Rhizopus stolonifer*

Horticultural Commodity	Type of Chitosan/ Concentration	Average Efficiency	Reference
Papaya	2 mg L^{-1}	73%	Hernández-Lauzardo et al., 2010
Peach	2 mg L^{-1}	66%	Hernández-Lauzardo et al., 2010
Raspberry	1.0, 2.0% Combination: chitosan 2% + TBZ	1.0% = 50% 2.0% = 52 % Combination = 54%	Zhang and Quantick, 1998
Strawberry cv. 'Chandler'	1. 10 mg L^{-1} 2. 15 mg L^{-1} (7 and 14 days)	1. 63–90% 2. 68–90%	El-Ghaouth et al., 1992
Strawberry	1.0, 2.0% Combination: chitosan 2% + TBZ	1.0% = 50% 2.0% = 55 % Combination = 60%	Zhang and Quantick, 1998
Strawberry	Alone (2.0%) 2.0% + potassium sorbate 0.3%	Alone = 48% Chitosan 2.0% + potassium sorbate 0.3% = 45%	Park et al., 2005
Strawberry cv. 'Camarosa'	1. Water-soluble commercial chitosan 1%, w/v 2. Grade chitosan 1%, v/v 3. Chitosan acetate 4. Chitosan chloride 5. Chitosan formate 6. Chitosan glutamate Resistance inducers: 7. Benzothiadiazole (0.2% w/v) 8. Oligosaccharides (1% v/v) 9. Calcium and organic acids (1% v/v)	1. Water-soluble commercial chitosan = 80.7% 2. Grade chitosan = 91.4% 3. Chitosan acetate = 86% 4. Chitosan chloride = 70% 5. Chitosan formate = 74% 6. Chitosan glutamate = 74% 7. Benzothiadiazole = 79% 8. Oligosaccharides = 63.7% 9. Calcium and organic = 87%	Romanazzi et al., 2013

Table 1.9 Summary of Chitosan Treatments Applied during the Postharvest Handling of Various Horticultural Commodities and their Level of Control on *Rhizopus stolonifer*—cont'd

Horticultural Commodity	Type of Chitosan/ Concentration	Average Efficiency	Reference
Table grapes cv. 'Isabella'	1. 5 mg mL^{-1} + oregano essential oil 2.5 μL^{-1} 2. 5 mg mL^{-1} + oregano essential oil 5 μL^{-1}	1. 5 mg ml^{-1} + oregano essential oil 2.5 μL^{-1} = 75% after 12 days storage 2. 5 mg mL^{-1} + oregano essential oil 5 μL^{-1} = 100% at 12 and 25°C	Dos Santos et al., 2012
Tomato Type 'Saladette'	Low, medium and high molecular weight (1.0, 1.5, 2.0%)	There was no difference due to chitosan Mw and fungicide. *Rhizopus* rot symptoms were only delayed.	Bautista-Baños and Bravo Luna, 2004
Tomato	2 mg L^{-1}	66%	Hernández-Lauzardo et al., 2010

at 0.25, 0.50 and 0.75 atm for 4 h reduced disease incidence by 95% in strawberries stored in the dark at 20°C for 4 days after treatment (Romanazzi et al., 2001). These authors also suggested that hypobaric treatment of fruit could trigger a stress response within the plant tissues resulting in the enhancement of disease resistance. In a recent study, the efficacy of a short-term hypobaric treatment prior to storage in order to control natural rot development in strawberries was evaluated (Hashmi et al., 2013a). The strawberries were treated with hypobaric pressures of 25 kPa_a, 50 kPa_a and 75 kPa_a for 4 h at 20°C and then stored at 20°C or 5°C. The treatment at 50 kPa_a showed consistent delayed rot development at both temperatures and did not result in either weight or firmness loss in the strawberries. In a further study carried out by the same authors, the possible modes of action of prestorage hypobaric treatments on decay reduction were studied (Hashmi et al., 2013b). They demonstrated that hypobaric treatment (50 kPa, 4 h) reduced decay in strawberries subsequently stored at 20°C and this fact was not attributable to the low O_2 partial pressure achieved during treatment but to an increase in some activities of defense-related enzymes such as phenylalanine ammonia-lyase (PAL), chitinase and peroxidase (POD), indicating stimulation of induced

resistance in non-inoculated hypobaric-treated fruits. In fact, this was confirmed when hypobaric-treated strawberries were inoculated with *R. stolonifer* immediately and 12 h after treatment resulting in 45% rot reduction, showing the response of the fruit to the fungal attack.

Generally Regarded as Safe (GRAS) Compounds

Salts and Minerals

Other methods still under experimentation for the control of *R. stolonifer* include the use of salts (additives) and minerals. Nunes et al. (2001b) and Pallou et al. (2012) have highlighted the efficacy of ammonium molybdate for controlling *R stolonifer* on apples cv. 'Golden Delicious' and peaches cvs. Flavorcrest', 'O Henry', and 'Last Chance'. In those studies, the ammonium molybdate salt totally inhibited the development of *R. stolonifer* during storage. Other salts with apparent fungicidal potential were sodium molybdate, potassium carbomate, potassium propanate and sodium benzoate. Among the minerals tested, boron products such as Clawbor® and Great® have shown a good ability to control storage rots on peaches cv. 'Andross' (Karagiannidis and Thomidis, 2008). Clawbor® and Great® reduced *Rhizopus* rot by 60% and 80%, respectively at a concentration of 6 mg L^{-1}.

Various

Other less studied methods reported for controlling *R. stolonifer* are the benzoic acid derivative named 2,5-dimethoxybenzoic acid (DMB), isothiocyanates, hot drench water, waxes and salicylic acid. Studies carried out by Lattanzio et al. (1996) demonstrated that postharvest application of the plant phenolic compound DMB delayed *R. stolonifer* appearance and notably reduced rot incidence when sprayed or dipped onto two cvs. of strawberry fruits, 'Pajaro' and 'Chandle'. Results also showed that the efficacy of this compound was dependent on storage temperature. For the isothiocyanates, Mari et al. (1996) tested various natural glucosinolates on pear fruit cv. 'Conference' reporting little or no effect on the control of *R. stolonifer.* In other evaluations on sweet potato, it was reported that the application of hot water drenches at various temperatures (50, 52.5, 55 and 57.5°C) resulted in complete inhibition of rots, including *R. stolonifer*, on this commodity, as also happened with the application of the fungicide Dicloran (Brash et al. 2009). A recent study confirmed the antifungal effect of wax (Gonçalves et al., 2010). The protective effects of carnauba wax when applied on plums and nectarines (obtained from a market in São

Paulo, Brazil) at concentrations of 4.5% and 9%, significantly reduced the incidence of *R. stolonifer* on their surface by 90%; however, when the wax was applied post-infection, the treatment was inefficient, indicating that the wax acts before the pathogens infect the fruit. In a further report, the postharvest application of up to 1.0 mM salicylic acid also resulted in a notable performance since the reduction of *Rhizopus* lesions on pre- and post-inoculated peach fruits (cv. 'Hale') was significantly lower than in the untreated fruits (Panahirad et al., 2012).

CONCLUDING REMARKS

Rhizopus stolonifer is a common member of the fungal phylum Zygomycota. It plays a key role in the carbon cycle because it works as a decomposer in soil, dung and discarded foods, it has also been reported to have medicinal properties. However, for the wholesale and retail business market, *R. stolonifer* is considered one of the most destructive fungi. The importance of *R. stolonifer* lies in its fast growth (it is considered the fastest-growing fungus) and its wide array of hosts, as it affects numerous horticultural commodities worldwide. Generally, *R. stolonifer* is acquired during the first steps of the postharvest chain; harvest and handling in the packing house. Other commodities such as strawberries and tomatoes may be infected before harvest. It is also known that *R. stolonifer* inoculum may be present on plant organs such as leaves and fruitlets, but few studies about what direct preharvest actions might be implemented to reduce the incidence of this fungus at this stage have been carried out.

There are numerous alternative methods for controlling *Rhizopus* rot during postharvest. Among the conventional ones, the most predominant action is the application of synthetic fungicides. In this respect, fungicides such as Benlate®, Rovral® and Botran® have serious limitation on their application in some countries, while others labeled as reduced-risk fungicides, including Boscalid, as well as Fluodixonil and Fenhexamid, have obtained clearance from the US Environmental Protection Agency (EPA). Currently, most fungicides are listed as 'slightly hazardous' to human health, and they may also enter the water, soil and air; therefore, care must be taken with their selection. In addition, the decision on whether to use fungicide is taken based on the target market.

Today, there is strong public resistance to consuming chemically treated horticultural commodities. As a result, the postharvest control of diseases has taken on new aims, i.e. developing products that are less harmful to human health and the environment. For controlling *R. stolonifer*, research on the development of alternative methods has also been intensified. The most extensive studies have been conducted on the use of antagonists although their registration and acceptance is difficult since the mode of action is unclear. To our knowledge, there are only a few commercial products available for controlling *R. stolonifer*, namely Biosave®, 10LP, 11LP and Shemer®.

Based on the information gathered, other extensively tested alternative methods include plant extracts and essential oils. Although most published information has been focused on *in vitro* studies, those evaluations carried out with these methods on horticultural commodities resulted in adequate control of this fungus. Further evaluations at a commercial level should be considered with thyme, oregano and cinnamon essential oils. Unfortunately, information about commercial products based on plant derivatives and used against *R. stolonifer* has not been published. With respect to UV-C irradiation, results are encouraging since significant control has been achieved with this technology; however, due to its high costs and unknown side effects, which have led to public concern, its future remains in doubt. Future research with the trend focused on better combined formulations based not only on the individual properties of the substances but also on their proportion and chemical compatibilities with chitosan and waxes is needed. To date, these technologies, together with UV-C irradiation and hypobaric atmospheres, not only may have fungicidal potential but also enhance the induction of resistance mechanisms in fruits and vegetables against *R. stolonifer*. Enzymatic defense activities including PAL, chitinase, POD and a β-1, 3-glucanase are reported to be involved in the reduction of *R. stolonifer* during the infection process.

Studies looking at combining fungicides at lower concentrations with other non-chemical alternatives such as essential oils, waxes, and antagonists have been undertaken, resulting in substantial control of *R. stolonifer*. According to the published results, they may soon be commercially available.

The reported postharvest technologies to control *R. stolonifer* seem to hold considerable promise in terms of reducing postharvest losses caused by this fungus. And, as with any other new technology, the overall impact should be assessed not only in the fruit itself and, in this case, on *R. stolonifer*, but also on an application system for harvesting, transport, packaging and store retail operations.

REFERENCES

Abd-Alla, M.A., El-Mougy, N.S., El-Gamal, N.G., 2009. Formulation of essential oils yeast for controlling postharvest decay of tomato fruits. Plant Pathol. Bull. 18, 23–33.

Abdel-Mallek, A.Y., Hemida, S.K., Bagy, M.M.K., 1995. Studies on fungi associated with tomato fruits and effectiveness of some commercial fungicides against three pathogens. Mycopathologia 130, 109–116.

Aboagye-Nuamah, F., Offei, S.K., Cornelius, E.W., Bancroft, R.D., 2005. Severity of spoilage storage rots on white yam (*Dioscorea rotundata* Poir.). Annals Appl. Biol. 147, 183–190.

Adaskaveg, J.E., 2009. Epidemiology and management of pre- and post-harvest diseases of peach, plum and nectarine. Calif. Tree Fruit Agreement. Annu. Res. Rep., 29–48.

Adaskaveg, J.E., Conn, K., Mauk, P.A., 1992. Evaluation and control of postharvest decay of fresh market pear fruit. Postharvest Res. Rep., 1–5.

Adaskaveg, J.E., Föster, H., Gubler, W.D., Teviotdale, B.L., Thompson, D.F., 2005. Reduced-risk fungicides help manage brown rot and other fungal diseases of stone fruit. Calif. Agriculture April–June, 109–114.

Adikaram, N.K.B., 1993. A survey of post-harvest losses in some fruits and vegetables and the fungi associated with them. Ceylon J. Sci. 19-20, 10.

Akinmusire, O.O., 2011. Fungal species associated with the spoilage of some edible fruits in Maiduguri northern eastern Nigeria. Adv. Environ. Biol. 5, 157–161.

Al-Hindi, R.R., Al-Najada, A.R., Mohamed, S.A., 2011. Isolation and identification of some fruit spoilage fungi: Screening of plant cell wall degrading enzymes. Afr. J. Microbiol. Res. 5, 443–448.

Alvarez, A.M., Nishijima, W.T., 1987. Postharvest diseases of papaya. Plant Dis. 71, 681–686.

Amusa, N.A., Ashaye, O.A., Oladapo, M.O., Kafaru, O.O., 2003. Pre-harvest deterioration of sour sop (*Annona muricata*) at Ibadan southwestern Nigeria and its effect on nutrient composition. Afri. J. Biotechnol. 2, 23–25.

Amienyo, C.A., Ataga, A.E., 2007. Use of indigenous plant extracts for the protection of mechanically injured sweet potato [*Ipomea batatas* (L.) Lam] tubers. Sci. Res. Essay 2, 167–170.

Avis, T.J., Martinez, C., Tweddell, R.J., 2006. Effect of chlorine atmospheres on the development of Rhizopus rot [*Rhizopus stolonifer*] and gray mold [*Botrytis cinerea*] on stored strawberry fruits. Can. J. Plant Pathol. 28, 526–532.

Bartz, J.A., Eayre, C.G., Mahovic, M.J., Concelmo, D.E., Brecht, J.K., Sargent, S.A., 2001. Chlorine concentration and the inoculation of tomato fruit in packinghouse dump tanks. Plant Dis. 85, 885–889.

Bashar, M.A., Shamsi, S., Hossain, M., 2012. Fungi associated with rotten fruits in Dhaka metropolis. Bangladesh J. Botany 41, 115–117.

Batta, Y.A., 2007. Control of postharvest diseases of fruit with an invert emulsion formulation of *Trichoderma harzianum* Rifai. Postharvest Biol. Technol. 43, 143–150.

Bautista-Baños, S., Hernández-López, M., Díaz-Pérez, J.C., Cano-Ochoa, C.F., 2000. Evaluation of the fungicidal properties of plant extracts to reduce *Rhizopus stolonifer* of 'ciruela' fruit (*Spondias purpurea* L.) during storage. Postharvest Biol. Technol. 20, 99–106.

Bautista-Baños, S., García-Dominguez, E., Barrera-Necha, L.L., Reyes-Chilpa, R., Wilson, C.L., 2003. Seasonal evaluation of the postharvest fungicidal activity of powders and extracts of huamuchil (*Pithecellobium dulce*): action against *Botrytis cinerea, Penicillium digitatum* and *Rhizopus stolonifer* of strawberry fruit. Postharvest Biol. Technol. 29, 81–92.

Bautista-Baños, S., Bravo-Luna, L., 2004. Evaluación del quitosano en el desarrollo de la pudrición blanda del tomate durante el almacenamiento. Rev. Iberoam. de Tecnología Postcosecha 1, 63–67.

Bautista-Baños, S., Velázquez-del Valle, M.G., Hernández-Lauzardo, Ait Barka, E., 2008. The *Rhizopus stolonifer*-tomato interaction. In: Ait Barka, E., Clement, C. (Eds.). Plant Microbe Interaction Research Signpost, Kerala, India, pp. 269–289.

Bayewu, R.A., Amusa, N.A., Ayoola, O.A., Babalola, O.O., 2007. Survey of the post harvest diseases and aflatoxin contamination of marketed pawpaw fruit (*Carica papaya* L) in South Western Nigeria. Afri. J. Agri. Res. 2, 178–181.
Ben-Aire, R., Droby, S., Zutkhi, J., et al., 1991. Preharvest and postharvest biological control of Rhizopus and Botrytis bunch rots of table grapes with antagonistic yeasts. In: Wilson, C., Chalutz, E. (Eds.), Biological Control of Postharvest Diseases of Fruits and Vegetables. Workshop Proceedings. Shepherdstown, West Virginia, USDA, pp. 100–113.
Bengston Frye, C., Reinhardt, D.J., 1993. Characterization of groups of the zygomycete genus *Rhizopus*. Mycopathologia 124, 139–147.
Bonaterra, A., Mari, M., Casalini, L., Montesinos, E., 2003. Biological control of *Monilinia laxa* and *Rhizopus stolonifer* in postharvest of stone fruit by *Pantoea agglomerans* EPS125 and putative mechanisms. Int. J. Food Microbiol. 84, 93–104.
Bosquez-Molina, E., Ronquillo-de, J.,E., Bautista-Baños, S., Verde-Calvo, J.R., Morales-López, J., 2010. Inhibitory effect of essential oils against *Colletotrichum gloeosporioides* and *Rhizopus stolonifer* in stored papaya fruit and their possible application in coatings. Postharvest Biol. Technol. 57, 132–137.
Brash, D.W., Cheah, L.-H., Hedderley, D.I., Olsson, S., Woolf, A.B., 2009. Hot water for control of soft rots on washed kumara (sweet potato). Acta Horticulturae 880, 455–462.
Bruton, B.D., Conway, W.S., Gross, K.C., Zhang, J.X., Biles, C.L., Sams, C.E., 1998. Polygalacturonase of a latent and wound postharvest fungal pathogen of muskmelons fruit. Postharvest Biol. Technol. 13, 2015–2214.
Caccioni, D.R.L., Tonini, G., Guizzardi, M., 1995. Antifungal activity of stone fruit aroma compounds against *Monilinia laxa* (Aderh. et Ruhl.) Honey and *Rhizopus stolonifer* (Ehrenb.): *In vivo* trials. J. Plant Dis. Prot. 102, 518–525.
Camele, I., De Feo, V., Altieri, L., Mancini, E., De Martino, L., Rana, G.L., 2010. An attempt of postharvest orange fruit rot control using essential oils from Mediterranean plants. J. Med. Food 13, 1515–1523.
Ceponis, M.J., Butterfield, J.E., 1979. Losses in fresh tomatoes at the retail and consumer levels in the Greater New York Area. J. Am. Soc. Horticultural Sci. 104, 751–754.
Chalutz, E., Droby, S., Cohen, L., et al., 1991. Biological control of Botrytis, Rhizopus and Alternaria rots of tomato fruit by *Pichia guilliermondii*. In: Wilson, C., Chalutz, E. (Eds.), Biological Control of Postharvest Diseases of Fruits and Vegetables. Workshop Proceedings. Shepherdstown, West Virginia, USDA, pp. 71–85.
Chuku, E.C., Osakwe, J.A., Daddy-West, C., 2010. Fungal spoilage of tomato (*Lycopersicon esculentum* Mill), using garlic and ginger. Sci. Afri. 9, 42–50.
Clark, C.A., Hoy, M.W., 1994. Identification of resistance in sweetpotato to Rhizopus soft rot using two inoculation methods. Plant Dis. 78, 1078–1082.
De Carvalho, V.L., Da Cunha, R.L., Jorge, C.N.J., Abreu, M.P.H., 2009. Alternatives e controle pós-colheita da podridão-parda e da podridão-mole em frutos de pessegueiro. Rev. Bras. Fruticola Jaboticabal 3, 78–83.
Dos Santos, N.S.T., Athayde Aguiar, A.J.A., Vasconcelos de Oliveira, C.E., et al., 2012. Efficacy of the application of a coating composed of chitosan an *Origanum vulgare* L. essential oil to control *Rhizopus stolonifer* and *Aspergillus niger* in grapes (*Vitis labrusca* L.). Food Microbiol. 32, 345–353.
Droby, S., Wisniewsky, M., El Ghaouth, A., Wilson, C., 2003. Influence of food additives on the control of postharvest rots of apple and peach and efficacy of the yeast-based biocontrol Aspire. Postharvest Biol. Technol. 27, 127–135.
Edmunds, B.A., Holmes, G.J., 2009. Evaluation of alternative decay control products for control of postharvest Rhizopus soft rot of sweetpotatotes. Plant Manag. Netw. February, 1–10.
El Ghaouth, A., Arul, J., Grenier, J., Asselin, A., 1992. Antifungal activity of chitosan on two post-harvest pathogens of strawberry fruits. Phytopathology 82, 398–402.

El-Katatny, M.H., Emam, A.S., 2012. Control of postharvest tomato rot by spore suspension and antifungal metabolites of *Trichoderma harzianum*. J. Microbiol. Biotechnol. Food Sci. 1, 1505–1528.
Fajola, O., 1979. The postharvest fruit rots of tomato (*Lycopersicum esculentum*) in Nigeria. Nahrung 23, 105–109.
Feliziani, E., Santini, M., Landi, L., Romanazzi, G., 2013. Pre- and postharvest treatment with alternatives to synthetic fungicides to control postharvest decay of sweet cherry. Postharvest Biol. Technol. 78, 133–138.
Ferreira, M.D., Bartz, J.A., Sargent, S.A., Brecht, J.K., 1996. An assessment of the decay hazard associated with hydrocooling strawberries. Plant Dis. 80, 117–1122.
Föster, H., Driever, G.F., Thompson, D.C., Adaskaveg, J.E., 2007. Postharvest decay management for stone fruit crops in California using the 'Reduced. Risk' fungicides fluodoxonil and fenhexamid. Plant Dis. 91, 209–215.
Gonçalves, F.P., Martins, M.C., Silva Junior, G.J., Lourenço, S.A., Amorim, L., 2010. Postharvest control of brown rot and Rhizopus rot in plums and nectarines using carnauba wax. Postharvest Biol. Technol. 58, 211–217.
Harris, J., Dennis, C., 1980. Distribution of *Mucor piriformis, Rhizopus sexualis* and *R. stolonifer* in relation to their spoilage disease. Transactional British Mycolo. Soc. 75, 445–450.
Harveson, R., 2012. Managing rhizopus head rot in sunflower. University of Nebraska-Lincoln. Nebraska Crop Production & Pest Management Information. CropWatch July 1–4.
Hashmi, M.S., East, A.R., Palmer, J.S., Heyes, J.A., 2013a. Pre-storage hypobaric treatments delay fungal decay of strawberries. Postharvest Biol. Technol. 77, 75–79.
Hashmi, M.S., East, A.R., Palmer, J.S., Heyes, J.A., 2013b. Hypobaric treatment stimulates defence-related enzymes in strawberry. Postharvest Biol. Technol. 85, 77–82.
Hernández-Lauzardo, A.N., Bautista-Baños, S., Velázquez-del Valle, M.G., Trejo-Espino, J.L., 2005. Identification of *Rhizopus stolonifer* (Ehrenb.: Fr.) Vuill. causal agent of Rhizopus rot disease of fruits and vegetables. Mexican J. Phytopathol. 24, 65–69.
Hernández-Lauzardo, A.N., Velázquez-Del Valle, M.G., Veranza-Castelán, L., Melo-Giorgana, G.E., Guerra-Sánchez, M.G., 2010. Effect of chitosan on three isolates of *Rhizopus stolonifer* obtained from peach, papaya and tomato. Fruits 65, 245–253.
Holmes, G.J., Stange, R.R., 2002. Influence of wound type and storage duration on susceptibility of sweetpotatoes to Rhizopus soft rot. Plant Dis. 86, 345–348.
Hossain, M.T., Hossain, S.M.M., Bakr, M.A., Matiar Rahman, A.K.M., Uddin, S.N., 2010. Survey on major diseases of vegetables and fruit crops in Chittagong region. Bangladesh J. Agricul. Res. 35, 423–429.
Jacobsen, B.J., 2006. Root rot diseases of sugar beet. Matika SRPSKA Proc. Nat. Sci. 110, 9–19.
Karagiannidis, K., Thomidis, T., 2008. Effect of boron products on development of Rhizopus rots of peaches. Analele Univ. din Oradea Fascicula Proteactia Mediului XIII, 197–199.
Kwon, J.-H., Kang, S.-W., Kim, J.-S., Park, Ch-S., 2001. Rhizopus soft rot on cherry tomato caused by *Rhizopus stolonifer* in Korea. Korean Soc. Mycol. 29, 176–178.
Kwon, J.-H., Lee, Ch-J., 2006. Rhizopus soft rot on pear (*Pyrus serotina*) caused by *Rhizopus stolonifer* in Korea. Mycobiology 34, 151–153.
Kwon, J.-H., Ryu, J.-S., Phuong Chi, T.T., Shen, S.-S., Choi, O., 2012. Soft rot of *Rhizopus oryzae* as a postharvest pathogen of banana fruit in Korea. Mycobiology 40, 214–216.
Lattanzio, V., Di Venere, D., Linsalata, V., Lima, G., Ippolito, A., Salerno, M., 1996. Antifungal activity of 2,5-dimethoxybenzoic acid on postharvest pathogens of strawberry fruits. Postharvest Biol. Technol. 9, 325–334.
Latorre, B.A., Viertel, S.C., Spadaro, I., 2002. Severe outbreaks of bunch rots caused by *Rhizopus stolonifer* and *Aspergillus niger* on table grapes in Chile. Plant Dis. 86, 815.

Lisker, N., Keren-Shacham, Z., Sarig, P., Zutkhi, Y., Ben-Aire, R., 1996. The biology and pathology of the fungus *Rhizopus stolonifer*, cause of black mould disease of table grapes in Israel. Plant Pathol. 45, 1099–1109.

Mahovic, M., Steven, S.A., Bartz, J.A., 2006. Identifying and controlling postharvest tomato diseases in Florida. Series Horticultural Sci. Department, Florida Cooperative Extension Service, Institute of Florida. website http://edis.ifas.ufl.edu/HS131.

Marchi, de A.F., Afonseca, L.S., Basetto, E., Packer, G.F., Cagnim, M.M., Amorim, L., 2008. Efeito de sanificantes no controle pós-colheita da podridão parda (*Monilinia fructicola*) e da podridão mole (*Rhizopus stolonifer*) em pêssegos. Summa Phytopathol. 34, 83–85.

Margosan, D.A., Smilanick, J.L., Simmons, G.F., 1994. Hot ethanol for the postharvest control of gray mold and black rot of strawberries. Biol. Cultural Tests 10, 60.

Margosan, D.A., Smilanick, J.L., Simmons, G.F., Henson, D.J., 1997. Combination of hot water and ethanol to control postharvest decay of peaches and nectarines. Plant Dis. 81, 1405–1409.

Mari, M., Iori, R., Leoni, O., Marchi, A., 1996. Bioassays of glucosinolate-derived isothiocyanates against postharvest pear pathogens. Plant Pathol. 45, 753–760.

Mari, M., Gregori, R., Donate, I., 2004. Postharvest control of *Monilinia laxa* and *Rhizopus stolonifer* in stone fruit by peracetic acid. Postharvest Biol. Technol. 33, 319–325.

Mass, J.L., 1998. Compendium of Strawberry Diseases. The American Phytopathological Society. APS Press, St Paul, Minnesota.

Matan, N., 2010. Postharvest control of Rhizopus stolonifer on tomato by citronella oil. Proceedings of the 8th International Symposium of Biocontrol and Biotechnology, Pattaya, Thailand. Oct. 4-6 pp. 197–202.

Mathew, S., 2010. The prevalence of fungi on the postharvested guava (*Psidium guava* L.) in Askum. Int. J. Pharmaceutical Sci. Res. 1, 145–149.

McLaughlin, R.J., Wilson, C.L., Droby, S., Ben-Aire, R., Chaluts, E., 1992. Biological control of postharvest diseases of grape, peach and, apple with the yeast *Kloeckera apiculata* and *Candida guilliermondii*. Plant Dis. 76, 470–473.

McMillan Jr, R.T., 1986. Serious diseases of tropical fruits in Florida. Proc. Florida State Horticultural Sci. 99, 224–227.

Mena-Nevarez, G., Valencia-del Toro, G., Piña-Guzmán, A.B., Villanueva-Arce, R., Durán-Páramo, E., Robles-Martínez, F., 2012. Degradation capacity of fungi (*Colletotrichum* sp., *Penicillium* sp. and *Rhizopus* sp.) on mangoes and oranges. Afri. J. Agricultural Res. 21, 4564–4574.

Meon, S., Nik, W.Z.W., 1988. Seed-borne infection and development of *Colletotrichum capsici* in naturally infected chilli seed. Pertanika 11, 341–344.

Miles, W.G., Shaker, F.M., Nielson, A.K., Ames, R.R., 1977. A laboratory study on the ability of fungicides to control beet rotting fungi. J. ASSBT 19, 288–293.

Nabigol, A., Morshedi, H., 2011. Evaluation of the antifungal activity of the Iranian thyme essential oils on the postharvest pathogens of strawberry fruits. Afri. J. Biotechnol. 10, 9864–9869.

Nahunnaro, H., 2008. Effects of different extracts in the control of yam rot induced by *Rhizopus stolonifer* on stored yam (*Dioscorea* sp). in Yola, Adamawa state of Nigeria. Agricultural J. 3, 382–387.

Narciso, J.A., Baldwin, E.A., Plotto, A., Ference, C.M., 2007. Postharvest peroxyacetic acid sprays slow decay and extend shelf life of strawberries. HortScience 42, 617–621.

Navarro, D., Díaz-Mua, H.M., Guillén, F., et al., 2011. Reduction of nectarine decay caused by *Rhizopus stolonifer, Botrytis cinerea* and *Penicillium digitatum* with *Aloe vera* gel alone or with the addition of thymol. Int. J. Food Microbiol. 151, 241–246.

Nelson, S., 2005. Rhizopus rot of jackfruit. Cooperative Extension Service. College of Tropical Agriculture and Human Resources. University of Hawaii. Plant Disease PD-29.

Nguyen The, C., Hugueney, C., Chamel, A., Arnoux, M., 1989. Penetration of *Monilinia laxa* and *Rhizopus stolonifer* through nectarine fruit epidermis. Acta Horticulturae 254, 193–198.

Nicoué, E.E., Émond, J.-P., Vuillermard, J.-C., do Nascimento, N.C., 2004. Destruction de *Rhizopus stolonifer* et de *Botrytis cinerea* par des traitments ozone/ions. Phytoprotection 85, 81–87.

Northover, J., Zhou, T., 2002. Control of Rhizopus rot of peaches with postharvest treatments of tebuconazle, fluodoxonil, and *Pseudomona syringae*. Can. J. Plant Pathol. 24, 144–153.

Nunes, C., Usall, J., Teixedó, N., Viñas, I., 2001a. Biological control of postharvest diseases using bacterium *Pantoea agglomerans* CPA-2. Int. J. Food Microbiol. 70, 52–61.

Nunes, C., Usall, J., Teixidó, N., Ochoa de, E.X., Viñas, I., 2001b. Control of post-harvest decay of apples by pre-harvest and post-harvest application of ammonium molybdate. Pest Manage. Sci. 57, 1093–1099.

Nunes, C., Usall, J., Teixidó, N., Fons, E., Viñas, I., 2002. Post-harvest biological control by *Pantoea agglomerans* (CPA) on Golden Delicious apples. J. Appl. Microbiol. 92, 247–255.

Ogawa, J.M., Zehr, E.I., Bird, G.W., Ritchie, D.F., Uriu, K., Uyemoto, J.K., 2000. Plagas y enfermedades de los frutales de hueso. The American Phytopathological Society. Ediciones Mundi-Prensa, España.

Okigbo, R.N., Ogbonaya, U.O., 2006. Antifungal effects of two tropical plant leaf extracts (*Ocimum gratissimum* and *Aframomum melegueta*) on postharvest yam (*Dioscorea* spp.) rot. Afri. J. Biotechnol. 5, 727–731.

Ooka, J.J., 1980. Guava fruit rot caused by *Rhizopus stolonifer* in Hawaii. Plant Dis. 64, 412–413.

Pallou, L., Criososto, C.H., Smilanick, J.L., 2012. Evaluation of food additives and low toxicity compounds as nonpolluting means to control the main postharvest diseases of California peaches. Acta Horticulturae 962, 539–547.

Pan, J., Vicente, A.R., Martínez, G.A., Chaves, A.R., Civello, P.M., 2004. Combined use of UV-C irradiation and heat treatment to improve postharvest life of strawberry fruit. J. Sci. Food Agri. 84, 1831–1838.

Panahirad, S., Zaare-Nahandi, F., Safaralizadeh, R., Alizadeh-Salteh, S., 2012. Postharvest control of *Rhizopus stolonifer* in peach (*Prunus persica* L. Batsch) fruits using salicylic acid. J. Food Saf. 32, 502–507.

Park, S., Stan, S.D., Daeschel, M.A., Zhao, Y., 2005. Antifungal coatings on fresh strawberries (*Fragaria* × *ananassa*) to control mold growth during cold storage. J. Food Sci. 70, 202–207.

Perera, G.S., Pérez, H.E., Lobo, R.G., et al., 2010. Evaluación de la eficacia de fungicidas naturales y químicos para el control de enfermedades postcosecha sobre distintas variedades de papaya. Cabildo Insular de Tenerife. Servicio Técnico de Agricultura y Desarrollo Rural, 25.

Pierson, C.F., 1966. Effect of temperature on the growth of *Rhizopus stolonifer* on peaches and on agar. Phytopathology 56, 276–278.

Plotto, A., Roberts, R.G., Roberts, D.D., 2003. Evaluation of plant essential oils as natural postharvest disease control of tomato (*Lycopersicon esculentum*). Acta Horticulturae 628, 737–745.

Pratella, G.C., Mari, M., Guizzardi, M., Folchi, A., 1993. Preliminary studies on the efficacy of endophytes in the biological control of the postharvest pathogens *Monilinia laxa* and *Rhizopus stolonifer* in stone fruit. Postharvest Biol. Technol. 3, 361–368.

Qin, G., Tian, S., Xu, Y., 2004. Biocontrol of postharvest diseases on sweet cherries by four antagonistic yeasts in different storage conditions. Postharvest Biol. Technol. 31, 51–58.

Qing, F., Shiping, T., 2000. Postharvest biological control of Rhizopus rot of nectarine fruits by *Pichia membranefaciens*. Plant Dis. 84, 212–1216.

Rathod, G.M., 2010. Survey of post-harvest fungal diseases of some fruits from Marathwada regions of Maharashtram India. J. Ecobiotechnol. 2/6, 7–10.

Reddy, B.M.V., Angers, P., Gosseli, A., Arul, J., 1998. Chracterization and use of essential oil from *Thymus vulgaris* against *Botrytis cinerea* and *Rhizopus stolonifer* in strawberry fruits. Phytochemistry 47, 1515–1520.

Romanazzi, G., Nigro, F., Ippolito, A., Salerno, M., 2001. Effect of short hypobaric treatments on postharvest rots of sweet cherries, strawberries and tablegrapes. Postharvest Biol. Technol. 22, 1–6.
Romanazzi, G., Feliziani, E., Santini, M., Landi, L., 2013. Effectiveness of postharvest treatment with chitosan and other resistance inducers in the control of storage decay of strawberry. Postharvest Biol. Technol. 75, 24–27.
Saini, S., Sattra, A., Alam, M., 1996. Occurrence of stem rot disease caused by *Rhizopus stolonifer* on periwinkle. Indian J. Plant Pathol. 14, 1–4.
Sallato, B.V., Torres, R., Zoffoli, J.P., Latorre, B.A., 2007. Effect of boscalid on postharvest decay of strawberry caused by *Botrytis cinerea* and *Rhizopus stolonifer*. Spanish J. Agri. Res. 5, 67–78.
Sarig, P., Zutkhi, Y., Monjauze, A., Lisker, N., Ben-Aire, R., 1996. Phytoalexin elicitation in grae berries and their susceptibility to *Rhizopus stolonifer*. Physiological Mol. Plant Pathol. 50, 337–347.
Serey, R.A., Torres, R., Latorre, B.A., 2007. Pre- and post-infection activity of new fungicides against *Botrytis cinerea* and other fungi causing decay of table grapes. Ciencia e Investigación Agraria 34, 215–224.
Shao, X., Wang, H., Xu, F., Cheng, S., 2013. Effects and possible mechanisms of tea tree oil vapor treatment on the main diseases in postharvest strawberry fruit. Postharvest Biol. Technol. 77, 94–101.
Shehu, K., Muhammad, S., 2011. Fungi associated with storage rots of onion bulbs in Sokoto. Nigeria. Int. J. Modern Bot. 1, 1–3.
Shena, L., Ippolito, A., Zahavi, T., Cohen, L., Droby, S., 2000. Molecular approaches to assist the screening and monitoring of postharvest biocontrol yeast. Eur. J. Plant Pathol. 106, 681–691.
Shipper, M.A.A., 1984. A revision of the genus Rhizopus. I. The *Rhizopus stolonifer*-group and *Rhizopus oryzae*. Studies Mycol. 25, 1–34.
Sholberg, P.L., 1998. Fumigation of fruit with short-chain organic acids to reduce the potential of postharvest decay. Plant Dis. 82, 689–693.
Sholberg, P., 2009. Control of postharvest decay by fumigation with acetic acid or plant volatile compounds. Fresh produce 3 (Special Issue 1), 80–86. Global Science Books.
Sholberg, P.L., Gaunce, A.P., 1996. Fumigation of stone fruit with acetic acid to control postharvest decay. Crop Prot. 15, 681–686.
Shukla, R.S., Alam, M., Sattar, A., Abdul-Khaliq., Singh, H.N., 2006. First report of *Rhizopus stolonifer* causing inflorescence and fruit rot of *Rauvolfia serpentine* in India. EPPO Bulletin 36, 11–13.
Singh, P., Mishra, A.K., Tripathi, N.N., 2012. Assessment of mycoflora associated with postharvest losses of papaya fruits. J. Agricultural Technol. 8, 961–968.
Snowdon, L.A., 1991. A Colour Atlas of Post-Harvest Diseases & Disorders of Fruits and Vegetables, vol. 2, Vegetables. Wolfe Scientific Ltd, Aylesbury, UK.
Stange, Jr. R.R., Midland, S.L., Holmes, G.J., Sims, J.J., Mayer, T.R., 2001. Constituents from the periderm and outer cortex of *Ipomea batatas* with antifungal activity against *Rhizopus stolonifer*. Postharvest Biology and Technology 23, 85–92.
Stevens, C., Khan, V.A., Lu, J.Y., et al., 1997. Integration of ultraviolet (UV-C) light with yeast treatment for control of postharvest storage rots of fruits and vegetables. Biol. Control 10, 98–103.
Stevens, C., Liu, J., Khan, V.A., et al., 1998. Application of hermetic UV-C for delayed ripening and reduction of *Rhizopus* soft rot in tomatoes: The effect of tomatine on storage rot development. J. Phytopathol. 146, 211–221.
Stevens, C., Liu, J., Khan, V.A., et al., 2004. The effects of low-dose ultraviolet light-C treatment on polygalacturonase activity, delay ripening and Rhizopus soft rot development of tomatoes. Postharvest Biol. Technolol. 23, 551–554.

Swingle, D.B., 1903. Formation of the spores in the sporangia of *Rhizopus nigricans* and of *Phycomices nitens*. US Department of Agriculture. Bureau of Plant Industry-Bulletin No. 37, Washington.

Syngenta, 2004. Postharvest disease protection for pome fruit. Syngenta Crop Protection.

Taiga, A., 2009. Efficacy of selected plant extracts in the control of fungal dry rot of white yam (*Dioscorea rotundata*) tubers in Kogi state. Am. Eurasian J. Sustainable Agriculture 3, 310–313.

Takeda, F., Janisiewskicz, W.J., Roitman, J., Mahoney, N., Abeles, F.B., 1990. Pyrrolnitrin delays postharvest fruit rot in strawberries. HortScience 25, 20–322.

Teng-fei, M., Bo, Y., Yue, Y., et al., 2009. Market diseases pathogens detection of imported fruits in Shangai. Agricultural Sci. China 8, 1087–1096.

Teviotdale, B.L., Michailides, T.J., Goldhamer, D.A., Viveros, M., 1995. Reduction of almond hull rot disease caused by *Rhizopus stolonifer* by early termination of preharvest irrigation. Plant Dis. 79, 402–405.

Teviotdale, B.L., Goldhamer, D.A., Viveros, M., 2001. Effects of deficit irrigation on hull rot disease of almond trees caused by *Monilinia fructicola* and *Rhizopus stolonifer*. Plant Dis. 85, 399–403.

Thind, T.S., Jhooty, J.S., 1985. Relative prevalence of fungal disease on chilli fruit in Punjab. J. Mycol. Plant Pathol. 15, 305–307.

Tian, S.P., Fan, Q., Xu, Y., Jiang, L.A., 2002. Effects of calcium on biocontrol activity of yeast antagonist against the postharvest fungal pathogen *Rhizopus stolonifer*. Plant Pathol. 51, 352–358.

Vicente, L.M.L., 2004. Estudio de las pectin metalmetil esterasas fúngicas causantes de la pudrición blanda en el fruto de jitomate (Lycopersicon esculentum M.). Thesis. Universidad Veracruzana, Veracruz, Mexico.

Vicente, A.R., Civello, P.M., Martínez, G.A., Powell, A.L.T., Labavitch, J.M., Chaves, A.R., 2005. Control of postharvest spoilage in soft fruit. Stewart Postharvest 1, 1–11.

Villanueva, C.B., 2004. Determinación de la actividad enzimática de PG en el proceso de maceración causado por *Rhizopus stolonifer* en jitomate (*Lycopersicon esculentum* M.). Thesis. Universidad Veracruzana, Veracruz, Mexico.

Vigneault, C., Bartz, J.A., Sargent, S.A., 2000. Postharvest decay risk associated with hydrocooling tomatoes. Plant Dis. 84, 1314–1318.

Viñas, I., Usall, J., Teixidó, N., Sanchis, V., 1998. Biological control of major postharvest pathogens on apple with *Candida sake*. Int. J. Food Microbiol. 40, 9–16.

Wang, X., Wang, J., Jin, P., Zheng, Y., 2013a. Investigating the efficacy of *Bacillu subtilis* SM21 on controlling Rhizopus rot in peach fruit. Int. J. Food Microbiol. 164, 141–147.

Wang, X., Xu, F., Wang, J., Jin, P., Zheng, Y., 2013b. *Bacillus cereus* AR156 induces resistance against Rhizopus rot through priming of defense in peach fruit. Food Chem. 136, 400–406.

Washington, W.S., Shanmuganathan, N., Forbes, C., 1992. Fungicide control of strawberry fruit rots, and the field occurrence of resistance of *Botrytis cinerea* to iprodione, benomyl and dichlofluanid. Crop protection 11, 355–360.

Welch, N.C., Paulus, A.O., Scheuerman, R.W., 1966. Control of Rhizopus soft rot. California Agriculture November, 14–15.

Wells, J.M., 1972. Heated wax-emulsions with benomyl and 2, 6.dichloro-4-nitroniline for control of postharvest decay of peaches and nectarines. Phytopathology 62, 129–133.

Wilson, C.L., Franklin, J.D., Pusey, P.L., 1987. Biological control of Rhizopus rot of peach with *Enterobacter cloacae*. Phytopathology 77, 303–305.

Yildirim, I., Turhan, H., Osgen, B., 2010. The effects of head rot disease (*Rhizopus stolonifer*) on sunflower genotypes at two different growing stages. Turkish J. Field Crops 15, 94–98.

Yoshida, S., Tsuyumu, S., Tsukiboshi, T., 2003. Macerating enzymes produced by *Rhizopus oryzae* in infected mulberry roots. J. Phytopathol. 151, 436–441.

Zacharia, R.M., Philip, S., 2010. Biological control of postharvest diseases of brinjal. Academic Review 1& 2. 101–112.
Zahavi, T., Cohen, L., Weiss, B., et al., 2000. Biological control of *Botrytis, Aspergillus* and *Rhizopus* rots on table and wine grapes in Israel. Postharvest Biol. Technol. 20, 115–124.
Zhang, D., Quantick, P.C., 1998. Antifungal effects of chitosan coating on fresh strawberries and raspberries during storage. J. Horticultural Sci. Biotechnol. 73, 763–767.
Zhang, H., Fu, Ch., Zheng, X., Xi, Y., Jiang, W., Wang, Y., 2004. Control of postharvest *Rhizopus* rot of peach by microwave treatment and yeast antagonist. Eur. Food Res. Technol. 218, 568–572.
Zhang, H., Zheng, X., Yu, T., 2007a. Biological control of postharvest diseases of peach with *Cryptoccoccus laurentii*. Food Control 18, 287–291.
Zhang, H., Wang, L., Zheng, X., Dong, Y., 2007b. Effects of yeast antagonist in combination with heat treatment on postharvest mold decay and Rhizopus decay of peaches. Int. J. Food Microbiol. 115, 53–58.
Zhang, H., Zheng, X., Wang, L., Li, S., Liu, R., 2007c. Effect of yeast antagonist in combination with hot water dips on postharvest Rhizopus rot of strawberries. J. Food Eng. 78, 281–287.
Zhang, H., Ma, L., Turner, M., et al., 2010. Salycilic acid enhances biocontrol efficacy of *Rhodotorula glutinis* against postharvest *Rhizopus* rot of strawberries and the possible mechanisms involved. Food Chem. 122, 577–583.
Zhao, Y., Tu, K., Shao, X.F., Jing, W., Yang, L., Su, Z.P., 2008. Biological control of the postharvest pathogens *Alteranria solani, Rhizopus stolonifer*, and *Botrytis cinerea* on tomato fruit by *Pichia gulliermondii*. J. Horticultural Sci. Biotechnol. 83, 132–136.
Zhao, Y., Tu, K., Liu, M., Su, J., Hou, Y.-P., 2010. A combination of heat treatment and *Pichia guilliermondii* prevents cherry tomato spoilage by fungi. Int. J. Food Microbiol. 137, 106–110.
Zhou, X., Lu, Z., Zhao, H., Wang, Y., Bie, X., 2011. Antagonistic action of *Bacillus subtilis* strain fmbj on the postharvest pathogen *Rhizopus stolonifer*. J. Food Sci. 76, 254–259.
Ziedan, El-S.H.E., Farrag, E.S.H., 2008. Fumigation of peach fruits with essential oils to control postharvest decay. Res. J. Agriculture Biol. Sci. 4, 512–519.
Zitter, T.A., Hopkins, D.L., Thomas, C.E., 1998. Compendium of Cucurbit Diseases. The American Phytopathological Society. APS Press, St Paul, Minnesota.

CHAPTER 2

Penicillium digitatum, *Penicillium italicum* (Green Mold, Blue Mold)

Lluís Palou
Laboratori de Patologia, Centre de Tecnologia Postcollita (CTP), Institut Valencià d'Investigacions Agràries (IVIA), Montcada, València, Spain

Contents

Postharvest Decay
http://dx.doi.org/10.1016/B978-0-12-411552-1.00002-8

OVERALL IMPORTANCE OF FUNGI

Importance in Citrus Fruits

Citrus spp. (Rutaceae) are the most widely produced fruits for human consumption and they are grown in over one hundred countries. The total production of citrus fruits has been increasing over the last decades and exceeded 115 million tons in 2011. Oranges (*Citrus sinensis* L.), with a world production of 71 million tons in 2011; mandarins or tangerines (*Citrus reticulata* Blanco), including clementines (*Citrus clementina* hort. ex Tanaka), satsumas (*Citrus unshiu* Marcow.), and a variety of hybrid mandarins (26 million tons); lemons [*Citrus limon* (L.) Burm. f.] and limes [*Citrus aurantiifolia* (Christm.) Swingle] (13 million tons); and grapefruits (*Citrus paradisi* Macfad.) (6 million tons), are the largest cultivated citrus species and cultivars. In 2011, the most important producing countries were China, Brazil, the USA, India, Mexico, Spain, Egypt, Italy, Turkey, Argentina, Iran, Pakistan, Indonesia, South Africa, and Morocco, with 22.9, 22.7, 10.4, 8.2, 6.7, 6.6, 3.6, 3.2, 3.1, 2.5, 2.3, 2.2, 2.1, 1.9, and 1.7 million tons of total citrus, respectively. In terms of international trade, citrus are the highest value fruit crop and Spain is the leading country, with 3.6 million tons of exports of fresh produce in 2011 (FAO, 2012).

Postharvest handling in citrus packing houses is intended to commercialize fruit of maximum quality, increase their postharvest life, and reduce produce losses. Among postharvest losses, those of pathological origin are typically of considerable economical importance. Green and blue molds, caused by *Penicillium digitatum* (Pers.: Fr.) Sacc. and *P. italicum* Wehmer, respectively, are the most economically important postharvest diseases of citrus in all production areas that, like Spain or California, are characterized by a Mediterranean-type climate with low summer rainfall (Eckert and Eaks, 1989). Both pathogens have been isolated not only from common commercial citrus species and cultivars, but also from other citrus-related genera such as *Fortunella*, *Poncirus*, or *Citrofortunella* (Farr and Rossman, 2013). The symptoms of both diseases in the orange cv. Valencia are shown in Figure 2.1A. The geographical distribution of both species includes all citrus-producing areas in the world and they have been also described in citrus fruits imported to many countries (Frisvad and

Samson, 2004). Both *P. digitatum* and *P. italicum* are strict wound pathogens that only infect the fruit through peel injuries produced in the field, the packing house, or during the fruit commercialization chain. A third *Penicillium* species that has been described as pathogenic in citrus fruits, but with much less economical importance, is *P. ulaiense* (Hsieh et al., 1987; MycoBank #126489). It was found in California in 1987 causing a citrus postharvest disease that was characterized by Holmes et al. (1994) and named 'whisker mold'. Later, it was also comprehensively described by Frisvad and Samson (2004). The fungus resembles *P. italicum*, especially in colony color, but it grows more slowly and shows paler reverse colors on all media. It can be found in packing houses in mixed infections with *P. digitatum* in stored citrus fruits and it is typically more resistant to postharvest fungicides such as imazalil (IMZ) or thiabendazole (TBZ) than *P. italicum*. It is considerably less aggressive than *P. digitatum* and *P. italicum*, which decay citrus fruit about three to five times faster. Conidia are also less efficient in causing infection than those of *P. digitatum* and *P. italicum*. In contrast to these species, *P. ulaiense* has never been collected in citrus groves in California (Holmes et al., 1994).

Actual losses due to penicillium decay are variable and depend upon climate and orchard factors, citrus cultivar, the extent of physical injury to the fruit during harvest and subsequent handling, the effectiveness of antifungal treatments, and the postharvest environment (Smilanick et al., 2006a). In Spain, a study by Tuset (1988) estimated that fruit rots caused by *Penicillium* spp. accounted for 55–80% of total postharvest decay observed in oranges and mandarins during the entire commercialization season, and for 30–55% of decay observed in storage rooms in citrus packing houses. It was found in inspections in New York of citrus from California and Florida that green and blue molds were present in 30% of the inspected shipments (Ceponis et al., 1986). Early work by Pelser (1977) showed that penicillium molds accounted for about 75% of total decay present in South African 'Valencia' oranges shipped to London; they accounted for more than 50% on lemons and grapefruits. Green mold typically causes larger losses during commercialization because it is predominant at ambient temperatures, but blue mold becomes more important when citrus fruit are cold-stored for long periods because *P. italicum* grows faster than *P. digitatum* below 10°C (Plaza et al., 2003a).

In general, the incidence of postharvest decay is higher in production areas with abundant rainfall, such as Brazil, Florida or southeastern Asia. Wound pathogens such as *Penicillium* spp. are very important in all areas

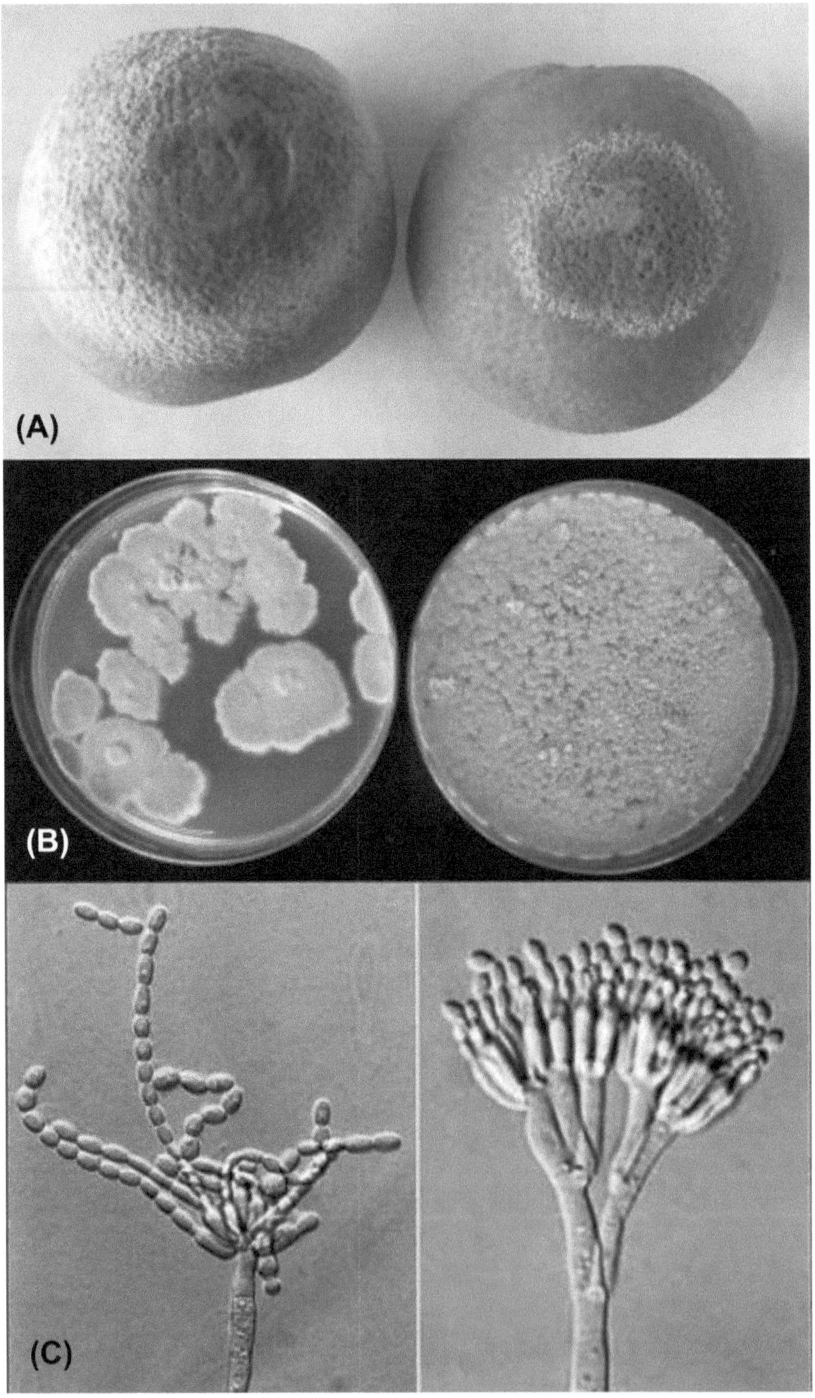
(A)
(B)
(C)

because they reproduce very rapidly and their spores are ubiquitous in the atmosphere and on fruit surfaces and are readily disseminated by air currents. Therefore, the source of fungal inoculum in citrus orchards and packing houses is practically continuous during the season and the fruit can become contaminated and infected in the grove, the packing house, and during distribution and marketing. In packing house facilities and storage rooms, the inoculum may build up to high levels if appropriate sanitation measures are not adopted. In addition, citrus fruit in containers can become soiled with conidia of *Penicillium* spp. that are loosened in handling of diseased fruit. The impact of postharvest decay is not only restricted to fruit losses. When rotten fruit is found in export shipments, even though the incidence may be relatively low, wholesale buyers typically reject the load and charge the producer for the transport and handling costs. Furthermore, they can abandon the affected producer and seek other sources in the market (Smilanick et al., 2006a).

Other Hosts

According to the USDA–ARS Systematic Botany and Mycology Laboratory Fungal Databases (Farr and Rossman, 2013), both *P. digitatum* and *P. italicum* have been reported in soil debris in citrus orchards (Domsch et al., 1980) and in plant hosts other than citrus fruits. Other hosts reported for *P. digitatum* include papaya, beet, corn kernels, Arabian coffee, melon, abisin, iris, tomato, apple, goldenberry (*Physalis* spp.), tamarind, and persimmon (Farr and Rossman, 2013). In addition, *P. digitatum* has been occasionally isolated from other food sources such as hazelnuts, pistachio, kola nuts, rice, peanuts, soybeans, sorghum, and even meats (Pitt and Hocking, 2009). In the case of *P. italicum*, other hosts listed in the USDA–ARS Fungal Database include avocado, mango, sweet potato, persimmon, melon, tomato, *Prunus* spp., *Pyrus* spp., wheat, and grape (Farr and Rossman, 2013). However, no economically important plant diseases caused by *P. digitatum* and *P. italicum*

Figure 2.1 (A) Symptoms of green mold caused by *Penicillium digitatum* (left) and blue mold caused by *Penicillium italicum* (right) on oranges cv. Valencia artificially inoculated and incubated at 20°C for 7 days. (B) Colonies of *P. digitatum* (left) and *P. italicum* (right) after incubation in potato dextrose agar (PDA) medium at 25°C for 7 days. (C) Conidiophores of *P. digitatum* (left) and *P. italicum* (right) showing branchlet, metule, phialides and conidia (× 1000). *(C source: Mycobank (*www.mycobank.org*; from records #6152, 6158, filed from: Samson, R.A., Hoekstra, E.S., Frisvad, J.C., 2004. Introduction to Food- and Airborne Fungi, 7th edn, Centraalbureau voor Schimmelcultures, Utrecht, The Netherlands.)*

in these hosts have been reported and it can be assumed that citrus fruits may be their only real habitat (Frisvad and Samson, 2004).

TAXONOMY, MORPHOLOGY AND GENOMICS

Penicillium digitatum

Penicillium digitatum, the cause of citrus green mold, was described and classified by Saccardo in 1881 (Saccardo, 1881). The currently accepted scientific name is *P. digitatum* (Pers.: Fr.) Sacc. (MycoBank #169502). According to the MycoBank database, there are three more legitimate taxon names: *P. digitatum* var. *californicum* Thom (1930), *P. digitatum* var. *digitatum*, and *P. digitatum* var. *latum* S. Abe (1956). The known nomenclatural synonyms (obligate or homotypic synonyms) are *Aspergillus digitatus* Pers.:Fr. (1794), which is the basionym or original name, *Monilia digitata* (Pers.: Fr.) Pers. (1801), and *Mucor digitata* (Pers.: Fr.) Mérat (1821). The taxonomic synonyms (facultative or heterotypic synonyms) are *P. olivaceum* Wehmer (1895), *P. olivaceum* Sopp (1912), *P. olivaceum* var. *italicum* Sopp (1912), *P. olivaceum* var. *norvegicum* Sopp (1912), *P. digitatoides* Peyronel (1913), and *P. lanosogrisellum* Biourge (1923). The species is classified in the class Fungi, division Ascomycota, subdivision Pezizomycotina, class Eurotiomycetes, subclass Eurotiomycetidae, order Eurotiales, family Trichocomaceae, and genus *Penicillium*.

Accurate descriptions of *P. digitatum* have been provided by Raper and Thom (1949), Onions (1966a), Frisvad and Samson (2004) and Pitt and Hocking (2009) among others. Briefly, colonies are plane and grow rapidly on malt extract agar (MEA) and potato dextrose agar (PDA; see Fig. 2.1B left), but poorly on Czapek agar and similar synthetic media. The colony obverse is olive green and the reverse colorless to cream yellow or pale dull brown. Colony texture is velutinous with no exudate droplets. The fungus is able to germinate in artificial media at 5°C and, in some cases, can produce colonies of up to 3 mm in diameter. There is no growth at 37°C. The odor can be strong, as volatile metabolites such as limonene, valencene, ethylene, ethyl alcohol, ethyl acetate, or methyl acetate have been detected.

The conidial apparatus is very fragile and tends to break up into many cellular elements. Conidiophores are terverticillate, borne from subsurface or aerial hyphae, irregularly branched and consist of short stipes with few metulae and branches that terminate in whorls of three to six phialides, which are often solitary, cylindrical with a short neck. Conidia are smooth-walled, ellipsoidal to cylindrical, variable in size, but mostly 3.5–8.0 × 3.0–4.0 μm (see Fig. 2.1C left).

Penicillium digitatum is the first phytopathogenic *Penicillium* species whose complete genome has been entirely sequenced (Marcet-Houben et al., 2012). This recent and important result was achieved through the collaboration effort of research teams in Catalonia and Valencia. Two strains of *P. digitatum* differing in their resistance to common chemical fungicides applied after harvest for green mold control were sequenced and it was found that few mutations were responsible for such differences. Earlier, the complete mitochondrial genome of *P. digitatum* was reported for the first time in a phytopathogenic *Penicillium* spp. by a Chinese group (Sun et al., 2011). Comparative analysis by Marcet-Houben et al. (2012) revealed identical mitochondrial genome sequences in Spanish and Chinese strains, suggesting a recent worldwide expansion of the species, probably in parallel to the industrialization of citrus agriculture. Further, a comparison with the closely related but non-phytopathogenic *P. chrysogenum* revealed a much smaller gene content in *P. digitatum*, consistent with a more specialized lifestyle. The analysis was also indicative of heterothallic sexual reproduction and revealed the molecular basis for the inability of *P. digitatum* to assimilate nitrate or produce the metabolites patulin and penicillin. The authors also identified the predicted secretome, which can provide tools for understanding the mechanisms underlying the virulence and host-specificity of the pathogen. The new fungal phylomes *P. digitatum* and *P. chrysogenum* were uploaded to the public database PhylomeDB (www.phylomedb.org). Prior to the complete genome sequence, the overall response of citrus fruit to *P. digitatum* infection was described from a genomic perspective (González-Candelas et al., 2010a, b). Subtracted and regular cDNA libraries were constructed and genes upregulated as a response to infection were identified using a cDNA macroarray generated from the subtracted library. Moreover, a 12k citrus cDNA microarray was used to study transcriptional changes in flavedo and albedo of the peel of citrus fruits whose disease resistance mechanisms had been elicited by an intense postharvest heat treatment. Work by Wang and Li (2008) showed that *Agrobacterium tumefaciens*-mediated transformation could be used as a genetic tool for conducting insertional mutagenesis in *P. digitatum* to study functional genomics. Recently, chitin synthase genes of *P. digitatum* were isolated and characterized by Gandía et al. (2012). Zhang et al. (2013) identified and cloned the gene that regulates the sucrose non-fermenting protein kinase in *P. digitatum*. They observed that disruption of this gene in mutants resulted in impaired conidiation and caused malformation of the conidiophore structures. Further, appearance of disease symptoms on fruit artificially inoculated with mutant strains was significantly delayed.

Penicillium italicum

Penicillium italicum, the cause of citrus blue mold was described by Wehmer in 1894. The currently accepted scientific name is *P. italicum* Wehmer (MycoBank #162660). According to the MycoBank database, there are three more legitimate taxon names: *P. italicum* var. *album* CT Wei (1940), *P. italicum* var. *avellaneum* Samson and Y. Gutter (1976), and *P. italicum* var. *italicum*. This species has no obligate synonyms and the facultative synonyms are *Oidium fasciculatum* Berk. (1836), *P. aeruginosum* Dierckx (1901), *P. digitatum* var. *latum* S. Abe (1956), *P. italicum* var. *album* C.T. Wei (1940), *P. italicum* var. *avellaneum* Samson & Y. Gutter (1976), *P. italicum* var. *italicum*, and *P. ventruosum* Westling (1911). The complete classification of the species is obviously the same as that of *P. digitatum*.

Penicillium italicum has been thoroughly described by Raper and Thom (1949), Onions (1966b), Samson et al. (2004), and Frisvad and Samson (2004) among other authors. At 25°C, colonies grow restrictedly on Czapek agar but more rapidly on MEA and PDA (see Fig. 2.1B right). Colonies are plane, heavy sporing, blue or gray-green colored and often appear granular due to the presence of bundles of conidiophores and conidial heads. The reverse is uncolored or gray to yellow-brown, although it can turn to brownish orange or red brown on media such as Czapek's Dox + yeast extract agar (CYA) or yeast extract sucrose (YES). The texture is velutinous to fasciculate, crustose, with exudates absent or very limited. On CYA, colony diameters after 7 days of incubation at temperatures of 5, 15, 30 and 37°C are 2–4, 17–34, 0–12 and 0 mm, respectively. The odor is caused by volatile metabolites such as ethyl acetate, isopentanol, linalool, isobutanol, 1-octene, ethyl butanoate, 1-nonene, styrene, or citronellene. The conidial apparatus consists of asymmetric penicilli bearing tangled chains of conidia. Conidiophores originate from the substratum or occasionally from superficial hyphae and are terverticillate, hyaline, usually with the branches appressed, with 100–250 × 3.5–5.0 μm stipes and metulae more or less cylindrical, smooth-walled, bearing three to six phialides each. The phialides are slender, cylindrical with short but distinct necks. Conidia are cylindrical at first, but often become elliptical or subglobose, smooth, 4.0–5.0 × 2.5–3.5 μm in size, greenish, smooth-walled (see Fig. 1.C right). Colorless to light brown sclerotia, measuring 200–500 μm, have been occasionally observed in fresh isolates.

BIOLOGY OF *PENICILLIUM DIGITATUM* AND *P. ITALICUM* INFECTION PROCESS

Disease Triangle

A graphical representation of the disease triangle for citrus postharvest green and blue molds is shown in Figure 2.2. This triangle represents the relationships between the pathogen, the fruit host and the environmental conditions that determine the occurrence of disease. Citrus postharvest diseases have been classically classified into two different groups according to the predominant time of infection: preharvest infections, caused generically by the so-called latent pathogens, and postharvest infections, caused generically by the so-called wound pathogens (Eckert and Eaks, 1989). Citrus penicillium molds belong to the second group.

Penicillium digitatum and *P. italicum* cause citrus fruit disease only through the infection of rind wounds. Usually, these wounds are inflicted during harvesting and subsequent handling of fruit in the packing house or during commercialization, but some infections can occur before harvest through injuries, cracks, or wounds made by insects. In this case, fruit infected long before harvest often drops from the tree, but fruit infected less than 3 days before harvest cannot be detected and may be harvested as sound (Eckert and Eaks, 1989). Fungal spores from fruit rotting on the ground in the orchard, in packing house facilities and storage rooms, or in any place during transportation and marketing are massively transported by air currents

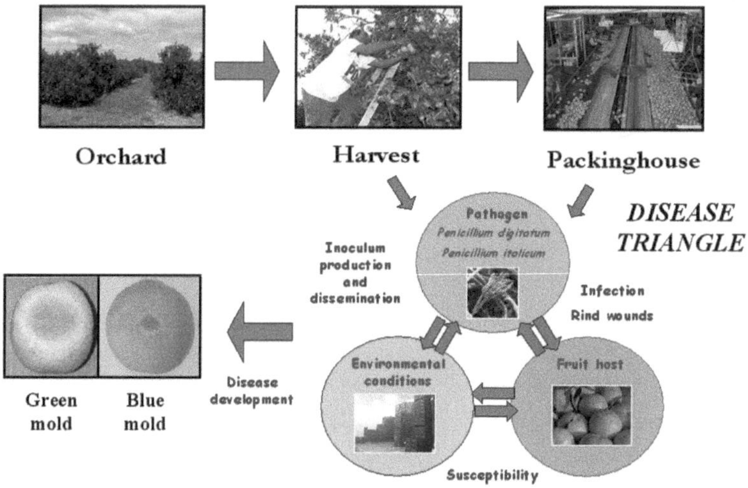

Figure 2.2 Disease triangle for citrus postharvest green and blue molds.

and can easily contaminate the surrounding fruit. No infection occurs if the fruit rind is intact because free conidia located on the peel surface are not able to germinate. In contrast, the conidia situated in injuries that rupture oil glands or penetrate into the albedo of the peel usually cause irreversible infection within 48 h at 20–25°C (Green, 1932; Kavanagh and Wood, 1967).

Symptomatology

Incipient infections of *P. digitatum* and *P. italicum* are usually seen by the naked eye only after about 3 days of incubation at room temperature. A circular area surrounding the infection site (rind wound) appears water-soaked, soft and decolorized, and can be easily penetrated with the finger. As necrotrophic pathogens, both fungi produce hydrolytic enzymes, mostly polygalacturonases and cellulases, which appear responsible for the maceration of the tissue during disease development (Eckert and Eaks, 1989; Barkai-Golan and Karadavid, 1991). As the fungus grows, an aerial white mycelium develops in the center of the lesion and expands radially. Depending on the inoculum load, sporulation begins after 3–5 days at room temperatures (15–28°C) and also expands radially forming a colored layer of velutinous texture. In the case of green mold, after 7–8 days, the central area of the lesion is olive green surrounded by a broad band of dense, non-sporulating white mycelium limited by fairly firm decaying peel. In the case of blue mold, the central sporulating area is blue or bluish-green surrounded by a very narrow band of non-sporulating white mycelium limited by a broad band of soft, water-soaked peel. With time, the entire surface of the fruit is completely covered with spores, the fruit then begins to shrink and, if exposed to air, becomes a hollow mummified shell in the case of green mold and a slimy shapeless mass in the case of blue mold. Although it is not uncommon in packing houses or markets to find symptoms of both diseases in the same fruit, usually green mold overgrows blue mold in mixed infections on fruits kept at room temperature.

Factors Determining Host–Pathogen Interaction

Disease development is mediated by complex interactions between pathogen virulence mechanisms and host defense responses. Extensive research work is being conducted to analyze and understand such interactions at either the biochemical or molecular level.

All citrus commercial species and cultivars have been found to be susceptible to green and blue molds. In this sense, genetic characteristics have lower impact on host susceptibility than, for example, the physical and physiological condition of the rind. Nonetheless, different degrees of susceptibility are commonly observed among different cultivars. According to Eckert and Eaks (1989), general susceptibility to postharvest diseases typically decreases in this order: mandarins, lemons, oranges, and grapefruits. Research is currently ongoing in our laboratory at the IVIA CTP to rank the most representative commercial Spanish citrus cultivars according to their susceptibility to green and blue molds. We believe that such information could be useful in integrated disease management (IDM) programs to discriminate among fruit sets in order to tailor the control strategies to be applied in the packing house. Furthermore, we are seeking potential relationships between susceptibility to disease and rind quality attributes. Strains of *P. digitatum* and *P. italicum* can, of course, vary in aggressiveness, but there are no reports of avirulent non-genetically-modified isolates. Viable infection and disease development are dependent on the amount of conidia of *Penicillium* spp. that reach the infection court (rind wound). The relationship between this amount and the incidence of infection is practically linear, provided that the fruit is highly susceptible (not immature) and the temperature is suitable for infection (20–25°C) (Eckert and Eaks, 1989). In these conditions, 50 and 500 spores of *P. digitatum* inoculated into rind wounds on oranges led to 10 and 65% of green mold incidence, respectively. Inoculum densities of 10^6 spores/mL of both *P. digitatum* and *P. italicum* have been recommended to obtain acceptable disease levels after artificial inoculation of citrus fruit in experiments to evaluate the control ability of postharvest antifungal treatments (Eckert and Brown, 1986). In contrast to green mold, blue mold can spread by contact from infected to sound fruit. Barmore and Brown (1982) suggested that the highest mycelial density of *P. digitatum* may prevent the healthy fruit from contacting the corrosive juices of the decaying fruit containing the polygalacturonases and the galacturonic acid that degrade the peel tissue, causing injuries through which the hyphae of *P. italicum* can infect the healthy fruit. It has been reported that *Penicillium* spp. are often synergistic with another important citrus postharvest wound pathogen, *Geotrichum citri-aurantii* (Ferraris) E.E. Butler, the cause of sour rot (Morris, 1982).

The germination of conidia of *Penicillium* spp. inside rind wounds and the subsequent hyphal development requires free water, nutrients, and specific temperature and pH conditions. While green mold is more frequent on fruit held at room temperature because it grows faster in these conditions, blue mold can be more prevalent on fruit stored at 3–5°C, the usual cold storage temperatures for oranges or mandarins. Different *in vitro* studies (Plaza et al., 2003a, 2004c) showed that *P. italicum* germinated and grew faster than *P. digitatum* at low temperature and under drier conditions. The optimum temperature for germination and growth of both species was 25°C and they were active in the studied range of 4–30°C. However, they were not able to grow at 37°C. *Penicillium italicum* has also been reported as able to grow at 0°C. While *P. italicum* germinated and grew at water activities (a_W) of 0.87, *P. digitatum* failed to grow at 0.90. It was found in additional *in vivo* work that on oranges stored at 4°C blue and green mold symptoms were visible after 16 and 23 days, respectively. In contrast, *P. digitatum* grew faster at different temperatures above 10°C and when the two pathogens were inoculated into the same wound in the rind of oranges maintained at 25°C, blue mold was practically inhibited. Regarding the effect of pH, results by Prusky et al. (2004) suggested that *P. digitatum* and *P. italicum* are able to enhance their virulence by acidifying the ambient environment in citrus rind wounds.

Extensive evidence has been found that volatiles emitted from wounded host tissue play a significant role in the pathogenicity of *P. digitatum* and *P. italicum*. The major component of the peel oil of a variety of citrus species including oranges, mandarins, lemons and grapefruits is the monoterpene limonene (4-isopropenyl-1-methylcyclohexene), which accounts for 60–95% of the total volatiles present in the oil. Other components are other minor monoterpenes, sesquiterpenes, aliphatic aldehydes and alcohols, and esters (Caccioni et al., 1998; Droby et al., 2008). The profile of volatile compounds in the atmosphere surrounding citrus fruit is different in the case of intact fruit, fruit only wounded, and fruit infected by *P. digitatum*. On oranges, Ariza et al. (2002) found that sesquiterpenes such as valencene, together with relatively small amounts of monoterpenes such as limonene, were the major volatile metabolites released from undamaged oranges, while in the case of wounded fruit high amounts of limonene and other known citrus monoterpenes such as myrcene or 3-carene were released instead of sesquiterpenes. In the case of *P. digitatum*-infected oranges, the pattern resembled that of wounded fruit, but a number of more volatile compounds such as ethanol, methyl acetate, and ethyl acetate was also found. These researchers also reported the production of mycotoxins by *P. digitatum*. Limonene and other oil compounds such as the terpenes myrcene, α-pinene, β-pinene, sabinene (Eckert and Ratnayake, 1994; Droby et al., 2008), or prangolarin (Arimoto et al., 1995) showed a stimulatory effect on the germination and germ tube elongation of both *P. digitatum* and *P. italicum*, acting as a mechanism of host recognition by the pathogenic fungi. Comparative studies with other non-host pathogens like *P. expansum* or *P. sclerotiorum* indicated no effect or even an inhibitory effect (Droby et al., 2008; Wang et al., 2012a). While germ tube elongation in *P. digitatum* responded most strongly to limonene, germination and germ tube elongation in *P. italicum* responded most strongly to myrcene, the second most abundant compound in the peel oil (Droby et al., 2008). Furthermore, recent research work conducted at the IVIA demonstrated the role of limonene in pathogenicity by introducing an antisense construct of a limonene synthase gene in orange plants (Rodríguez et al., 2011). When transgenic oranges, which showed a reduced accumulation of limonene in the peel, were challenged with *P. digitatum*, the incidence and severity of green mold were markedly reduced, indicating an effective induction of disease resistance. It has been reported, on the other hand, that *P. digitatum* is able to biotransform limonene to α-terpineol, which is an oxygenated derivative very valuable in the industry of flavors and fragrances (Badee et al., 2011).

As a response to the stress caused by the pathogen attack, the fruit host can trigger at the same time and to different extents several biochemical mechanisms in an attempt to overcome the action of the pathogen and inhibit decay. The partial contribution of each mechanism in this complex multifaceted response and the final success or failure will depend on all factors determining the disease triangle. An important line of defense against *Penicillium* spp. and other pathogens is the presence of constitutive or preformed antifungal compounds in the fruit peel. The oxygenated monoterpene citral (3,7-dimethyl 2,6-octadienal) (Rodov et al., 1995b) and different flavonoids are among the most important that have been identified in citrus fruits. The latter include *p*-coumaric acid, a precursor of coumarins, polymethoxyflavones such as tangeretin, nobiletin, heptamethoxyflavone, or sinensetin, located in the flavedo and considered as phytoanticipins, and flavanones such as naringin, hesperidin, narirutin, or didymin, located in the albedo (Ortuño et al., 2006, 2011). The synthesis of these compounds is primarily regulated by the activity of the enzyme phenylalanine ammonia lyase (PAL), as part of the phenylpropanoid pathway in citrus fruits (González-Candelas et al., 2010b). An aging-associated decline of the concentration of preformed flavonoids has been reported that may explain why only young immature citrus fruits can be completely resistant against green or blue molds (Del Río and Ortuño, 2004).

Another line of defense comprises induced resistance mechanisms that are elicited by rind wounding (abiotic stress) and/or fungal infection (biotic stress). One of the most studied is the accumulation of lignin or lignin-like polymers in the cell walls at sites of wounding or pathogen inoculation. The process, catalyzed by the enzymes PAL and peroxidase, creates a physical barrier in rind wounds that impedes the penetration or development of the pathogens. Significant increases in PAL activity have been observed after rind wounding (Ismail and Brown, 1979), infection by *P. digitatum* (Ballester et al., 2006), and application of some postharvest treatments like heat or UV-C light (Nafussi et al., 2001; Ben-Yehoshua, 2005). Numerous studies reported that the biosynthesis of phytoalexins, which are secondary metabolites with antifungal activity, can be triggered by rind wounding and, particularly, by the infection by *P. digitatum* or *P. italicum*. The best-known citrus phytoalexins are the coumarins scoparone (6,7-dimethoxy coumarin) and scopoletin (7-hydroxy, 6-dimethoxy coumarin) (Kim et al., 1991). These compounds can be effectively induced by some physical or chemical postharvest treatments via the enhancement of the activity of enzymes such as PAL (Ben-Yehoshua, 2005; Venditti et al., 2005; Ben-Yehoshua et al., 2008; Ballester et al., 2010; Rojas-Argudo et al., 2012). Another fruit host defense mechanism extensively documented in citrus fruit is the production of pathogenesis-related proteins (PRP). Chitinases or β-1,3-glucanases are well-characterized proteins that inhibit mycelial growth by damaging fungal cell walls (Pavoncello et al., 2001). Gene expression and PRP activity in citrus rind following treatment with different elicitors have been studied (Porat et al., 2002; Ballester et al., 2010).

Another important investigation area related to host–pathogen interaction refers to the ability of *P. digitatum* or *P. italicum* to suppress the burst of reactive oxygen species (ROS) with accumulation of hydrogen peroxide (H_2O_2) that occur in citrus fruit tissue as the precursor step of most of the resistance mechanisms just described above. It is considered that this oxidative burst leads to disease resistance in incompatible interactions like, for instance, *P. expansum* and citrus fruit (Ballester et al., 2006; Macarisin et al., 2007). The enzyme superoxide dismutase (SOD) controls the metabolism of ROS by dismutation of superoxide radicals with the subsequent production of H_2O_2. The enzyme catalase (CAT) and other antioxidant enzymes such as ascorbate peroxidase (APX), or glutathione reductase (GR) are known for their contribution to the elimination of H_2O_2 in host–pathogen interactions (De Gara et al., 2003). In *P. digitatum* infected fruit, the activities of all these enzymes, and also those of PAL and peroxidase, were higher in the flavedo than in the albedo (Ballester et al., 2006, 2013). The activity of ROS-related enzymes in the citrus peel can be significantly enhanced by the application of postharvest treatments such as some biological control agents (Macarisin et al., 2010; Lu et al., 2013), heat treatments (Perotti et al., 2011), or oxidants like ozone (Boonkorn et al., 2012). These responses are clearly regulated by the fruit host maturity and the environmental conditions. Vilanova et al. (2012) found that even a non-host pathogen like *P. expansum* was able to infect citrus fruit under favorable conditions, namely very high inoculum concentration (10^7 spores/mL) and mature or over-mature oranges. When wound response in oranges was studied in both compatible and incompatible interactions, Vilanova et al. (2013) found that *P. digitatum*, in contrast to *P. expansum*, was able to suppress or counteract the host resistance mechanisms.

POSTHARVEST FACTORS INFLUENCING *PENICILLIUM DIGITATUM* AND *P. ITALICUM* INFECTION

Both *P. digitatum* and *P. italicum* are strict wound pathogens that cause disease after harvest. Infection of immature fruit by these fungi is very rarely followed by disease development because of a range of complex mechanisms triggered by the fruit host that lead to natural disease resistance. Conversely, infection of mature fruit is ordinarily followed by fruit decay. However, factors other than those intrinsic to pathogen and host can influence infection and/or disease development. In contrast to other postharvest diseases that mainly initiate in the field as latent infections, in the case of citrus penicillium molds most of these factors are harvest or postharvest factors. The role of crucial parameters such as temperature, inoculum density, rind maturity and condition, etc. has been discussed in the previous section and can explain the influence of important commercial fruit handling operations such as cleaning, sorting or prolonged cold storage. In this section, the influence of other important handling procedures like harvest and degreening is described.

Harvest and Transportation

Being the only feasible infection sites, it is imperative to pick and handle citrus fruits very carefully to minimize the production of rind punctures, wounds, bruises, compression damage, and general mechanical injuries. There is no doubt about the direct relationship between the degree of rind injury and the incidence of penicillium molds. Picking by pulling the fruit should be always avoided if the production is for fresh consumption. Moreover, the stems should be clipped at the shortest distance from the button in order to minimize subsequent punctures on nearby fruit. Unfortunately, the large amounts of fruit to be moved, the high labor costs and, in many cases, the lack or low availability of conscientious, skilled and well-trained teams of pickers are factors that may impede the adoption of optimum harvesting practices. On the other hand, harvest should not be conducted after rainfall or when free water is present on the fruit surface. In these conditions, pathogen sporulation is favored and excessive turgidity of the fruit peel makes it more susceptible to mechanical damage and subsequent infection (Eckert and Eaks, 1989).

Degreening

Early season citrus fruits, particularly mandarins and oranges, are usually exposed to low doses of exogenous ethylene gas (2–3 ppm C_2H_4) at high relative humidity (RH) for 2–4 days. This procedure is conducted to remove the rind green color of fruit that internally has already reached their commercial maturity. This removal is accomplished by two independent processes, the degradation of chlorophyll and the accumulation of carotenoids (Sdiri et al., 2012). Adequate fruit presorting before degreening based on initial peel color is highly recommendable for treatment rationalization and minimization of the required doses of ethylene. Although several research works have been conducted to date to clarify the relationship between ethylene applications and postharvest decay by *P. digitatum* and *P. italicum*, a well-determined trend has not been found. Discrepancies observed in research results seem to be related to the amount of ethylene applied, treatment duration, and environmental conditions.

In Florida and other humid citrus production areas, degreening is typically performed at temperatures around 30°C and RH higher than 90% and a significant reduction of green mold in these conditions has been reported (Brown, 1973). It was observed that exposure to such a high temperature exerted a curing effect to the fruit peel that reduced decay by wound lignification. Nevertheless, fruit resistance induction to penicillium molds has also been reported when citrus have been degreened at 20–22°C (El-Kazzaz et al., 1983a, b; Porat et al., 1999), which is the standard commercial degreening temperature in Spain and other Mediterranean countries and is near the optimum temperature for development of *P. digitatum* and *P. italicum*. These authors discussed that ethylene treatment induced an increase in total phenolic compounds of the fruit rind. Later, molecular work by Marcos et al. (2005) confirmed the involvement of ethylene in the expression of genes that regulate the activity of enzymes related to the induction of resistance in the peel of citrus in response to *P. digitatum* infection. It was also observed that treatment with 1-methylcyclopropene (1-MCP) significantly increased the fruit susceptibility to green mold.

However, on the other hand, other research results have suggested no significant effect (Plaza et al., 2004a) or an increase in the incidence of penicillium decay in citrus following ethylene exposure at 20°C (McGlasson and Eaks, 1972; Wardowski et al., 2006). In this case, such an increase has been attributed to a negative effect of ethylene on fruit susceptibility by accelerating the biochemical processes that lead to the senescence of the fruit peel. Thus, dip or drench fungicide applications, usually of TBZ or IMZ, to field fruit bins prior to the degreening operation have been recommended. In some cases, drenching is conducted in containers still loaded in field trucks (Smilanick et al., 2006c). Although being non-climacteric fruits, many reports stated detrimental effects of prolonged ethylene exposure on the condition of citrus rind, especially the induction of calyx alterations or even the loss of calyxes (Sdiri et al., 2013). Conversely, no negative effects have been reported in the amount of phenolic compounds present in the peel or in the internal fruit quality (Mayuoni et al., 2011; Sdiri et al., 2012). Applied research is being conducted in our laboratory at the IVIA CTP to clarify the effects on the most important early-season orange and mandarin cultivars of the standard commercial degreening practice conducted in most of Spanish citrus packing houses (Moscoso-Ramírez and Palou, 2014a).

CONTROL

Due to their high relative importance as cause of economical losses, effective control of green and blue molds is the main goal of postharvest disease management programs for fresh citrus fruits in all production areas with low summer rainfall. Typically, these postharvest diseases have been controlled worldwide for many years solely by the application of conventional fungicides after harvest.

Postharvest Treatments with Conventional Fungicides

Postharvest treatments with synthetic chemicals typically have a reasonable cost, are easy to apply, and provide a curative effect against pre-existing or established infections, a persistent preventive effect against potential new infections that can occur after the application of the treatment in the packing house, and also a satisfactory inhibition of sporulation on decaying fruit that breaks the infection cycles. Among fruit treated with conventional fungicides, losses are typically 2–4%, while without postharvest treatment or refrigeration, losses of 15–30% occur within 1–3 weeks after harvest (Naqvi,

Table 2.1 Fungicides Permitted in the European Union for Postharvest Treatment of Citrus Fruits

Fungicide	Directives	Inclusion	Expiration	MRL[a] (mg/kg)
Included Annex I Council Directive 91/414/EEC				
Imazalil	2010/57/EU	1/8/2011	31/7/2021	5.0
Fosetyl-Al	2006/64/EC	1/5/2007	30/4/2017	75.0[b]
Thiophanate methyl	2005/53/EC	1/3/2006	28/2/2016	6.0
Myclobutanil	2011/2/EU	1/6/2011	31/5/2021	3.0
Ortho-phenyl phenol (and salts)	2009/160/EC 2010/81/EU	1/1/2010	31/12/2019	5.0
Pyrimethanil	2006/74/EC	1/6/2007	31/5/2017	10.0
Thiabendazole	2010/77/EU	1/1/2002	31/12/2015	5.0
Excluded Annex I Council Directive 91/414/EEC (voluntary withdrawal)				
Guazatine			30/6/2012	5.0
Prochloraz			30/6/2012	10.0

[a]Maximum residue limit
[b]temporary MRL.
Sources: Council Directive 91/414/EEC, amended 1 August 2011; Spanish Ministry of Agriculture, Food and Environment (MAGRAMA), Madrid, 24 April 2013; Commission Decision of 5 December 2008 (2008/934/EC).

2004; Smilanick et al., 2006a). Factors that affect the effectiveness and deposition of fungicides and their dissipation rate in fruit include fungicide concentration, treatment mode (spray, drench, or dip), type of mixture (aqueous- or wax-based mixtures), species, cultivar, fruit age, treatment duration, temperature, and pH of the fungicide mixture. These and other aspects related to the use of conventional fungicides and their synergy with heat for the control of citrus penicillium molds have been recently reviewed (Schirra et al., 2011). The fungicides currently approved for postharvest use on citrus fruits in the European Union (EU), with their correspondent inclusion and expiration dates, EU Directives and maximum residue limits (MRL) are presented in Table 2.1.

Imazalil (IMZ)

IMZ [(*RS*)-1-(β-allyloxy-2,4-dichlorophenethyl)imidazole] is nowadays the postharvest fungicide most commonly employed by the citrus industry worldwide. Its status in the EU was recently reviewed and its inclusion in the Annex I of the Council Directive 91/414/EEC (current list of permitted agrochemicals, revised in 2011) was approved for 10 more years until 2021 with an MRL of 5 mg kg^{-1} (ppm) (see Table 2.1). This imidazole is a sterol 14-α-demethylation inhibitor (DMI) that is commercialized worldwide by several companies under a variety of trade names, e.g. Deccozil, Fungaflor, Fruitgard, Fecundal, Magnate, Bromazil, Fungazil, Panpan, etc. IMZ is typically applied in water at 500–1000 ppm as a dip, drench or spray, or in amended citrus waxes at 2000–3000 ppm. It generally shows curative, protective and antisporulant effects against penicillium molds, but its activity greatly depends on the formulation (IMZ sulfate or IMZ emulsifiable concentrate) and application mode, being in general higher in water than in waxes, and higher in dips than in drenches or sprays. IMZ moves into the rind of citrus fruit during treatment and most is absorbed by the epicuticular wax and cuticle. IMZ residues in fruit resist removal by washing and decline very slowly during storage (Smilanick et al., 1997, 2006a; Erasmus et al., 2013). The fungicidal activity of IMZ increases as the pH increases and it is compatible and highly synergistic with heat and GRAS (generally regarded as safe) compounds such as sodium carbonate (SC), SBC, PS, or paraben sodium salts (Smilanick et al., 1999, 2005, 2008; Montesinos-Herrero et al., 2009a; Dore et al., 2010; Moscoso-Ramírez et al., 2013a, b). For example, recent research at the IVIA CTP has evidenced, if compared with each treatment alone, a considerable reduction of the incidence and severity of green and blue molds and the sporulation of *P. digitatum* and *P. italicum* on 'Valencia' oranges dipped in a mixture of 200 mM sodium methylparaben (SMP) with 25 ppm IMZ and cold-stored at 5°C for 2 months (Fig. 2.3). In general, the combination with food additives allows the use of lower doses of IMZ, which is currently very important for producers to satisfy the increasing demand of export markets for fruit with very low chemical residue levels. Continuous and sometimes incorrect use of IMZ in citrus packing houses has led to the proliferation worldwide of resistant strains of *P. digitatum* and *P. italicum* (Holmes and Eckert, 1999; Kinay et al., 2007; Pérez et al., 2011). One of the main problems that often results in ineffective decay control and sporulation inhibition on IMZ-treated citrus fruit is the suboptimal residue loading (Erasmus et al., 2013). Molecular studies have been performed to explain the mechanisms of DMI and IMZ resistance and also to develop effective, rapid methods for the detection of resistant genotypes of *P. digitatum* or *P. italicum* (Sánchez-Torres and Tuset, 2011; Sun et al., 2013).

Thiabendazole (TBZ)

TBZ [2-(1,3-thiazol-4-yl)benzimidazole] is a benzimidazole used for many years for the control of citrus green and blue molds. Current brand names

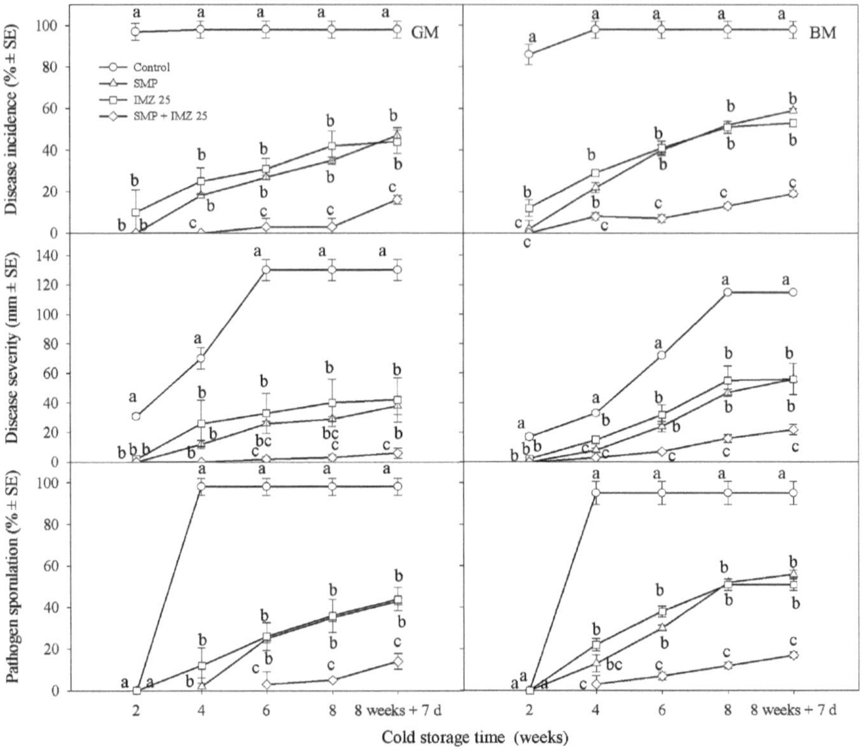

Figure 2.3 Incidence, severity and sporulation of green (GM) and blue (BM) molds on 'Valencia' oranges artificially inoculated with *Penicillium digitatum* or *Penicillium italicum*, dipped 24 h later in water (control), 200 mM sodium methylparaben alone (SMP), 25 ppm fungicide imazalil (IMZ 25), or 200 mM SMP combined with 25 ppm IMZ (SMP + IMZ 25) for 60 s at 20°C, and cold stored at 5°C and 90% RH for 8 weeks followed by 7 days of shelf-life at 20°C. For each mold and evaluation date, means with different letters are significantly different, according to Fisher's protected LSD test (*P* = 0.05) applied after an ANOVA. Disease incidence and pathogen sporulation were arcsine-transformed. Non-transformed means are shown. *Reprinted with permission from: Moscoso-Ramírez, P.A., Montesinos-Herrero, C., Palou, L. 2013. Characterization of postharvest treatments with sodium methylparaben to control citrus green and blue molds. Postharvest Biology and Technology 77, 128-137.*

include Tecto, Textar, Mintezol, Mycozol, Tresaderm, etc. Although its solubility in water is limited, it is commonly applied as a suspension of very fine particles that shows an acceptable systemic activity. The status of TBZ in the EU will be revised in 2015; current MRL is 5 mg/kg (see Table 2.1). Typical modes of application include water drench, dip or spray over rotating brushes at concentrations of 500–2000 ppm, and also wax formulations at higher concentrations. Dips and drenches are the most effective application modes because fruit wetting and product infiltration is improved; thus better curative and preventive activities are provided (Smilanick et al., 2006a). TBZ is compatible with heat and antifungal salts such as SBC or PS (Smilanick et al., 2006c; Schirra et al., 2008; D'Aquino et al., 2013a). The use of heated and/or pH-increased solutions increases the TBZ residue level in fruit, which allows the use of doses lower than 500 ppm. Since *P. italicum* is usually more tolerant than *P. digitatum* to the benzimidazole fungicides, blue mold can be prevalent on TBZ-treated citrus fruits (Gutter et al., 1981). Resistance is a major problem related to TBZ use for many years. Molecular studies showed that TBZ-resistance is characterized by a unique point mutation in the β-tubulin gene sequence corresponding to amino acid 200 (Schmidt et al., 2006; Sánchez-Torres and Tuset, 2011).

Sodium Ortho-Phenylphenate (SOPP)

SOPP [sodium (1,1'-biphenyl)-2-olate] was revisited by EU regulators in 2009 and extended for postharvest treatment of citrus fruits with an MLR of 5 mg/kg (see Table 2.1). It had been included for many years on the list of approved food additives with E-number 232. Although the legislation generically refers to ortho-phenyl phenol (OPP) and salts, only the sodium salt is currently in use. SOPP has been extensively used to wash, disinfect and protect fruit in citrus packing houses worldwide. Usual applications are at room temperature in soak tanks or foamer washes at concentrations of 0.5–2.0% (5000–20000 ppm) and pH 11.2–12.0, with a final rinsing of treated fruit with water (Smilanick et al., 2006a). Wax formulations are also available, but they are not as common as water applications.

Reduced Risk Fungicides

For the first time in over 25 years, three new fungicides, pyrimethanil (PYR), fludioxonil (FLU) and azoxystrobin (AZX), all belonging to different mode of action classes and classified as 'reduced-risk' fungicides, were registered for postharvest use to control penicillium decays of citrus fruits in the USA and other countries. Currently, only PYR is fully registered for this use in the EU (see Table 2.1).

Pyrimethanil (PYR)

In trials where citrus fruit were inoculated with *P. digitatum* 24 h before treatment, PYR (4,6-dimethyl-*N*-phenyl-2-pyrimidinamine; Penbotec™

400SC, Janssen PMP, Beerse, Belgium) at 500 ppm reduced green mold incidence by 90% applied by immersing or drenching the fruit for 30 s. PYR at 1000 or 2000 ppm applied in wax over rotating brushes reduced green mold incidence by about 65% (Smilanick et al., 2006b). Effective PYR treatments left fruit residues of 1–2 mg/kg. TBZ and IMZ-resistant PD isolates were controlled by PYR. The addition of SBC (Smilanick et al., 2006b; Kanetis et al., 2008b) or PS (Smilanick et al., 2008) improved PYR performance. PYR was not compatible with chlorine, but it was with a mixture of H_2O_2 and PAA (Kanetis et al., 2008b). PYR residues in citrus fruit were very persistent and greatly increased by increasing the solution temperature (D'Aquino et al., 2006; Smilanick et al., 2006b). A commercial formulation containing 20% PYR and 20% IMZ (Philabuster™ 400SC; Janssen PMP) has been introduced to several countries all over the world.

Fludioxonil (FLU)

FLU [4-(2,2-difluoro-1,3-benzodioxol-4-yl)-1*H*-pyrrole-3-carbonitrile; Scholar® 50WP, Syngenta Crop Protection, Greensboro, NC, USA] is a phenylpyrrole derived from the natural antibiotic pyrrolnitrin produced by several *Pseudomonas* spp. In general, postharvest applications of FLU (usual doses are 500–1200 ppm) are less effective than IMZ for the control of green mold, particularly because deep penetration of the active ingredient into the fruit rind is limited and its antisporulation activity is lower (Smilanick et al., 2008). According to Schirra et al. (2005), FLU and heat are synergistic for the control of green mold and the residue concentrations were notably higher in fruit treated at 50°C than in fruit treated at 20°C. However, Kanetis et al. (2008b) found no enhancement of green mold control when drench solutions of FLU at 300 ppm were heated to 50°C. FLU also showed synergistic activity when applied in combination with TBZ or SBC (D'Aquino et al., 2013b). Isolates of *P. digitatum* resistant to FLU and PYR have been already collected in California citrus packing houses (Kanetis et al., 2008a, 2010). Resistance levels to phenylpyrrole fungicides are related to the ability of mutated isolates to impede the synthesis of glycerol to counteract osmotic stress (Kanetis et al., 2008c).

Azoxystrobin (AZX) and Trifloxystrobin (TFX)

AZX (methyl (2*E*)-2-{2-[6-(2-cyanophenoxy)pyrimidin-4-yloxy]phenyl}-3-methoxyacrylate; Abound®, Syngenta Crop Protection) and TFX (methyl

(*E*)-methoxyimino-{(*E*)-α-[1-(α,α,α-trifluoro-*m*-tolyl) ethylideneaminooxy]-*o*-tolyl}acetate); Flint®, Bayer CropScience, Monheim, Germany) are strobilurins that have been tested as postharvest fungicides against citrus penicillium molds. AZX showed high activity against either *P. digitatum in vitro* (Zhang et al., 2009) or natural decay in grapefruits (Schirra et al., 2002), but poor efficacy on wound inoculated oranges, probably due to poor systemic activity (Schirra et al., 2010). AZX applied at ambient temperature was less effective than when heated to 50°C (Schirra et al., 2010). UV mutagenesis studies showed that it is highly likely that *P. digitatum* will evolve high levels of resistance if AZX enters common use in citrus packing houses (Zhang et al., 2009). Unlike AZX, TFX was highly effective in controlling decay caused by green and blue mold in fruit artificially inoculated with *P. digitatum* or *P. italicum*, when applied at 100 ppm and 50°C (Schirra et al., 2006). Residues of TFX were significantly correlated with dip temperature.

Other Fungicides

Other fungicides evaluated for postharvest use on citrus fruit include myclobutanil [MYB; (*RS*)-2-(4-chlorophenyl)-2-(1*H*-1,2,4-triazol-1-ylmethyl)hexanenitrile; Rally® 40W SP, Dow AgroSciences LLC, Indianapolis, IN, USA], the benzimidazole thiophanate methyl [TPM; dimethyl 4,4′-(*o*-phenylene)bis(3-thioallophanate); Topsin®-M 70 WP, United Phosphorus, Inc., King of Prussia, PA, USA], cyprodinil (CYP, 4-cyclopropyl-6-methyl-*N*-phenylpyrimidin-2-amine; Unix®; Schirra et al., 2009), and propiconazole (PCZ, 1-[2-(2,4-dichlorophenyl)-4-propyl-1,3-dioxolan-2-ylmethyl]-1*H*-1,2,4-triazole; Mentor® 45WP; Syngenta Crop Protection). MYB and TMP are both currently approved for postharvest use in the EU with MRLs of 3.0 and 6.0 mg kg^{-1}, respectively (see Table 2.1), but their actual usage in Spanish packing houses is not extended. PCZ is a triazole that shares its mode of action with IMZ, which is an important handicap because *Penicillium* spp. isolates resistant to IMZ can be also resistant to PCZ. The main interest of PCZ lies in its capacity to control sour rot, caused by *G. citri-aurantii* (McKay et al., 2012), especially after the withdrawal of guazatine in the EU in 2012. The use of the active ingredient prochloraz also expired in the EU in 2012. New experimental fungicides for the control of green and blue molds are also being evaluated. For instance, very recently, a series of new experimental 2-imidazolyl-3,4-dihydroquinazolines exhibited good fungicidal activity against *P. digitatum* (Li et al., 2013).

Integrated Disease Management (IDM) Strategies

Despite the effectiveness of conventional chemical fungicides, concerns about environmental contamination and human health risks associated with fungicide residues periodically lead to regulatory reviews and potential restrictions or cancellations of available active ingredients. Likewise, traditional citrus export markets are increasingly demanding products with lower levels of pesticides in order to satisfy the safety demands from the general public. In addition, new higher-value markets based on organically grown, sustainable, environmentally friendly, ecological, or green agricultural produce are currently arising and becoming more popular. Furthermore, the widespread and continuous use of these synthetic compounds has led to a build-up of resistant biotypes of the pathogens in commercial packing houses that seriously compromise the effectiveness of control treatments. Isolates of grove origin, where these chemicals are not used, are typically all fungicide sensitive (Gutter et al., 1981; Holmes and Eckert, 1999; Schmidt et al., 2006; Kinay et al., 2007; Kanetis et al., 2008a). There is, therefore, a clear and increasing need to find and implement control methods alternative to conventional fungicides for the control of penicillium molds of citrus fruits. If conventional chemicals are not used, the goal is to accomplish satisfactory mold control by adopting IDM programs (Palou et al., 2008). The purpose of such strategies, based on the knowledge of pathogen biology and epidemiology and the consideration of all preharvest, harvest, and postharvest factors that may influence disease incidence, is to minimize decay losses with no adverse effects on fruit quality by taking cost-effective action on every one of those factors at the right moment. Approaches to IDM strategies for citrus postharvest penicillium decay control are discussed in this section.

Preharvest Operations

Their relative importance is low if compared with their influence on the final incidence of citrus postharvest diseases caused by latent pathogens. However, a variable proportion of total citrus fruit decayed by *P. digitatum* and *P. italicum* is infected in the grove. Hence, field treatments could be of use either to reduce inoculum levels or protect the fruit. Satisfactory fruit protection might be achieved through either effective field antifungal treatments, if direct action against the pathogen persists long enough after harvest, or field treatments that indirectly induce higher levels of disease resistance to the fruit host. Among different fungicides and mixtures applied before harvest, treatments with thiophanate methyl (TPM) have shown the best results to control postharvest green mold (Ritenour et al.,

2004; Smilanick et al., 2006c; Zhang and Timmer, 2007). The efficacy of this chemical was acceptable even when applied 7 weeks before harvest, but it was especially high when applied 1 week before harvest. Low-toxicity chemicals like food additives or generally regarded as safe (GRAS) compounds (Youssef et al., 2012b) and biocontrol agents (Cañamás et al., 2008) have also been successfully applied before harvest to reduce green and blue molds. Trials by Yildiz et al. (2005) showed that preharvest applications with the growth regulators 2,4-dichlorophenoxyacetic acid (2,4-D) or gibberellic acid (GA3) were not effective in reducing postharvest green mold in mandarins. The removal of fallen fruit in the orchard is a cultural practice that may help to reduce inoculum levels in the field; however, the costs involved usually hinder the adoption of this practice (Smilanick et al., 2006a).

Early Detection of Infection

In the last few years, extensive research work has been conducted on new vision technologies able to discriminate, in an accurate and fast manner, citrus fruit with incipient infections by *Penicillium* spp. from sound fruit, and also from fruit with other peel decays or defects. The value of potential commercial implementation of such technologies is multiple. Machine vision systems would provide means to detect infected fruit automatically, thus preventing the drawbacks and high costs related to human inspection. Early detection of field infections not visible by the naked eye in conjunction with automatic sorting and elimination systems would prevent all posterior problems related to fruit decay and sporulation in the packing house and would also save the costs of treating and handling fruit that will be discarded later. In this sense, these detection systems should be ideally applied within the first presorting operation before degreening. Besides the technological development of these systems, the most important challenge for commercial implementation is the suitable integration of the systems into standard citrus packing line operations. Machine vision systems that have been tested for early detection of *P. digitatum* or *P. italicum* include chlorophyll fluorescence imaging (Nedbal et al., 2000), hyperspectral computer vision systems (Gómez-Sanchis et al., 2013), and laser-light backscattering imaging (Lorente et al., 2013).

Another research field that is being examined for early detection of mold infections on citrus packing houses is olfactometric analysis. Very recently, Pallottino et al. (2012) and Gruber et al. (2013) developed electronic devices able to distinguish, through the detection of specific volatile biomarkers, between *P. digitatum*-infected and non-infected citrus fruit, even when the levels of infection were very low or the infections were only 24 h old.

Fruit and Packing House Sanitation

As it has been previously stated, there is a linear relationship between the amount of conidia of *P. digitatum* or *P. italicum* that reach the infection court (rind wound) and the incidence of successful infections. Therefore, measures adopted to reduce the level of inoculum present on the fruit surface can contribute to disease control. On the one hand, fruit arriving at the packing house from the field should be conveniently washed and/or surface disinfected and, on the other hand, effective sanitation practices are needed to minimize the amount of inoculum available for fruit surface contamination in packing house facilities, including storage rooms. Further, an appropriate design of packing house facilities that conveniently separates 'clean' (waxing and packaging packing line segments, storage rooms, etc.) from 'dirty' areas (fruit reception, degreening rooms, washing and selection packing line segments, etc.) is worthwhile (Bancroft et al., 1984). Hoods with air aspiration systems to trap airborne fungal spores have been installed in the ceiling of field bin downloading areas in some Spanish citrus packing houses. It is also important to establish, in parallel to the fruit packing line, an additional automatic line for washing and disinfection of empty field bins or containers. Although obvious, problems related to inadequate location of discarded fruit containers can be of importance in some cases. The outdoor location of these containers must impede the entrance of spores inside the packing house via air currents or insects and they must be replaced with a reasonable frequency (Palou, 2011).

Chlorine, particularly sodium hypochlorite (NaClO), is the most widely used sanitizer for surface disinfection of citrus fruits and prevention of contamination of dip or drench solutions (Smilanick et al., 2002). Nevertheless, its use greatly depends on production areas; for instance, while it is of very common use in California or Australia, it is barely used in Spain. Because of rising problems associated with chlorine and derivatives like chlorine dioxide (ClO_2) or calcium hypochlorite [$Ca(ClO)_2$], namely pH dependence, deficient performance with high levels of organic matter, nitrosamine formation, or corrosive activity, research work to seek for alternatives is being conducted. Active ingredients such as peracetic acid (syn.: peroxiacetic acid, PAA), H_2O_2, iodine, ethanol, bromo-chloro-dimethylhydantoin, or some mixtures have been evaluated. The latter is also a chlorine derivative but can be used at much lower doses because of the synergistic activity between bromine and chlorine (Cunningham, 2007). Numerous studies are being conducted in South Australia in this matter, not only testing the efficacy of active ingredients, but also evaluating their compatibility with common citrus postharvest fungicides or salts and also conducting surveys of commercial usage (Cunningham, 2007, 2009; Taverner, 2013). Kanetis et al. (2008b) evaluated the performance of a mixture of H_2O_2 with PAA. Dao et al. (2008) examined the inactivation of spores of *P. digitatum* and *P. italicum* treated with ethanol solutions and vapors. Recently, Cerioni et al. (2012) developed a sequential oxidative treatment, using NaClO and H_2O_2 in the presence of cupric sulfate ($CuSO_4$), that was very effective in killing spores of *P. digitatum*, and also reduced green mold in lemons when combined with heat and antifungal salts such as sodium bicarbonate (SBC), potassium sorbate (PS) or potassium phosphite (Cerioni et al., 2012, 2013b). It is known that sanitizers, mostly compounds with a very high oxidant activity, are ineffective in controlling postharvest diseases because their mode of action is by contact and they cannot reach the fungal structures established within rind wounds or developing under the peel of the fruit. Therefore, they cannot be used as substitutes for IMZ or other conventional fungicides, although they can show a synergistic activity when applied in combination. On the other hand, if applied at excessive doses or for too long an exposure time, they can be highly phytotoxic to the fruit rind (Palou et al., 2007; Kanetis et al., 2008b). Chlorine and derivatives and other compounds such as quaternary ammonia, isopropyl alcohol, formaldehyde, or ozone can be also used to disinfect packing house facilities and equipment, floors and walls of storage rooms, and field containers. The application of these chemicals should be preceded by thorough washes because only previously cleaned surfaces are effectively disinfected. Steam or very hot water are also used in some cases. Smilanick and Mansour (2007) proposed a physical treatment based on the use of heat and moisture as an alternative to chemical sanitation. They found that good disinfection of *P. digitatum* and *P. italicum* was achieved after heating storage rooms for 1–2 days at 50°C or more with RH exceeding 75%. High humidity was very important because the sensitivity of dry conidia to heat or other treatments is considerably lower than that of well-hydrated conidia. Ozone (O_3) is a residue-free very potent oxidizer with very high toxicity against free fungal spores and hyphae. If applied correctly, both gaseous and aqueous ozone are effective sanitizers, but they cannot replace the use of synthetic fungicides in the citrus industry. Continuous exposure to ozone gas at 0.3 ppm at the usual citrus storage temperature of 5°C did not reduce disease incidence in oranges, although it delayed the development of both *P. digitatum* and *P. italicum* inoculated in rind wounds (Fig. 2.4). Furthermore, gaseous ozone inhibited aerial mycelial growth and sporulation of the fungi, which can help to reduce the proliferation of fungicide-resistant strains of the pathogens during cold storage (Palou et al., 2001a). However, these effects were transitory and limited to infected citrus fruit stored in highly vented packages or open-top containers that allowed direct contact with the gas (Palou et al., 2003). Storage of citrus fruits in ozonated atmospheres and general ozone applications for sanitation and control of

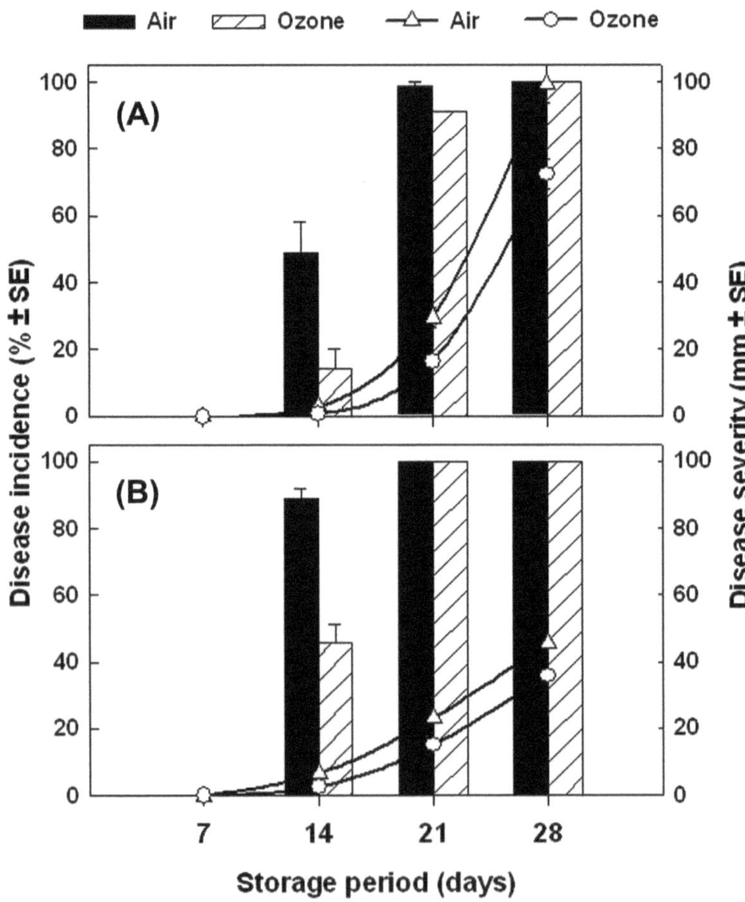

Figure 2.4 Green (A) and blue (B) molds incidence (bars) and severity (lines) on artificially inoculated 'Valencia' oranges continuously exposed for 4 weeks at 5°C and 90% RH to ambient air or 0.3 ppm ozone. *Reprinted with permission from: Palou, L., Smilanick, J.L., Crisosto, C.H., Mansour, M.F. 2001. Effect of gaseous ozone exposure on the development of green and blue molds on cold stored citrus fruit. Plant Disease 85, 632-638.*

postharvest diseases including penicillium molds have been comprehensively reviewed (Palou et al., 2007; Karaca, 2010). Ozone gas has also been assayed against *P. digitatum* in combination with electrolyzed oxidizing water (Whangchai et al., 2010). More recently, the effect of very low concentrations of ozone gas on the germination of spores of *P. digitatum* during long-term cold storage has been reported (Rubio Ames et al., 2013). Ozone can be also of use to eliminate fungicide residues on fruit and deplete ethylene in rooms. Being highly reactive, ozone is potentially harmful to humans and phytotoxic to fruit. Hence, dimension, monitoring, and safety issues must be addressed before commercial installation of ozonation systems in citrus packing houses. On the other hand, novel technologies such as high-density non-equilibrium atmospheric pressure plasma have been also evaluated for inactivation of *P. digitatum* spores (Iseki et al., 2010).

Control Methods Alternative to Conventional Fungicides

Besides preharvest, harvest, and other postharvest considerations, the basis of successful IDM strategies to control citrus postharvest green and blue molds is the commercial adoption of suitable non-polluting postharvest antifungal treatments to replace the use of conventional fungicides. According to their nature, these alternative methods can be physical, chemical, or biological. Significant advances in the evaluation of these control treatments, either alone or in combination with other treatments of the same or different nature, have been accomplished over the last few years.

Physical Treatments

Heat Treatments

Despite their limitations, postharvest heat treatments increasingly play a key role in integrated strategies for non-polluting penicillium decay control because they are relatively effective, simple, cheap, easy to apply and easy to combine with other physical, chemical, or biological control methods. Antifungal heat treatments for fresh citrus fruits include curing, hot water dips, and hot water rinsing and brushing.

Curing of citrus typically employs exposure of fruit for 2–3 days to an air atmosphere heated to temperatures higher than 30°C at RH higher than 90%. Numerous studies have demonstrated the intense curative activity of curing treatments against postharvest penicillium molds in different citrus cultivars stored at 20°C (Ben-Yehoshua, 2005; Kinay et al., 2005; Zhang and Swingle, 2005). However, the efficacy of the treatment, especially against blue mold, is lower on citrus fruit cold-stored for long periods (Plaza et al.,

2003b). In spite of their good efficacy, commercial implementation of curing is rare, first because of the expense of heating and immobilizing large amounts of fruit for relatively long periods and, secondly, because excessive or uncontrolled treatments may increase weight loss or cause rind phytotoxicity (Palou, 2009). A new experimental approach is the integration of curing in the degreening process (Plaza et al., 2004a).

Relatively brief immersions (2–5 min) in water at 45–55°C have repeatedly shown value in reducing citrus penicillium molds in a wide variety of citrus species and cultivars (Rodov et al., 1995a; Palou et al., 2001b; Schirra et al., 2004; Ben-Yehoshua, 2005). In general, lower and higher temperatures were ineffective and phytotoxic, respectively. Hot water dips are a technology easier, cheaper, and more feasible for heat application than curing but, because of their limitations, they are only commercially applied as stand-alone treatments to small fruit like kumquat, whose peel is also eaten, or some organically grown citrus fruits (Ben-Yehoshua, 2005). This is primarily due to the lack of persistence and preventive activity of the treatment. Also, the range of effective yet non-phytotoxic temperatures is very narrow (Palou, 2009). Another important reason is that the curative activity of hot water is greatly influenced by the fruit host and its condition. In general, it is lower on fruit naturally more susceptible to disease; thus, it is lower on mandarins than on oranges and decreases with fruit maturity at harvest (Schirra et al., 1998; Palou et al., 2002a). The use of heated chemical antifungal solutions is more common in citrus packing houses and they are usually applied through continuous systems in which the treated fruit move slowly from one end of the water tank to the other. Other limitations to the commercial use of hot water dips are the energy costs and the need of large high-volume tanks.

Hot water rinsing and brushing (HWRB) consists basically of packing line machinery that applies hot water over rotating brushes at high temperature (55–65°C) for a very short time (10–30 s). This technology was first developed in Israel, where it was found that HWRB at 56°C for 20 s reduced decay by 45–55% on organically-grown tangerines, oranges, and grapefruits with no rind injuries or adverse influence on fruit weight loss or internal quality parameters (Porat et al., 2000a). It was also observed that an indirect mode of action of HWRB in grapefruits was the induction of fruit resistance against *P. digitatum* (Porat et al. 2000b).

Irradiation and Illumination Treatments

Treatments that have been tested against citrus green or blue molds include irradiation at wavelengths (λ) corresponding to very different regions in the electromagnetic spectrum, i.e. blue light ($390<\lambda<540$ nm), far ultraviolet light (UV-C, $100<\lambda<280$ nm) and ionizing radiations ($\lambda<100$ nm). Obviously, the technological needs for the application of these irradiation treatments differ absolutely from one other, and also from those needed for the application of heat.

It has been recently reported that blue light at wavelengths from 390 to 500 nm and intensity of 40 μmol/m^2 reduced decay caused by *P. digitatum* in tangerines (Alférez et al., 2012). Further work showed that blue light significantly reduced the activity of polygalacturonase in *P. digitatum* and *P. italicum* and induced the accumulation of octanal in the flavedo (Liao et al., 2013). Exposure of citrus fruit to UV-C doses ranging from 0.5 to 8 kJ/m^2 has been reported as effective in reducing postharvest green or blue molds in different citrus species and cultivars (Palou, 2009). Although direct effects on the fungi have been described, in general, the prevalent mode of action of UV-C light for the control of postharvest diseases is the induction of beneficial responses in the fruit host, a phenomenon known as hormesis (Droby et al., 1993). UV-C exposure should be optimized for each type of fruit and application case in order to maximize effectiveness with no production of rind damage (Ben-Yehoshua, 2005). For feasible commercial implementation, illumination devices should be integrated into citrus packing lines with the appropriate safety measures. Ionizing radiation sources for food treatment include radioactive (^{60}Co or ^{137}Cs, γ-rays) and machine sources [electron beams (β particles) and X-rays (bremsstrahlung)]. The maximum dose permitted by the US Food and Drug Administration (US FDA) for irradiation of fruits and vegetables is 1000 Gy (100 krad). According to early research (Sommer et al., 1964; Barkai-Golan, 1992), effective control of established infections of *P. digitatum* or *P. italicum* required γ-ray or electron irradiation doses higher than 1000 Gy and such doses often induced apparent fruit damage, mostly rind pitting and browning. Because of this negative impact on fruit quality, ionizing radiation as a single treatment for decay suppression cannot be commercially adopted and lower doses should be evaluated in combination with other physical or chemical treatments (Palou, 2009). Additional handicaps for the use of this technology are the costs required for implementation and operation of radiation treatment plants and limited consumer acceptance, especially in EU countries.

Controlled Atmospheres

Long-term cold storage in conventional controlled atmospheres (CA) (5–10% O_2 + 0–5% CO_2) is not commercially feasible for citrus fruits because potential benefits do not compensate the high installation and maintenance costs (Smilanick et al., 2006a). Brief gaseous exposure to CO_2- or O_2-enriched atmospheres at curing temperatures has recently been tested in our laboratory to control green and blue molds avoiding the risks and costs of conventional curing. For instance, treatments with 15 kPa CO_2 or 30 kPa O_2 at 33°C for 24 or 48 h significantly reduced green and blue molds on artificially inoculated 'Ortanique' hybrid mandarins after 4 and 7 days of incubation at 20°C (Fig. 2.5). Similar results were obtained with other mandarin cultivars and 'Valencia' oranges (Montesinos-Herrero et al., 2012). However, control of both diseases was minimal after 15 days. Therefore, these treatments were not persistent and showed a fungistatic rather than fungicidal effect. In an attempt to overcome these limitations, the antifungal atmospheres were satisfactorily combined with previous dips in heated PS solutions (Montesinos-Herrero and Palou, 2013).

Low-Toxicity Chemical Treatments

Chemicals alternative to conventional fungicides for postharvest disease control should be natural or synthetic compounds with known and very low toxicological effects on mammals and impact on the environment. Chemicals tested as postharvest treatments against citrus green and blue molds can be classified into food additives and GRAS substances, natural compounds (essential oils, plant extracts, peptides, chitosan, etc.), resistance inducers, and other low-toxicity chemicals. In general, these compounds have been typically evaluated as aqueous solutions, vapors, or ingredients of edible coatings.

Food Additives and GRAS Compounds

The sodium carbonate salts SC and SBC are food additives that are in use in citrus packing houses because of their effectiveness, relatively low cost and lack of restrictions for many applications including organic agriculture. Dip treatments in 2–3% SC or SBC aqueous solutions for 60–150 s showed curative activity against penicillium molds and their performance was significantly improved by heating the solutions to 45–50°C (Smilanick et al., 1999, 2005; Palou et al., 2001b, 2002a). For instance, the influence of SC concentration, solution temperature and immersion period on the efficacy of SC dips against blue mold on 'Valencia' oranges is shown in Figure 2.6. The mode of action of carbonates has been related to direct (ion toxicity) and indirect (pH increase and phytoalexin accumulation in rind wounds) effects. Therefore, fruit host genotype and initial condition play an

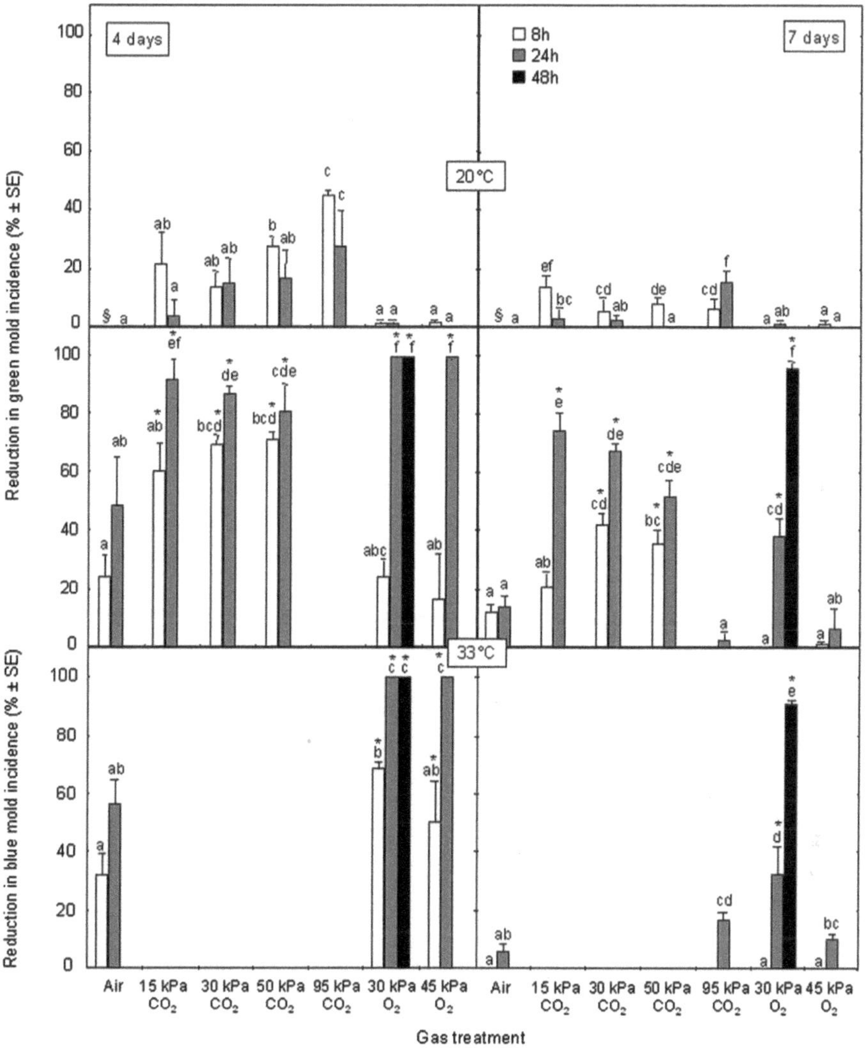

Figure 2.5 Reduction in green and blue molds incidence with respect to the control treatment (§, air at 20°C for 8 h; 100% incidence of both molds after 7 days of incubation) on 'Ortanique' mandarins artificially inoculated with *Penicillium digitatum* and *Penicillium italicum*, respectively, treated 24 h later with air, 15, 30, 50, or 95 kPa CO_2 or 30 or 45 kPa O_2 at 20 or 33°C for 8, 24, or 48 h, and incubated at 20°C for 4 and 7 days. For each individual graph, values within columns with the same letters did not differ significantly according to Fisher's protected LSD test (P <0.05). Where no letters are shown, treatment conditions were not tested or readings were not performed. Asterisk indicates synergistic activity was present between temperature and gas treatment according to Limpel's formula. *Reprinted with permission from: Montesinos-Herrero, C., del Río, M.A., Rojas-Argudo, C., Palou, L. 2012. Short exposure to high CO_2 and O_2 at curing temperature to control postharvest diseases of citrus fruit. Plant Disease 96, 423-430.*

important role in treatment efficacy. Heated SC or SBC solutions were generally more effective on oranges than on mandarins (Palou et al., 2002a; Venditti et al., 2005; Torres et al., 2007; Dore et al., 2010).

The potassium salt of sorbic acid, PS, is a wide spectrum antimicrobial food additive that has recently gained commercial attention for the control of citrus green and blue molds. PS aqueous solutions at 2–3% are effective at different dip conditions, e.g. 2–3 min at room temperature or 30–60 s at 50–62°C (Hall, 1988; Palou et al., 2002b; Smilanick et al., 2008). Similarly to carbonates, the effectiveness of these treatments is clearly influenced by the host species, cultivar and condition (Montesinos-Herrero et al., 2009a). Rinsing the fruit with tap water at low pressure after dip treatment does not considerably reduce the effectiveness and is a good practice to remove salt residues that can adversely affect fruit quality, especially during long-term storage (Montesinos-Herrero et al., 2009b). The commercial implementation of SC, SBC or PS dips is primarily limited by the low persistence of the treatments and the lack of preventive activity or residual effect to protect the fruit against subsequent infections. For example, the high effectiveness of 3% PS dips at 62°C for 60 s against both green and blue molds on 'Ortanique' mandarins artificially inoculated 24 h before treatment, gradually declined during fruit storage at 5°C for 60 days. This decline was lower on 'Valencia' oranges and, in general, more important in the case of blue mold (Fig. 2.7). In addition, disposal of carbonate solutions can be a problem due

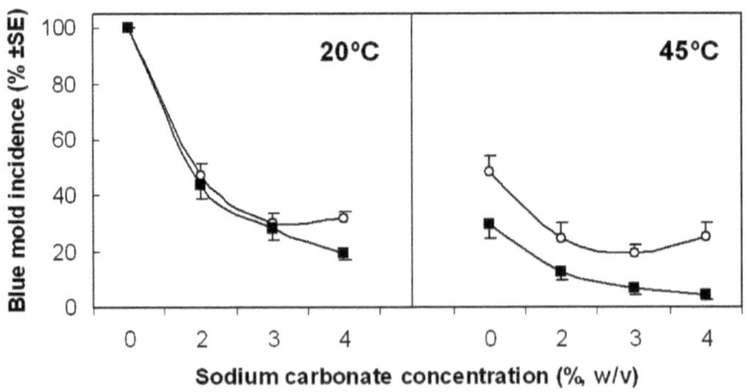

Figure 2.6 Influence of solution temperature, sodium carbonate concentration, and immersion period (○ = 60 s, ■ = 150 s) on the incidence of blue mold on 'Valencia' oranges artificially inoculated 24 h before treatment, rinsed at low pressure, and stored at 20°C and 90% RH for 7 days. Data are the means of two experiments with five replicates of 25 fruit each. *Reprinted with permission from: Palou, L., Smilanick, J.L., Usall, J., Viñas, I. 2001. Control of postharvest blue and green molds of oranges by hot water, sodium carbonate, and sodium bicarbonate. Plant Disease 85, 371-376.*

to high sodium content, pH and solution conductivity (Smilanick et al., 2008).

Other GRAS compounds that have been tested against citrus penicillium molds include sodium propionate (SP) and sodium benzoate (SB) (Palou et al., 2002b). Very recently, the use of aqueous solutions of sodium paraben salts for the control of both green and blue molds has been investigated in our laboratory. Considerable curative activity of SMP, sodium ethylparaben (SEP) and sodium propylparaben (SPP) at different concentrations

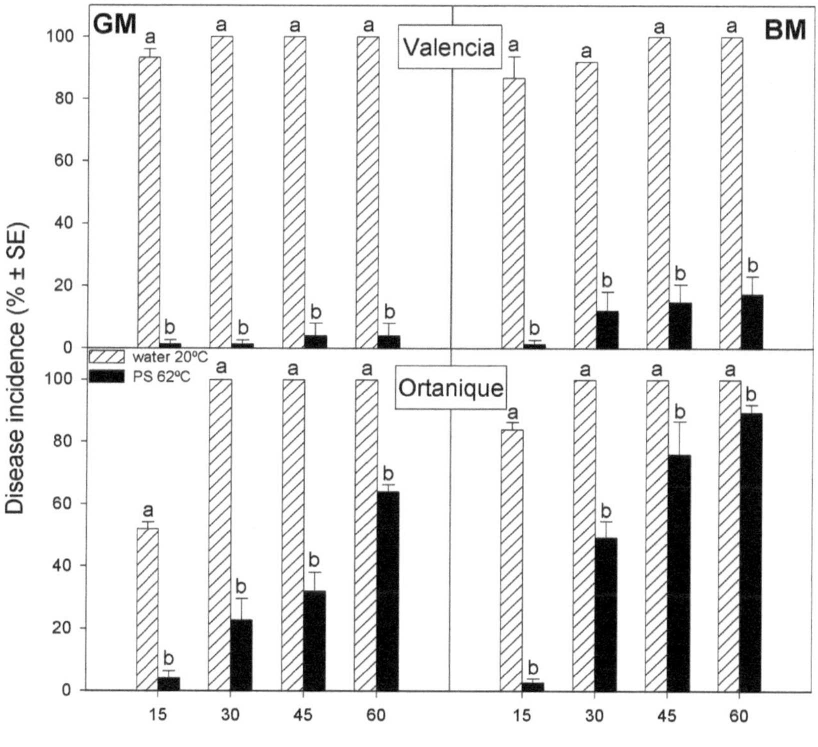

Figure 2.7 Incidence of green (GM) and blue (BM) molds on 'Valencia' oranges and 'Ortanique' mandarins artificially inoculated, dipped 24 h later in water at 20°C or aqueous solutions of 3% (w/v) potassium sorbate (PS) at 62°C for 60 s, and stored for 60 days at 5°C and 90 % RH. For each disease and cultivar, columns with unlike letters are significantly different according to Fisher's protected LSD test ($P \leq 0.05$) applied to arcsine-transformed data. Non-transformed means ± SE are shown. *Reprinted with permission from: Montesinos-Herrero, C., del Río, M.A., Pastor, C., Brunetti, O., Palou, L. 2009. Evaluation of brief potassium sorbate dips to control postharvest penicillium decay on major citrus species and cultivars. Postharvest Biology and Technology 52, 117-125.*

was obtained in *in vivo* primary screenings with 'Valencia' oranges artificially inoculated with *P. digitatum* and *P. italicum*, treated 24 h later, and incubated for 6 days at 20°C (Fig. 2.8). Dips of 200 mM SMP and 80 mM SEP at 20°C for 60 s were selected as the most interesting postharvest treatments, with efficacy comparable to that of solutions heated to 50°C. Paraben salts were compatible with the fungicide IMZ and their effectiveness was significantly higher on oranges than on mandarins (Moscoso-Ramírez et al., 2013a, b).

Natural Compounds

A wide range of substances with bioactive and antifungal properties can be extracted from plants or animals. Palou et al. (2008) listed a number of natural compounds with effective antifungal properties against citrus green and blue molds, and classified them into the following categories: volatiles and essential oils, plant extracts, peptides and proteins, and chitosan and derivatives. In this chapter, the latter will be discussed in the section referring to edible coatings.

Although a large list of essential oils or volatile compounds obtained from plants has been reported to have *in vitro* antifungal activity against *P. digitatum* or *P. italicum*, only a few of these compounds showed antifungal activity in *in vivo* tests (Palou et al., 2011). Plaza et al. (2004b) tested about 20 essential oils *in vitro* and selected thyme and cinnamon oils to be tested *in vivo* for the control of green and blue molds. However, applications of these essential oils in wax or volatile phase were ineffective and caused rind damage. Similarly, other studies found that natural compounds selected *in vitro* showed inefficacy and/or phytotoxicity when applied *in vivo* (Ameziane et al., 2007; Szczerbanik et al., 2007). In contrast, good inhibition of blue mold was obtained with the GRAS-registered essential oil MO-1 applied for 24 h in a gas phase (Ben-Arie et al., 2011). Likewise, oil vapors of thyme or clove applied inside polyethylene and nano-clay polyethylene films significantly reduced penicillium decay (Yahyazadeh et al., 2009). Plant extracts that showed a significant activity against *P. digitatum* or *P. italicum* include garlic, pomegranate, huamuchil, *Aloe vera*, *Thymus* spp., *Eucaliptus* spp., *Cistus* spp., *Juglans* spp., *Myrtus* spp., *Accacia* spp., *Whitania* spp., *Lippia* spp. and many others (Palou et al., 2008; Tayel et al., 2009; Askarne et al., 2013; Zapata et al., 2013). General barriers to aqueous or vapor application of volatile oils or plant extracts are the phytotoxicity risks and the production of off-odors and/or off-flavors in treated fruit. Some peptides and small proteins produced by plants or animals have been characterized and tested for the control of green and blue molds with promising results (Muñoz et al., 2007).

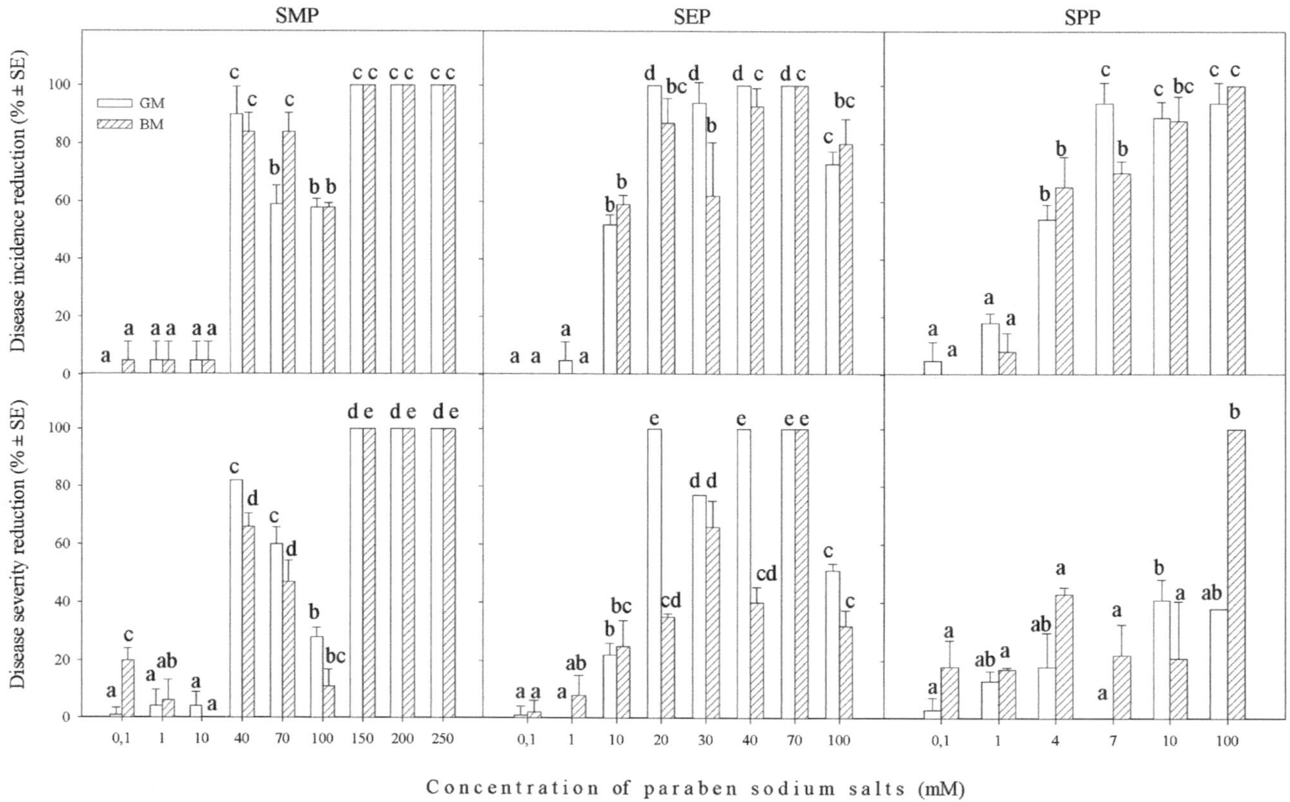

Figure 2.8 Curative activity of sodium methylparaben (SMP), sodium ethylparaben (SEP) and sodium propylparaben (SPP) at different concentrations against green (GM) and blue (BM) molds in *in vivo* primary screenings with 'Valencia' oranges artificially inoculated with *Penicillium digitatum* or *Penicillium italicum*, treated 24 h later, and incubated for 6 days at 20°C and 90% RH. Reductions of disease incidence and severity were determined with respect to control fruit treated with water. For each salt and mold, columns with different letters are significantly different according to Fisher's protected LSD test ($P = 0.05$) applied after an ANOVA. Incidence reduction values were arcsine-transformed. Non-transformed means are shown.

Chemical Resistance Inducers

Several chemicals are known by their general ability to induce resistance in treated plants and they are generically called resistance inducers. Recent research in our laboratory focused on the evaluation of postharvest dips in aqueous solutions of selected chemical resistance inducers to control green and blue molds on oranges. Acetylsalicylic acid (ASA) and harpin protein (Messenger®) were ineffective and discarded in preliminary trials. Among a wide range of concentrations evaluated in *in vivo* primary screenings, sodium silicate (SSi) at 1000 mM, potassium silicate (PSi) at 90 mM, 2,6-dichloroisonicotinic acid (INA) at 0.03 mM, salicylic acid (SA) at 0.25 mM, β-aminobutyric acid (BABA) at 0.3 mM, and benzothiadiazole (BTH) at 0.9 mM were selected and tested afterwards as dips at 20°C for 60 or 150 s with oranges artificially inoculated with *P. digitatum* or *P. italicum* before (curative activity) or after (preventive activity) the treatment and incubated for 7 days at 20°C. PSi dips provided significant preventive and curative activities without injuring the fruit (Moscoso-Ramírez and Palou, 2014b). This result was in agreement with previous work by Liu et al. (2010). SSi treatments were effective, but showed high potential for induction of rind phytotoxicity. In our experimental conditions, the rest of the treatments had no effect or only reduced mold development very slightly (Moscoso-Ramírez and Palou, 2013). According to previous reports (Porat et al., 2003; Iqbal et al., 2012), particularly unexpected results were the lack of activity of postharvest SA and BABA dips. We concluded that, with the exception of PSi, postharvest dip treatments with chemical resistance inducers cannot be recommended for inclusion in commercial decay management programs for citrus packing houses. Further research should focus on field applications of these substances.

Edible Coatings

In the citrus industry, fruit coating is a normal practice to replace the natural waxes removed during washing in order to reduce fruit weight loss, shrinkage and improve appearance. A very active research field is nowadays the replacement of commercial citrus waxes, which are often a vehicle for the application of synthetic fungicides like IMZ, TBZ or SOPP, for natural edible coatings with antifungal properties. Antifungal ingredients of such coatings can be food additives, natural compounds or microbial antagonists (Valencia-Chamorro et al., 2011a).

The application of chitosan and other derivatives from chitin present in crustacean shells, insect cuticles, and fungal cell walls resulted in significant reduction of postharvest penicillium decay and delayed fruit senescence in different citrus species and cultivars during long-term cold storage (Chien and Chou, 2006; Cháfer et al., 2012). Chitosan has also been used in recent studies as a carrier of additional antifungal compounds such as essential oils (Wang et al., 2011; Cháfer et al., 2012) or synthetic fungicides (El-Mougy et al., 2012) to enhance the activity against citrus green and blue molds.

Work by our research group showed that, in *in vitro* tests, hydroxypropyl methylcellulose (HPMC)-lipid edible composite films containing a variety of food additives or GRAS compounds exhibited antifungal properties against both *P. digitatum* or *P. italicum* (Valencia-Chamorro et al., 2008). In subsequent trials, coatings containing PS, SB, SP, SMP, and some mixtures of these preservatives were the most effective in reducing the incidence and the severity of both green and blue molds on 'Valencia' oranges and 'Ortanique' and 'Clemenules' mandarins coated 24 h after fungal inoculation and incubated at 20°C (Fig. 2.9) or cold-stored at 5°C (Valencia-Chamorro et al., 2009a, b, 2010, 2011b). The inhibitory activity of the coatings was strongly dependent on the susceptibility of each citrus cultivar to penicillium decay and it was higher on oranges than on mandarins. In general, all coatings significantly reduced weight loss and maintained the firmness of coated fruit. Although the coatings modified the gas composition in the internal atmosphere of coated oranges and mandarins, they did not affect the overall sensory quality of the fruit. GRAS substances have also been incorporated into commercial citrus waxes as substitutes for conventional fungicides (Youssef et al., 2012a).

Substantial research is recently being conducted on the development of new coatings amended with natural compounds as antifungal ingredients. Essential oils from *Mentha spicata*, *Lippia scaberrima* (du Plooy et al., 2009) or *Cinnamomum zeylanicum* (Kouassi et al., 2012), incorporated into different commercial citrus coatings such as shellac or carnauba, provided satisfactory control of green mold on oranges. This disease was also significantly reduced in lemons artificially inoculated with *P. digitatum* with the application of wax containing thymol and carvacrol (Pérez-Alfonso et al., 2012). HPMC-based films containing different concentrations of an ethanolic extract of propolis showed antifungal activity against *P. italicum* in *in vitro* tests (Pastor et al., 2010). Carboxymethyl cellulose coatings formulated with essential oil from *Impatiens balsamina* reduced natural decay on oranges stored at 5°C for up to 100 days (Zeng et al., 2013).

Biocontrol antagonists like *Candida* spp. or *Pseudomonas* spp. have been incorporated into shellac-, cellulose- or chitosan-based coatings for the

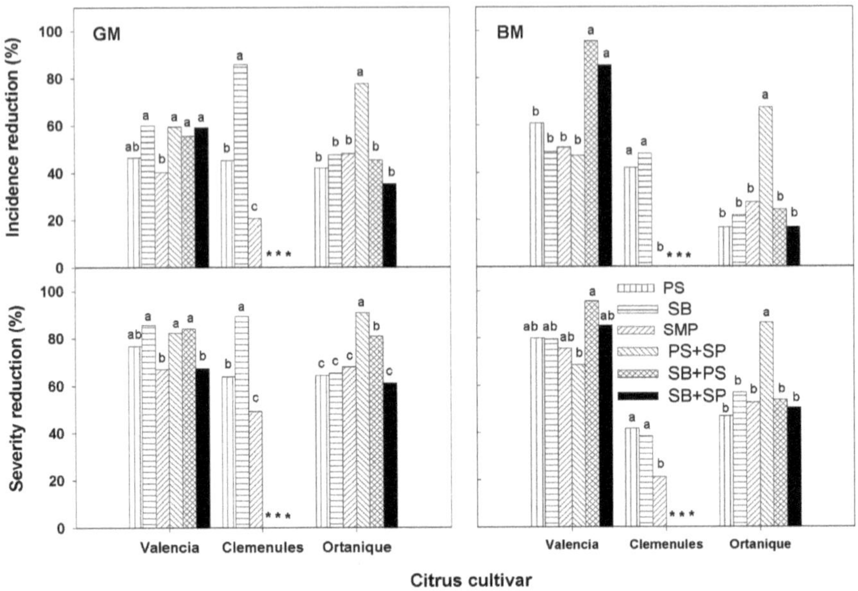

Figure 2.9 Curative activity of hydroxypropyl methylcellulose (HPMC)-lipid edible composite coatings, containing potassium sorbate (PS), sodium benzoate (SB), sodium propionate (SP), sodium methylparaben (SMP) or their mixtures, against green (GM) and blue (BM) molds on different citrus cultivars artificially inoculated with *Penicillium digitatum* and *Penicillium italicum*, coated 24 h later, and incubated for 7 days at 20°C. Disease incidence and severity reductions were determined with respect to control fruit (inoculated but uncoated). For each mold and cultivar, columns with different letters are different by Fisher's protected LSD test ($P < 0.05$). *No data available.

control of citrus penicillium molds (Potjewijd et al., 1995; El-Ghaouth et al., 2000; McGuire, 2000). The effectiveness of these coatings greatly depended on their ability to support populations of the antagonistic microorganism. Shellac and other commercial wax coatings can be toxic to the yeasts due to the addition of alcohols and bases that are used to dissolve the primary constituents.

Other Chemical Alternatives

The use of postharvest treatments with phosphite salts for the control of green and blue molds is currently arising great interest among citrus growers. Work done in California with potassium and calcium phosphites revealed satisfactory decay control on a variety of citrus species and cultivars and good synergistic effects with heat, conventional fungicides and GRAS substances such as SC, SBC or PS (Cerioni et al., 2013a). Phosphite salts are exempt from residue tolerances in the USA. Liquid lime sulfur solution, an inexpensive and widely available fungicide that contains calcium polysulfide, effectively controlled green mold in postharvest immersion treatments, but is not commercially used because of the objectionable sulfide odor it emits and its corrosiveness to some packing house equipment (Smilanick and Sorenson, 2001). The fertilizer ammonium molybdate has also shown activity against citrus penicillium molds (Palou et al., 2002b). In experimental trials, ammonia gas fumigation injected at 3000 ppm satisfactorily controlled green and blue molds on lemons and oranges without injuring the fruit peel (Montesinos-Herrero et al., 2011).

Biological Control Treatments

This denomination is usually restricted to the utilization of microbial antagonists. A comprehensive list of yeasts and bacteria that have been evaluated as biocontrol agents in postharvest treatments for the control of citrus green and blue molds can be found in Palou et al. (2008). Since then, several more microorganisms with antagonistic activity against *P. digitatum* or *P. italicum* have been identified: *Kluyveromyces marxianus* (Geng et al., 2011), *Metschnikowia andauensis* (Manso and Nunes, 2011), *Bacillus amyloliquefaciens* (Yu et al., 2012), *Paenibacillus polymyxa* (Lai et al., 2012), *Paenibacillus brasilensis* (Tu et al., 2013), *Streptomyces* spp. (Maldonado et al., 2010), *Saccharomyces cerevisiae*, *Wickerhamomyces anomalus* (Platania et al., 2012), *Pichia membranefaciens* (Luo et al., 2013), *Rhodosporidium paludigenum* (Lu et al., 2013), *Lactobacillus plantarum* (Wang et al., 2012b), and several strains of other acid lactic bacteria (Gerez et al., 2010). In general, microbial antagonists are applied as aqueous cell suspensions in postharvest spray, drench, or dip applications. However, some of them such as *Muscador albus* (Mercier and Smilanick, 2005), *Streptomyces globisporus* (Li et al., 2010) or *Nodulisporium* spp. (Suwannarach et al., 2013) have been tested as biofumigants. Others, as already discussed in this chapter, have been used as ingredients in different types of fruit coatings.

In spite of all the experimental work done with biocontrol agents, the commercial use of these products remains limited particularly due to inconsistency in performance when used as stand-alone postharvest treatments (Droby et al., 2009). Other handicaps are the excessive specificity, the lack of curative activity, the difficulties of developing stable, reliable, and economically suitable product formulations, the disinformation and subsequent limited acceptance by consumers, the limited market for developing companies and, in some cases (e.g. EU countries), the inappropriate and strict regulatory issues that prevent registration (Teixidó et al., 2013). Therefore, only a very few postharvest biological products are currently registered for use against citrus penicillium molds: Bio-Save® (JET Harvest Solutions, Longwood, FL, USA; based on the bacterium *P. syringae*), Shemer™ (Koppert B.V., Berkel en Rodenrijs, The Netherlands; based on the yeast *Metschnikowia fructicola*), and Pantovital® (BioDurcal S.L., Durcal, Spain; based on the bacterium *Pantoea agglomerans*). Approaches to enhance the biocontrol activity of the antagonists include the addition of nutrients, the combined use of two or more microorganisms, genetic manipulation of the antagonists, and the combination with other control methods (Nunes, 2012). Currently, molecular and biochemical research is devoted to exploring the induction of fruit resistance to disease by postharvest biological control treatments (Castoria and Wright, 2010). One of the most studied modes of action of several biological control agents to control green and blue molds is the enhancement in the citrus peel of the activity of ROS-related enzymes. Evidence has been provided for different citrus species and cultivars treated with antagonists such as *C. oleophila* (Macarisin et al., 2010), *P. agglomerans* (Torres et al., 2011), *M. fructicola* (Macarisin et al., 2010; Hershokovitz et al., 2012), *P. membranefaciens* (Luo et al., 2013), or *R. paludigenum* (Lu et al., 2013).

Combination of Treatments

For general acceptance by the industry, the efficacy of alternative treatments needs to be comparable to that provided by conventional fungicides. Unfortunately, alternative physical, chemical or biological stand-alone treatments cannot presently achieve such levels of efficacy, precisely due to their non-contaminant nature and consequent inherent limitations for decay control. Therefore, the use of an integrated rather than a single approach is advocated in order to offer a consistent level of cost-effective disease control. In general, three objectives may be pursued by the integration of two or more treatments: additive and/or synergistic effects to increase the effectiveness and/or the persistence of individual treatments, complementary effects to combine preventive and curative activities, and potential commercial implementation of effective treatments that are too unpractical, costly, or risky as single treatments (Palou et al., 2008). Most of the extensive research on integration of alternative treatments against citrus green and blue molds included postharvest heat or biocontrol treatments as major components of the combined treatments.

Combination of Heat with Other Alternative Treatments

The synergy between the postharvest application of curing treatments and CO_2 or O_2 antifungal atmospheres has already been discussed in this chapter. Other physical treatments that have been combined with curing or hot water for the control of green and blue molds include plastic packaging of fruit, ionizing radiation at low doses, and UV-C illumination (Montesinos-Herrero and Palou, 2010).

Synergistic effects from heating aqueous solutions of commercial synthetic fungicides have been previously described in this chapter. In general, heat allows the use of lower doses of fungicides, reducing costs and environmental risks. Likewise, heating aqueous solutions of low-toxicity alternative chemicals such as food additives or GRAS substances (e.g. SC, SBC, PS, SB, ethanol, sulfur dioxide, etc.), molybdate salts (Cunningham, 2010; Montesinos-Herrero and Palou, 2010) or phosphite salts (Cerioni et al., 2013a) significantly enhanced their effectiveness against penicillium molds. The most appropriate temperature solution should be specifically determined for each combination of active ingredient and fruit species and cultivar but, in general, if compared to hot water alone, similar effectiveness is obtained with lower solution temperatures, which considerably reduce the risk of production of heat damage. Curing treatments have also been combined with conventional fungicides and GRAS compounds such as SC, ethanol, or acetic acid (Venditti et al., 2009; Montesinos-Herrero and Palou, 2010).

Heat treatments and biological control are complementary methods that often show synergistic effects for the control of citrus green and blue molds. In some cases, both are components of complex integrated control strategies that also include other control means. Heat typically offers some curative activity against existing or incipient pathogenic infections but does not adequately protect the fruit. Biocontrol agents are able to colonize rind infection sites and offer effective preventive activity against pathogens that may reach the treated fruit during storage or commercialization (Janisiewicz and Korsten, 2002). Microbial antagonists that have been satisfactorily combined with thermal curing or hot water for citrus penicillium decay control include *C. oleophila*, *C. famata*, *Metschnikowia mulcherrima*, *Pseudomonas glathei*, *Bacillus subtilis*, or *P. agglomerans* (Palou, 2009).

Combination of Biocontrol Agents with Other Alternative Treatments

Besides heat treatments, other physical control means that have been combined with the application of antagonistic microorganisms for the control of citrus green or blue molds are UV-C illumination and storage in controlled atmospheres (Palou, 2009). Numerous literature references reported an improvement of citrus penicillium decay control by the integration of low-toxicity chemicals, especially food additives, with antagonistic microorganisms. The most frequent combination is the application of sodium carbonates, particularly SBC, with biocontrol antagonists such as *P. agglomerans*, *P. syringae*, *C. oleophila*, *B. subtilis*, or *Cryptococcus laurentii* (Palou et al., 2008). More recently, improved control of green mold was observed when SBC was combined with the yeast *K. marxianus* (Geng et al., 2011). Other natural compounds or low-toxicity chemicals examined to enhance biocontrol efficacy against *P. digitatum* or *P. italicum* are PAA, ethanol, oxalic acid, calcium chloride, 2-deoxy-D-glucose (Palou et al., 2008), or tea saponin (Hao et al., 2011).

CONCLUDING REMARKS

Postharvest diseases caused by the necrotrophic ascomycetes *P. digitatum* and *P. italicum* affect all citrus species and cultivars and are strictly originated from infections in peel wounds. Successful infection and disease development depend on intrinsic characteristics of the pathogen and the fruit host and on their interactions, modulated by the environmental conditions. Recent research on molecular and biochemical aspects of host–pathogen interaction has considerably increased the knowledge on the mechanisms of natural and induced fruit resistance to disease. On the one hand, the mitochondrial genome and, later, the complete genome of *P. digitatum* have been entirely sequenced for the first time for a phytopathogenic *Penicillium* species. On the other hand, the construction of cDNA libraries and subsequent transcriptomic analyses of fruit wounded, *P. digitatum*-infected or with resistance elicited by postharvest physical treatments have revealed genes that are involved in fruit responses to fungal attack. Mechanisms of resistance that have been studied include the biosynthesis of lignin and derivatives, the accumulation of antifungal flavonoids, and the production of PRPs like chitinases or β-1,3-glucanases. The ability of the pathogen in compatible interactions to suppress the burst of ROS and the accumulation of H_2O_2 that occurs in the peel tissue as a previous step to the activation of resistance mechanisms, as well as the role of antioxidant enzymes such as SOD, CAT, APX or GR have also been documented.

Traditional postharvest chemical fungicides such as IMZ, TBZ, SOPP and new reduced risk active ingredients such as PYR, FLU or AZX are the principal means used worldwide for effective control of citrus green and blue molds. Nevertheless, concerns about environmental contamination and human health risks, restricted markets, as well as the proliferation of resistant fungal strains increasingly lead to important usage restrictions.

Appropriate IDM strategies for effective control of green and blue molds should consider the influence on disease incidence of commercial fruit handling operations and procedures. Harvest should not be conducted under humid conditions and conscientious and well-trained pickers are required to minimize the production of rind wounds. The effect of ethylene degreening is controversial and greatly dependent on treatment characteristics and fruit condition. Preharvest field treatments could be of use to reduce inoculum levels or induce resistance in the fruit. Interesting research is being conducted on early detection of infections by *Penicillium* spp. through either machine vision systems or olfactometric analysis. Adequate procedures and technologies for fruit and packing house sanitation are important to reduce inoculum levels and potential reinfections.

Substantial progress has been accomplished in evaluating new physical, low-toxicity chemical, and biological alternative control methods as part of IDM programs. Curing, hot water, HWRB, irradiation and illumination (UV-C, blue light), CO_2- and O_2-based antifungal atmospheres, food additives and GRAS compounds, natural compounds (i.e. essential oils and other volatiles, plant extracts, antifungal peptides, chitosan), resistance inducers (e.g. PSi), antifungal edible coatings, or phosphite salts are non-polluting alternative control treatments that have shown variable value. However, the lack of either curative or preventive activity, low persistence, inconsistency, high variability and dependence on fruit condition, excessive specificity, risk of adverse effects on fruit quality, or technological problems for cost-effective application are general limitations associated with the nature of these alternatives that hinder their general commercial implementation as stand-alone treatments. Research work to find new and more effective alternatives and especially to integrate different treatments in a multifaceted approach is in progress in many public and private laboratories. As advances in the knowledge on the molecular and biochemical mechanisms underlying host–pathogen interactions and how they are influenced by potential postharvest treatments are achieved, more tools will be available for the postharvest citrus industry to design tailored cost-effective IDM strategies for the control of penicillium molds.

ACKNOWLEDGMENTS

The author thanks the Catalan, Spanish and EU public agencies that funded research in this topic. In memory of Dr Miguel Ángel del Río, for his unconditional friendship, guidance and support.

REFERENCES

Alférez, F., Liao, H.L., Burns, J.K., 2012. Blue light alters infection by *Penicillium digitatum* in tangerines. Postharvest Biol. Technol. 63, 11–15.

Ameziane, N., Boubaker, H., Msanda, F., Jilal, A., Benaoumar, A.A., 2007. Antifungal activity of Moroccan plants against citrus fruit pathogens. Agronomy Sustainable Dev. 27, 273–277.

Arimoto, Y., Sugarawa, F., Yoshida, S., Yamaguchi, I., 1995. Prangolarin is a chemical facilitator for the enhanced development of the infection process in the epicarp of *Citrus limon* by *Penicillium digitatum*. J. Agric. Food Chem. 43, 2283–2285.

Ariza, M.R., Larsen, T.O., Petersen, B.O., Duus, J.O., Barrero, A.F., 2002. *Penicillium digitatum* metabolites on synthetic media and citrus fruits. J. Agric. Food Chem. 50, 6361–6365.

Askarne, L., Talibi, I., Boubaker, H., et al., 2013. Use of Moroccan medicinal plant extracts as botanical fungicide against citrus blue mould. Lett. Appl. Microbiol. 56, 37–43.

Badee, A.Z.M., Helmy, S.A., Morsy, N.F.S., 2011. Utilisation of orange peel in the production of α-terpineol by *Penicillium digitatum* (NRRL 1202). Food Chem. 126, 849–854.

Ballester, A.R., Izquierdo, A., Lafuente, M.T., González-Candelas, L., 2010. Biochemical and molecular characterization of induced resistance against *Penicillium digitatum* in citrus fruit. Postharvest Biol. Technol. 56, 31–38.

Ballester, A.R., Lafuente, M.T., González-Candelas, L., 2006. Spatial study of antioxidant enzymes, peroxidase and phenylalanine ammonia-lyase in the citrus fruit-*Penicillium digitatum* interaction. Postharvest Biol. Technol. 39, 115–124.

Ballester, A.R., Lafuente, M.T., De Vos, R.C.H., Bovy, A.G., González-Candelas, L., 2013. Citrus phenylpropanoids and defence against pathogens. Part I: Metabolic profiling in elicited fruits. Food Chem. 136, 178–185.

Bancroft, M.N., Gardner, P.D., Eckert, J.W., Baritelle, J.L., 1984. Comparison of decay control strategies in California lemon packing houses. Plant Dis. 68, 24–28.

Barkai-Golan, R., 1992. Suppression of postharvest pathogens of fresh fruits and vegetables by ionizing radiation. In: Rosenthal, I. (Ed.), Electromagnetic Radiations in Fruit Science. Springer-Verlag, Berlin, Germany, pp. 155–193.

Barkai-Golan, R., Karadavid, R., 1991. Cellulolytic activity of *Penicillium digitatum* and *P. italicum* related to fungal growth and to pathogenesis in citrus fruits. J. Phytopathol. 131, 65–72.

Barmore, C.R., Brown, G.E., 1982. Spread of *Penicillium digitatum* and *Penicillium italicum* during contact between citrus fruits. Phytopathology 72, 116–120.

Ben-Arie, R., Kvitnitsky, E., Paluy, I., 2011. Control of blue mold rot on mandarins with gaseous phase of essential oil. Acta Horticulturae 905, 275–281.

Ben-Yehoshua, S. (Ed.), 2005. Environmentally Friendly Technologies for Agricultural Produce Quality. CRC Press, Taylor and Francis Group, Boca Raton, FL, USA. 534 pp.

Ben-Yehoshua, S., Rodov, V., Nafussi, B., et al., 2008. Involvement of limonene hydroperoxides formed after oil gland injury in the induction of defense response against *Penicillium digitatum* in lemon fruit. J. Agric. Food Chem. 56, 1889–1895.

Boonkorn, P., Gemma, H., Sugaya, S., Setha, S., Uthaibutra, J., Whangchai, K., 2012. Impact of high-dose, short periods of ozone exposure on green mold and antioxidant enzyme activity of tangerine fruit. Postharvest Biol. Technol. 67, 25–28.

Brown, G.E., 1973. Development of green mold in degreened oranges. Phytopathology 63, 1104–1107.

Caccioni, D.R.L., Guizzardi, M., Biondi, D.M., Renda, A., Ruberto, G., 1998. Relationship between volatile components of citrus fruit essential oils and antimicrobial action on *Penicillium digitatum* and *Penicillium italicum*. Int. J. Food Microbiol. 43, 73–79.

Cañamás, T.P., Viñas, I., Usall, J., Torres, R., Anguera, M., Teixidó, N., 2008. Control of postharvest diseases on citrus fruit by preharvest applications of biocontrol agent *Pantoea agglomerans* CPA-2: Part II. Effectiveness of different cell formulations. Postharvest Biol. Technol. 49, 96–106.
Castoria, R., Wright, S.A.I., 2010. Host responses to biological control agents. In: Prusky, D., Gullino, M.L. (Eds.). Postharvest Pathology, Plant Pathology in the 21st Century: Contributions to the 9th International Congress, vol. 2. SpringerScience + Business Media B.V, Dordrecht, The Netherlands, pp. 171–181.
Ceponis, M.J., Cappellini, R.A., Lightner, G.W., 1986. Disorders in citrus shipments to the New York market 1972–1984. Plant Dis. 70, 1162–1165.
Cerioni, L., Rodriguez-Montelongo, L., Ramallo, J., Prado, F.E., Rapisarda, V.A., Volentini, S.I., 2012. Control of lemon green mold by a sequential oxidative treatment and sodium bicarbonate. Postharvest Biol. Technol. 63, 33–39.
Cerioni, L., Rapisarda, V.A., Doctor, J., et al., 2013a. Use of phosphite salts in laboratory and semicommercial tests to control citrus postharvest decay. Plant Dis. 97, 201–212.
Cerioni, L., Sepulveda, M., Rubio-Ames, Z., et al., 2013b. Control of lemon postharvest diseases by low-toxicity salts combined with hydrogen peroxide and heat. Postharvest Biol. Technol. 83, 17–21.
Cháfer, M., Sánchez-González, L., González-Martínez, C., Chiralt, A., 2012. Fungal decay and shelf life of oranges coated with chitosan and bergamot, thyme, and tea tree essential oils. J. Food Sci. 77, E182–E187.
Chien, P.J., Chou, C.C., 2006. Antifungal activity of chitosan and its application to control post-harvest quality and fungal rotting ofTankan citrus fruit (*Citrus tankan* Hayata). J. Sci. Food Agric. 86, 1964–1969.
Cunningham, N.M., 2007. Getting to know your postharvest sanitizers. SARDI Packers Newsletter 86, 3–4.
Cunningham, N.M., 2009. Fungicide and sanitisers – compatibility is still an issue for citrus packers. SARDI Packers Newsletter 94, 6–7.
Cunningham, N., 2010. Combinations of treatments to replace the use of conventional fungicides for the commercial control of postharvest diseases of citrus fruit. Stewart Postharvest Rev. 1, 1–8.
Dao, T., Bensoussan, M., Gervais, P., Dantigny, P., 2008. Inactivation of conidia of *Penicillium chrysogenum, P. digitatum* and *P. italicum* by ethanol solutions and vapours. Int. J. Food Microbiol. 122, 68–73.
D'Aquino, S., Schirra, M., Palma, A., Angioni, A., Cabras, P., Migheli, Q., 2006. Residue levels and effectiveness of pyrimethanil vs imazalil when using heated postharvest dip treatments for control of *Penicillium* decay on citrus fruit. J. Agric. Food Chem. 54, 4721–4726.
D'Aquino, S., Fadda, A., Barberis, A., Palma, A., Angioni, A., Schirra, M., 2013a. Combined effects of potassium sorbate, hot water and thiabendazole against green mould of citrus fruit and residue levels. Food Chem. 141, 858–864.
D'Aquino, S., Palma, A., Angioni, A., Schirra, M., 2013b. Residue levels and efficacy of fludioxonil and thiabendazole in controlling postharvest green mold decay in citrus fruit when applied in combination with sodium bicarbonate. J. Agric. Food Chem. 61, 296–306.
De Gara, L., de Pinto, M.C., Tommasi, F., 2003. The antioxidant systems vis-a-vis reactive oxygen species during plant-pathogen interaction. Plant Physiol. Biochem. 41, 863–870.
Del Río, J.A., Ortuño, A., 2004. Biosynthesis of flavonoids in citrus and its involvement in the antifungal defense mechanisms. In: Dris, R., Niskanen, R., Jain, S.M. (Eds.), Crop Management and Postharvest Handling of Horticultural Products. Science Publishers, Enfield, NH, USA, pp. 185–221.

Domsch, K.H., Gams, W., Anderson, T.H., 1980. Compendium of Soil Fungi, vol. 1, Academic Press, London, UK.
Dore, A., Molinu, M.G., Venditti, T., D'hallewin, G., 2010. Sodium bicarbonate induces crystalline wax generation, activates host-resistance, and increases imazalil level in rind wounds of oranges, improving the control of green mold during storage. J. Agric. Food Chem. 58, 7297–7304.
Droby, S., Chalutz, E., Horev, B., et al., 1993. Factors affecting UV-induced resistance in grapefruit against the green mould decay caused by *Penicillium digitatum*. Plant Pathol. 42, 418–424.
Droby, S., Eick, A., Macarisin, D., et al., 2008. Role of citrus volatiles in host recognition, germination and growth of *Penicillium digitatum* and *Penicillium italicum*. Postharvest Biol. Technol. 49, 386–396.
Droby, S., Wisniewski, M., Macarisin, D., Wilson, C., 2009. Twenty years of postharvest biocontrol research: Is it time for a new paradigm? Postharvest Biol. Technol. 52, 137–145.
du Plooy, W., Regnier, T., Combrinck, S., 2009. Essential oil amended coatings as alternatives to synthetic fungicides in citrus postharvest management. Postharvest Biol. Technol. 53, 117–122.
Eckert, J.W., Brown, G.E., 1986. Evaluation of postharvest fungicide treatments for citrus fruits. In: Hickey, K.D. (Ed.), Methods for Evaluating Pesticides for Control of Plant Pathogens. APS Press, St Paul, MN, USA, pp. 92–97.
Eckert, J.W., Eaks, I.L., 1989. Postharvest disorders and diseases of citrus fruits. In: Reuter, W., Calavan, E.C., Carman, G.E. (Eds.), The Citrus Industry. DANR. University of California Press, Berkeley, CA, USA, pp. 179–260.
Eckert, J.W., Ratnayake, M., 1994. Role of volatile compounds from wounded oranges in induction of germination of *Penicillium digitatum* conidia. Phytopathology 84, 746–750.
El-Ghaouth, A., Smilanick, J.L., Brown, G.E., Ippolito, A., Wisniewski, M., Wilson, C.L., 2000. Application of *Candida saitoana* and glycolchitosan for the control of postharvest diseases of apple and citrus fruit under semi-commercial conditions. Plant Dis. 84, 243–248.
El-Kazzaz, M.K., Chordas, A., Kader, A.A., 1983a. Physiological and compositional changes in orange fruit in relation to modification of their susceptibility to *Penicillium italicum* by ethylene treatments. J. Am. Soc. Horticultural Sci. 108, 618–621.
El-Kazzaz, M.K., Sommer, N.F., Kader, A.A., 1983b. Ethylene effects on *in vitro* and *in vivo* growth of certain postharvest fruit-infecting fungi. Phytopathology 73, 998–1001.
El-Mougy, N.S., Abdel-Kader, M.M., Aly, M.H., 2012. Effect of a new chemical formula on postharvest decay incidence in citrus fruit. J. Plant Prot. Res. 52, 156–264.
Erasmus, A., Lennox, C.L., Smilanick, J.L., Lesar, K., Fourie, P.H., 2013. Imazalil residue loading and green mould control on citrus fruit as affected by formulation, solution pH and exposure time in aqueous dip treatments. Postharvest Biol. Technol. 77, 43–49.
FAO, Food and Agriculture Organization of the United Nations, 2012. Citrus Fruit, Fresh and Processed, Annual Statistics 2012. Available at: www.fao.org. Accessed July 27, 2013.
Farr, D.F., Rossman, A.Y., 2013. Fungal Databases, Systematic Mycology and Microbiology Laboratory, ARS, USDA. Available at: http://nt.ars-grin.gov/fungaldatabases. Accessed: July 12, 27, August 2, 7, 2013.
Frisvad, J.C., Samson, R.A., 2004. Polyphasic taxonomy of *Penicillium* subgenus *Penicillium*. A guide to identification of food and air-borne terverticillate Penicillia and their mycotoxins. Stud. Mycol. 49, 1–174.
Gandía, M., Harries, E., Marcos, J.F., 2012. Identification and characterization of chitin synthase genes in the postharvest citrus fruit pathogen *Penicillium digitatum*. Fungal Biol. 116, 654–664.

Geng, P., Chen, S., Hu, M., et al., 2011. Combination of *Kluyveromyces marxianus* and sodium bicarbonate for controlling green mold of citrus fruit. Int. J. Food Microbiol. 151, 190–194.

Gerez, C.L., Carbajo, M.S., Rollán, G., Torres Leal, G., Font de Valdez, G., 2010. Inhibition of citrus fungal pathogens by using lactic acid bacteria. J. Food Sci. 75, M354–M359.

Gómez-Sanchis, J., Blasco, J., Soria-Olivas, E., et al., 2013. Hyperspectral LCTF-based system for classification of decay in mandarins caused by *Penicillium digitatum* and *Penicillium italicum* using the most relevant bands and non-linear classifiers. Postharvest Biol. Technol. 82, 76–86.

González-Candelas, L., Alamar, S., Ballester, A.R., et al., 2010a. Global regulation of genes in citrus fruit in response to the postharvest pathogen Penicillium digitatum. In: Prusky, D., Gullino, M.L. (Eds.). Postharvest Pathology, Plant Pathology in the 21st Century: Contributions to the 9th International Congress, vol. 2. SpringerScience + Business Media B.V, Dordrecht, The Netherlands, pp. 57–67.

González-Candelas, L., Alamar, S., Sánchez-Torres, P., Zacarías, L., Marcos, J.F., 2010b. A transcriptomic approach highlights induction of secondary metabolism in citrus fruit in response to *Penicillium digitatum* infection. BMC Plant Pathol. 10, 194.

Green, F.M., 1932. The infection of oranges by *Penicillium*. J. Pomol. Horticultural Sci. 10, 184–215.

Gruber, J., Nascimento, H.M., Yamauchi, E.Y., et al., 2013. A conductive polymer based electronic nose for early detection of *Penicillium digitatum* in post-harvest oranges. Materials Science & Engineering. C, Mater. Biol. Appl. 33, 2766–2769.

Gutter, Y., Shachnai, A., Schiffmann-Nadel, M., Dinoor, A., 1981. Biological aspects of citrus molds tolerant to benzimidazole fungicides. Phytopathology 71, 482–487.

Hall, D.J., 1988. Comparative activity of selected food preservatives as citrus postharvest fungicides. Proc. Florida State Horticultural Soc. 101, 184–187.

Hao, W., Li, H., Hu, M., Yang, L., Rizwan-ul-Haq, M., 2011. Integrated control of citrus green and blue mold and sour rot by *Bacillus amyloliquefaciens* in combination with tea saponin. Postharvest Biol. Technol. 59, 316–323.

Hershokovitz, V., Ben-Dayan, C., Raphael, G., et al., 2012. Global changes in gene expression of grapefruit peel tissue in response to the yeast biocontrol agent *Metschnikowia fructicola*. Molecular Plant Pathol. 13, 338–349.

Holmes, G.J., Eckert, J.W., 1999. Sensitivity of *Penicillium digitatum* and *P. italicum* to postharvest citrus fungicides in California. Phytopathology 89, 716–721.

Holmes, G.J., Eckert, J.W., Pitt, J.I., 1994. Revised description of *Penicillium ulaiense* and its role as a pathogen of citrus fruits. Phytopathology 84, 719–727.

Hsieh, H.M., Su, H.J., Tzean, S.S., 1987. The genus *Penicillium* in Taiwan I. Two new taxa of synnematous *Penicillium*. Trans. Mycol. Soc. Republic China 2, 157–168.

Iqbal, Z., Singh, Z., Khangura, R., Ahmad, S., 2012. Management of citrus blue and green moulds through application of organic elicitors. Australas. Plant Pathol. 41, 69–77.

Ismail, M.A., Brown, G.E., 1979. Postharvest wound healing in citrus fruit: Induction of phenylalanine ammonia-lyase in injured 'Valencia' orange flavedo. J. Am. Soc. Horticultural Sci. 104, 126–129.

Iseki, S., Ohta, T., Aomatsu, A., et al., 2010. Rapid inactivation of *Penicillium digitatum* spores using high-density nonequilibrium atmospheric pressure plasma. Appl. Phys. Lett. 96, 153704.

Janisiewicz, W.J., Korsten, L., 2002. Biological control of postharvest diseases of fruits. Ann. Rev. Phytopathol. 40, 411–441.

Kanetis, L., Förster, H., Adaskaveg, J.E., 2008a. Baseline sensitivities for new postharvest fungicides against *Penicillium* spp. on citrus and multiple resistance evaluations in *P. digitatum*. Plant Dis. 92, 301–310.

Kanetis, L., Förster, H., Adaskaveg, J.E., 2008b. Optimizing efficacy of new postharvest fungicides and evaluation of sanitizing agents for managing citrus green mold. Plant Dis. 92, 261–269.

Kanetis, L., Förster, H., Jones, C.A., Borkovich, K.A., Adaskaveg, J.E., 2008c. Characterization of genetic and biochemical mechanisms of fludioxonil and pyrimethanil resistance in field isolates of *Penicillium digitatum*. Phytopathology 98, 205–214.

Kanetis, L., Förster, H., Adaskaveg, J.E., 2010. Determination of natural resistance frequencies in *Penicillium digitatum* using a new air-sampling method and characterization of fludioxonil- and pyrimethanil-resistant isolates. Phytopathology 100, 738–746.

Karaca, H., 2010. Use of ozone in the citrus industry. Ozone Sci. Eng. 32, 122–129.

Kavanagh, J.A., Wood, R.K.S., 1967. The role of wounds in the infection of oranges by *Penicillium digitatum* Sacc. Ann. Appl. Biol. 60, 375–383.

Kim, J.J., Ben-Yehoshua, S., Shapiro, B., Henis, Y., Carmeli, S., 1991. Accumulation of scoparone in heat-treated lemon fruit inoculated with *Penicillium digitatum* Sacc. Plant Physiol. 97, 880–885.

Kinay, P., Yildiz, F., Sen, F., Yildiz, M., Karacali, I., 2005. Integration of pre- and postharvest treatments to minimize *Penicillium* decay of Satsuma mandarins. Postharvest Biol. Technol. 37, 31–36.

Kinay, P., Mansour, M.F., Mlikota-Gabler, F., Margosan, D.A., Smilanick, J.L., 2007. Characterization of fungicide-resistant isolates of *Penicillium digitatum* collected in California. Crop Prot. 26, 647–656.

Kouassi, K.S., Bajji, M., Jijakli, H., 2012. The control of postharvest blue and green molds of citrus in relation with essential oil wax formulations, adherence and viscosity. Postharvest Biol. Technol. 73, 122–128.

Lai, K., Chen, S., Hu, M., et al., 2012. Control of postharvest green mold of citrus fruit by application of endophytic *Paenibacillus polymyxa* strain SG-6. Postharvest Biol. Technol. 69, 40–48.

Li, Q., Ning, P., Zheng, L., Huang, J., Li, G., Hsiang, T., 2010. Fumigant activity of volatiles of *Streptomyces globisporus* JK-1 against *Penicillium italicum* on *Citrus microcarpa*. Postharvest Biol. Technol. 58, 157–165.

Li, W.J., Li, Q., Liu, D.L., Ding, M.W., 2013. Synthesis, fungicidal activity, and sterol 14α-demethylase binding interaction of 2-azolyl-3,4-dihydroquinazolines on *Penicillium digitatum*. J. Agric. Food Chem. 61, 1419–1426.

Liao, H.L., Alférez, F., Burns, J.K., 2013. Assessment of blue light treatments on citrus postharvest diseases. Postharvest Biol. Technol. 81, 81–88.

Liu, J., Zong, Y., Qin, G., Li, B., Tian, S., 2010. Plasma membrane damage contributes to antifungal activity of silicon against *Penicillium digitatum*. Current Microbiol. 61, 274–279.

Lorente, D., Zude, M., Regen, C., Palou, L., Gómez-Sanchis, J., Blasco, J., 2013. Early decay detection in citrus fruit using laser-light backscattering imaging. Postharvest Biol. Technol. 86, 424–430.

Lu, L., Lu, H., Wu, C., et al., 2013. *Rhodosporidium paludigenum* induces resistance and defense-related responses against *Penicillium digitatum* in citrus fruit. Postharvest Biol. Technol. 85, 196–202.

Luo, Y., Zhou, Y., Zeng, K., 2013. Effect of *Pichia membranefaciens* on ROS metabolism and postharvest disease control in citrus fruit. Crop Prot. 53, 96–102.

Macarisin, D., Cohen, L., Eick, A., et al., 2007. *Penicillium digitatum* suppresses production of hydrogen peroxide in host tissue during infection of citrus fruit. Phytopathology 97, 1491–1500.

Macarisin, D., Droby, S., Bauchan, G., Wisniewski, M., 2010. Superoxide anion and hydrogen peroxide in the yeast antagonist-fruit interaction: A new role for reactive oxygen species in postharvest biocontrol? Postharvest Biol. Technol. 58, 194–202.

Maldonado, M.C., Orosco, C.E., Gordillo, M.A., Navarro, A.R., 2010. *In vivo* and *in vitro* antagonism of *Streptomyces* sp. RO3 against *Penicillium digitatum* and *Geotrichum candidum*. Afri. J. Microbiol. Res. 4, 2451–2456.

Manso, T., Nunes, C., 2011. *Metschnikowia andauensis* as a new biocontrol agent of fruit postharvest diseases. Postharvest Biol. Technol. 61, 64–71.
Marcet-Houben, M., Ballester, A.R., de la Fuente, B., et al., 2012. Genome sequence of the necrotrophic fungus *Penicillium digitatum*, the main postharvest pathogen of citrus. BMC Genomics 13, 646.
Marcos, J.F., González-Candelas, L., Zacarías, L., 2005. Involvement of ethylene biosynthesis and perception in the susceptibility of citrus fruits to *Penicillium digitatum* infection and the accumulation of defence-related mRNAs. J. Exp. Bot. 56, 2183–2193.
Mayuoni, L., Tietel, Z., Patil, B.S., Porat, R., 2011. Does ethylene degreening affect internal quality of citrus fruit? Postharvest Biol. Technol. 62, 50–58.
McGlasson, W.B., Eaks, I.L., 1972. A role for ethylene in the development of wastage and off-flavors in stored "Valencia" oranges. HortScience 7, 80–81.
McGuire, R.G., 2000. Population dynamics of postharvest decay antagonists growing epiphytically and within wounds on grapefruit. Phytopathology 90, 1217–1223.
McKay, A.H., Förster, H., Adaskaveg, J.E., 2012. Efficacy and application strategies for propiconazole as a new postharvest fungicide for managing sour rot and green mold of citrus fruit. Plant Dis. 96, 235–242.
Mercier, J., Smilanick, J.L., 2005. Control of green mold and sour rot of stored lemon by biofumigation with *Muscodor albus*. Biol. Control 32, 401–407.
Montesinos-Herrero, C., del Río, M.A., Pastor, C., Brunetti, O., Palou, L., 2009a. Evaluation of brief potassium sorbate dips to control postharvest penicillium decay on major citrus species and cultivars. Postharvest Biol. Technol. 52, 117–125.
Montesinos-Herrero, C., Smilanick, J.L., Hurley, J.M., Palou, L., 2009b. Potassium sorbate residue levels and persistence in citrus fruit as detected by a simple colorimetric method. J. Agric. Food Chem. 57, 3458–3463.
Montesinos-Herrero, C., Palou, L., 2010. Combination of physical and low-toxicity chemical postharvest treatments for integrated disease management of citrus fruit: a review. Stewart Postharvest Rev. 1, 1–11.
Montesinos-Herrero, C., Smilanick, J.L., Tebbets, J.S., Walse, S., Palou, L., 2011. Control of citrus postharvest decay by ammonia gas fumigation and its influence on the efficacy of the fungicide imazalil. Postharvest Biol. Technol. 59, 85–93.
Montesinos-Herrero, C., del Río, M.A., Rojas-Argudo, C., Palou, L., 2012. Short exposure to high CO_2 and O_2 at curing temperature to control postharvest diseases of citrus fruit. Plant Dis. 96, 423–430.
Montesinos-Herrero, C., Palou, L., 2013. Control of postharvest green mold of citrus fruit by combining potassium sorbate dip treatments with brief exposures to high CO2 or O2. Proceedings of the 10th International Congress of Plant Pathology. 25–30 August 2013, Beijing, China, in press.
Morris, S.C., 1982. Synergism of *Geotrichum candidum* and *Penicillium digitatum* in infected citrus fruit. Phytopathology 72, 1336–1339.
Moscoso-Ramírez, P.A., Palou, L., 2013. Evaluation of postharvest treatments with chemical resistance inducers to control green and blue molds on orange fruit. Postharvest Biol. Technol. 85, 132–135.
Moscoso-Ramírez, P.A., Palou, L., 2014a. Effect of ethylene degreening on the development of postharvest penicillium molds and fruit quality of early season citrus fruits. Postharvest Biol. Technol. 91, 1–8.
Moscoso-Ramírez, P.A., Palou, L., 2014b. Preventive and curative activity of postharvest potassium silicate treatments to control green and blue molds on orange fruit. Eur. J. Plant Pathol. 138, 721–732.
Moscoso-Ramírez, P.A., Montesinos-Herrero, C., Palou, L., 2013a. Characterization of postharvest treatments with sodium methylparaben to control citrus green and blue molds. Postharvest Biol. Technol. 77, 128–137.

Moscoso-Ramírez, P.A., Montesinos-Herrero, C., Palou, L., 2013b. Control of citrus postharvest penicillium molds with sodium ethylparaben. Crop Prot. 46, 44–51.

Muñoz, A., López-García, B., Marcos, J.F., 2007. Comparative study of antimicrobial peptides to control citrus postharvest decay caused by *Penicillium digitatum*. J. Agric. Food Chem. 55, 8170–8176.

Nafussi, B., Ben-Yehoshua, S., Rodov, V., Peretz, J., Ozer, B.K., D'hallewin, G., 2001. Mode of action of hot-water dip in reducing decay of lemon fruit. J. Agric. Food Chem. 49, 107–113.

Naqvi, S.A.M.H., 2004. Diagnosis and management of pre and post-harvest diseases of citrus fruit. In: Naqvi, S.A.M.H. (Ed.), Diseases of Fruits and Vegetables. Diseases and Management. Kluwer Academic Publishers, Dordrecht, The Netherlands, pp. 339–359.

Nedbal, L., Soukupová, J., Whitmarsh, J., Trtílek, M., 2000. Postharvest imaging of chlorophyll fluorescence from lemons can be used to predict fruit quality. Photosynthetica 38, 571–579.

Nunes, C.A., 2012. Biological control of postharvest diseases of fruit. Eur. J. Plant Pathol. 133, 181–196.

Onions, A.H.S., 1966a. *Penicillium digitatum*. C.M.I. Description Pathogenic Fungi Bacteria 96, 1–2.

Onions, A.H.S., 1966b. *Penicillium italicum*. C.M.I. Description Pathogenic Fungi Bacteria 99, 1–2.

Ortuño, A., Báidez, A., Gómez, P., et al., 2006. *Citrus paradisi* and *Citrus sinensis* flavonoids: Their influence in the defence mechanism against *Penicillium digitatum*. Food Chem. 98, 351–358.

Ortuño, A., Díaz, L., Álvarez, N., Porras, I., García-Lidón, A., Del Río, J.A., 2011. Comparative study of flavonoid and scoparone accumulation in different *Citrus* species and their susceptibility to *Penicillium digitatum*. Food Chem. 125, 232–239.

Pallottino, F., Costa, C., Antonucci, F., et al., 2012. Electronic nose application for determination of *Penicillium digitatum* in Valencia oranges. J. Sci. Food Agric. 92, 2008–2012.

Palou, L., 2009. Control of citrus postharvest diseases by physical means. Tree Forestry Sci. Biotechnol. 3, 127–142.

Palou, L., 2011. Control integrado no contaminante de enfermedades de poscosecha (CINCEP): nuevo paradigma para el sector español de los cítricos. Levante Agrícola 406, 173–183 (in Spanish with English summary).

Palou, L., Smilanick, J.L., Crisosto, C.H., Mansour, M.F., 2001a. Effect of gaseous ozone exposure on the development of green and blue molds on cold stored citrus fruit. Plant Dis. 85, 632–638.

Palou, L., Smilanick, J.L., Usall, J., Viñas, I., 2001b. Control of postharvest blue and green molds of oranges by hot water, sodium carbonate, and sodium bicarbonate. Plant Dis. 85, 371–376.

Palou, L., Usall, J., Muñoz, J.A., Smilanick, J.L., Viñas, I., 2002a. Hot water, sodium carbonate, and sodium bicarbonate for the control of postharvest green and blue molds of clementine mandarins. Postharvest Biol. Technol. 24, 93–96.

Palou, L., Usall, J., Smilanick, J.L., Aguilar, M.J., Viñas, I., 2002b. Evaluation of food additives and low-toxicity compounds as alternative chemicals for the control of *Penicillium digitatum* and *Penicillium italicum* on citrus fruit. Pest Manag. Sci. 58, 459–466.

Palou, L., Smilanick, J.L., Crisosto, C.H., Mansour, M., Plaza, P., 2003. Ozone gas penetration and control of the sporulation of *Penicillium digitatum* and *Penicillium italicum* within commercial packages of oranges during cold storage. Crop Prot. 22, 1131–1134.

Palou, L., Smilanick, J.L., Margosan, D.A., 2007. Ozone applications for sanitation and control of postharvest diseases of fresh fruits and vegetables. In: Troncoso-Rojas, R., Tiznado-Hernández, M.E., González-León, A. (Eds.), Recent Advances in Alternative Postharvest Technologies to Control Fungal Diseases in Fruits and Vegetables. Transworld Research Network, Trivandrum, Kerala, India, pp. 39–70.

Palou, L., Smilanick, J.L., Droby, S., 2008. Alternatives to conventional fungicides for the control of citrus postharvest green and blue molds. Stewart Postharvest Rev. 2 (2), 1–16.
Palou, L., Smilanick, J.L., Montesinos-Herrero, C., Valencia-Chamorro, S., Pérez-Gago, M.B., 2011. Novel approaches for postharvest preservation of fresh citrus fruits. In: Slaker, D.A. (Ed.), Citrus Fruits: Properties, Consumption and Nutrition. Nova Science Publishers, Inc., New York, USA, pp. 1–45.
Pastor, C., Sánchez-González, L., Cháfer, M., Chiralt, A., González-Martínez, C., 2010. Physical and antifungal properties of hydroxypropylmethylcellulose based films containing propolis as affected by moisture content. Carbohydr. Polymers 82, 1174–1183.
Pavoncello, D., Lurie, S., Droby, S., Porat, R., 2001. A hot water treatment induces resistance to *Penicillium digitatum* and promotes the accumulation of heat shock and pathogenesis-related proteins in grapefruit flavedo. Physiol. Plantarum 111, 17–22.
Pelser, P., du, T., 1977. Postharvest handling of South African citrus fruit. Proc. Int. Soc. Citriculture 1, 244–249.
Pérez, E., Blanco, O., Berreta, C., Dol, I., Lado, J., 2011. Imazalil concentration for *in vitro* monitoring of imazalil resistant isolates of *Penicillium digitatum* in citrus packinghouses. Postharvest Biol. Technol. 60, 258–262.
Pérez-Alfonso, C.O., Martínez-Romero, D., Zapata, P.J., Serrano, M., Valero, D., Castillo, S., 2012. The effects of essential oils carvacrol and thymol on growth of *Penicillium digitatum* and *P. italicum* involved in lemon decay. Int. J. Food Microbiol. 158, 101–106.
Perotti, V.E., Del Vecchio, H.A., Sansevich, A., et al., 2011. Proteomic, metabalomic, and biochemical analysis of heat treated Valencia oranges during storage. Postharvest Biol. Technol. 62, 97–114.
Pitt, J.I., Hocking, A.D., 2009. Fungi and Food Spoilage, third ed. Springer, Dordrecht, The Netherlands.
Platania, C., Restuccia, C., Muccilli, S., Cirvilleri, G., 2012. Efficacy of killer yeasts in the biological control of *Penicillium digitatum* on Tarocco orange fruits (*Citrus sinensis*). Food Microbiol. 30, 219–225.
Plaza, P., Usall, J., Teixidó, N., Viñas, I., 2003a. Effect of water activity and temperature on germination and growth of *Penicillium digitatum, P. italicum* and *Geotrichum candidum*. J. Appl. Microbiol. 94, 549–554.
Plaza, P., Usall, J., Torres, R., Lamarca, N., Asensio, A., Viñas, I., 2003b. Control of green and blue mould by curing of oranges during ambient and cold storage. Postharvest Biol. Technol. 28, 195–198.
Plaza, P., Sanbruno, A., Usall, J., et al., 2004a. Integration of curing treatments with degreening to control the main postharvest diseases of clementine mandarins. Postharvest Biol. Technol. 34, 29–37.
Plaza, P., Torres, R., Usall, J., Lamarca, N., Viñas, I., 2004b. Evaluation of the potential of commercial post-harvest application of essential oils to control citrus decay. J. Horticultural Sci. Biotechnol. 79, 935–940.
Plaza, P., Usall, J., Teixidó, N., Viñas, I., 2004c. Effect of water activity and temperature on competing abilities of common postharvest citrus fungi. Int. J. Food Microbiol. 90, 75–82.
Porat, R., Weiss, B., Cohen, L., Daus, A., Goren, R., Droby, S., 1999. Effects of ethylene and 1-methylcyclopropene on the postharvest qualities of 'Shamouti' oranges. Postharvest Biol. Technol. 15, 155–163.
Porat, R., Daus, A., Weiss, B., Cohen, L., Fallik, E., Droby, S., 2000a. Reduction of postharvest decay in organic citrus fruit by a short hot water brushing treatment. Postharvest Biol. Technol. 18, 151–157.
Porat, R., Pavoncello, D., Peretz, J., et al., 2000b. Induction of resistance to *Penicillium digitatum* and chilling injury in 'Star Ruby' grapefruit by a short hot-water rinse and brushing treatment. J. Horticultural Sci. Biotechnol. 75, 428–432.

Porat, R., McCollum, T.G., Vinokur, V., Droby, S., 2002. Effects of various elicitors on the transcription of a β-1,3-endoglucanase gene in citrus fruit. J. Phytopathol. 150, 70–75.
Porat, R., Vinokur, V., Weiss, B., et al., 2003. Induction of resistance to *Penicillium digitatum* in grapefruit by β-aminobutyric acid. Eur. J. Plant Pathol. 109, 901–907.
Potjewijd, R., Nisperos, M.O., Burns, J.K., Parish, M., Baldwin, E.A., 1995. Cellulose-based coatings as carriers for *Candida guillermondii* and *Debaryomyces* sp. in reducing decay of oranges. HortScience 30, 1417–1421.
Prusky, D., McEvoy, J.L., Saftner, R.A., Conway, W.S., Jones, R., 2004. Relationship between host acidification and virulence of *Penicillium* spp. on apple and citrus fruit. Phytopathology 94, 44–51.
Raper, K.B., Thom, C., 1949. A Manual of the Penicillia. Hafner Publishing Co., New York, USA.
Ritenour, M.A., Pelosi, R.R., Burton, M.S., Stover, E.W., Dou, H.T., McCollum, T.G., 2004. Assessing the efficacy of preharvest fungicide applications to control postharvest diseases of Florida citrus. HortTechnology 14, 58–62.
Rodov, V., Ben-Yehoshua, S., Albagli, R., Fang, D.Q., 1995a. Reducing chilling injury and decay of stored citrus fruit by hot water dips. Postharvest Biol. Technol. 5, 119–127.
Rodov, V., Ben-Yehoshua, S., Fang, D.Q., Kim, J.J., Ashkenazi, R., 1995b. Preformed antifungal compounds of lemon fruit: citral and its relation to disease resistance. J. Agric. Food Chem. 43, 1057–1061.
Rodríguez, A., San Andrés, V., Cervera, M., et al., 2011. Terpene down-regulation in orange reveals the role of fruit aromas in mediating interactions with insect herbivores and pathogens. Plant Physiol. 156, 793–802.
Rojas-Argudo, C., Palou, L., Bermejo, A., Cano, A., del Río, M.A., González-Mas, M.C., 2012. Effect of X-ray irradiation on nutritional and antifungal bioactive compounds of 'Clemenules' clementine mandarins. Postharvest Biol. Technol. 68, 47–53.
Rubio Ames, Z., Feliziani, E., Smilanick, J.L., 2013. Germination of fungal conidia after exposure to low concentration ozone atmospheres. Postharvest Biol. Technol. 83, 22–26.
Saccardo, P.A., 1881. *Penicillium digitatum*. Fungi Italici Autographice Delineati Fascs 17-28, tabs 641–1120, 894.
Samson, R.A., Hoekstra, E.S., Frisvad, J.C., 2004. Introduction to Food- and Airborne Fungi, seventh ed. Centralbureau voor Schimmelcultures, Utrecht, The Netherlands.
Sánchez-Torres, P., Tuset, J., 2011. Molecular insights into fungicide resistance in sensitive and resistant *Penicillium digitatum* strains infecting citrus. Postharvest Biol. Technol. 59, 159–165.
Schirra, M., D'hallewin, G., Cabras, P., Angioni, A., Garau, V.L., 1998. Seasonal susceptibility of Tarocco oranges to chilling injury as affected by hot water and thiabendazole postharvest dip treatments. J. Agric. Food Chem. 46, 1177–1180.
Schirra, M., Cabras, P., Angioni, A., Brandolini, V., 2002. Residue levels and storage decay control in cv. Star Ruby grapefruit after dip treatments with azoxystrobin. J. Agric. Food Chem. 50, 1461–1464.
Schirra, M., Mulas, M., Fadda, A., Cauli, E., 2004. Cold quarantine responses of blood oranges to postharvest hot water and hot air treatments. Postharvest Biol. Technol. 31, 191–200.
Schirra, M., D'Aquino, S., Palma, A., et al., 2005. Residue level, persistence and storage performance of citrus fruit treated with fludioxonil. J. Agric. Food Chem. 53, 6718–6724.
Schirra, M., D'Aquino, S., Palma, A., Angioni, A., Cabras, P., Migheli, Q., 2006. Residues of the quinone outside inhibitor fungicide trifloxystrobin after postharvest dip treatments to control *Penicillium* spp. on citrus fruit. J. Food Prot. 69, 1646–1652.
Schirra, M., D'Aquino, S., Palma, A., Angioni, A., Cabras, P., 2008. Factors affecting the synergy of thiabendazole, sodium bicarbonate, and heat to control postharvest green mold of citrus fruit. J. Agric. Food Chem. 56, 10793–10798.

Schirra, M., D'Aquino, S., Cabras, P., Angioni, A., 2009. Influence of post-harvest application rates of cyprodinil, treatment time and temperature on residue levels and efficacy in controlling green mould on 'Valencia' oranges. Food Additives & Contaminants. Part A, Chemistry, Analysis, Control. Expo. Risk Assess. 26, 1033–1037.

Schirra, M., Palma, A., Barberis, A., et al., 2010. Postinfection activity, residue levels, and persistence of azoxystrobin, fludioxonil, and pyrimethanil applied alone or in combination with heat and imazalil for green mold control on inoculated oranges. J. Agric. Food Chem. 58, 3661–3666.

Schirra, M., D'Aquino, S., Cabras, P., Angioni, A., 2011. Control of postharvest diseases of fruit by heat and fungicides: efficacy, residue levels, and residue persistence. A review. J. Agric. Food Chem. 59, 8531–8542.

Schmidt, L.S., Ghosoph, J.M., Margosan, D.A., Smilanick, J.L., 2006. Mutation at β-tubulin codon 200 indicated thiabendazole resistance in *Penicillium digitatum* collected from California citrus packinghouses. Plant Dis. 90, 765–770.

Sdiri, S., Navarro, P., Monterde, A., Ben Abda, J., Salvador, A., 2012. Effect of postharvest degreening followed by a cold-quarantine treatment on vitamin C, phenolic compounds and antioxidant activity of early-season citrus fruit. Postharvest Biol. Technol. 65, 13–21.

Sdiri, S., Navarro, P., Salvador, A., 2013. Postharvest application of a new growth regulator reduces calyx alterations of citrus fruit induced by degreening treatment. Postharvest Biol. Technol. 75, 68–74.

Smilanick, J.L., Sorenson, D., 2001. Control of postharvest decay of citrus fruit with calcium polysulfide. Postharvest Biol. Technol. 21, 157–168.

Smilanick, J.L., Michael, I.F., Mansour, M.F., et al., 1997. Improved control of green mold of citrus with imazalil in warm water compared with its use in wax. Plant Dis. 81, 1299–1304.

Smilanick, J.L., Margosan, D.A., Mlikota-Gabler, F., Usall, J., Michael, I.F., 1999. Control of citrus green mold by carbonate and bicarbonate salts and the influence of commercial postharvest practices on their efficacy. Plant Dis. 83, 139–145.

Smilanick, J.L., Aiyabei, J., Mlikota-Gabler, F., Doctor, J., Sorenson, D., Mackey, B., 2002. Quantification of the toxicity of aqueous chlorine to spores of *Penicillium digitatum* and *Geotrichum citri-aurantii*. Plant Dis. 86, 509–514.

Smilanick, J.L., Mansour, M.F., Margosan, D.A., Mlikota-Gabler, F., Goodwine, W.R., 2005. Influence of pH and $NaHCO_3$ on effectiveness of imazalil to inhibit germination of *Penicillium digitatum* and to control postharvest green mold on citrus fruit. Plant Dis. 89, 640–648.

Smilanick, J.L., Brown, G.E., Eckert, J.W., 2006a. The biology and control of postharvest diseases. In: Wardowski, W.F., Miller, W.M., Hall, D.J., Grierson, W. (Eds.), Fresh Citrus Fruits, second ed. Florida Science Source, Inc., Longboat Key, FL, USA, pp. 339–396.

Smilanick, J.L., Mansour, M.F., Mlikota-Gabler, F., Goodwine, W.R., 2006b. The effectiveness of pyrimethanil to inhibit germination of *Penicillium digitatum* and to control citrus green mold after harvest. Postharvest Biol. Technol. 42, 75–85.

Smilanick, J.L., Mansour, M.F., Sorenson, D., 2006c. Pre- and postharvest treatments to control green mold of citrus fruit during ethylene degreening. Plant Dis. 90, 89–96.

Smilanick, J.L., Mansour, M.F., 2007. Influence of temperature and humidity on survival of *Penicillium digitatum* and *Geotrichum citri-aurantii*. Plant Dis. 91, 990–996.

Smilanick, J.L., Mansour, M.F., Mlikota-Gabler, F., Sorenson, D., 2008. Control of citrus postharvest green mold and sour rot by potassium sorbate combined with heat and fungicides. Postharvest Biol. Technol. 47, 226–238.

Sommer, N.F., Maxie, E.C., Fortlage, R.J., Eckert, J.W., 1964. Sensitivity of citrus fruit decay fungi to gamma irradiation. Radiat. Bot. 4, 317–322.

Sun, X., Li, H., Yu, D., 2011. Complete mitochondrial genome sequence of the phytopathogenic fungus *Penicillium digitatum* and comparative analysis of closely related species. FEMS Microbiol. Lett. 323, 29–34.

Sun, X., Xu, Q., Ruan, R., Zhang, T., Zhu, C., Li, H., 2013. PdMLE1, a specific and active transposon acts as a promoter and confers *Penicillium digitatum* with DMI resistance. Env. Microbiol. Rep. 5, 135–142.
Suwannarach, N., Kumla, J., Bussaban, B., Nuangmek, W., Matsui, K., Lumyong, S., 2013. Biofumigation with the endophytic fungus *Nodulisporium* spp. CMU-UPE34 to control postharvest decay of citrus fruit. Crop Prot. 45, 63–70.
Szczerbanik, M., Jobling, J., Morris, S., Holford, P., 2007. Essential oil vapours control some common postharvest fungal pathogens. Austr. J. Exp. Agric. 47, 103–109.
Taverner, P.D., 2013. Compatibility of peracetic acid in fungicide and salt mixtures. SARDI Packers Newsletter 106, 3.
Tayel, A.A., El-Baz, A.F., Salem, M.F., El-Hadary, M.H., 2009. Potential applications of pomegranate peel extract for the control of citrus green mould. J. Plant Dis. Prot. 116, 252–256.
Teixidó, N., Torres, R., Viñas, I., Abadias, M., Usall, J., 2013. Biological and natural materials for postharvest disease management: success and failures. Proceedings of the 2nd International Symposium on Discovery and Development of Innovative Strategies for Postharvest Disease Management, 28 April-2 May 2013, Kusadasi, Turkey. Acta Horticulturae, in press.
Torres, R., Nunes, C., García, J.M., et al., 2007. Application of *Pantoea agglomerans* CPA-2 in combination with heated sodium bicarbonate solutions to control the major postharvest diseases affecting citrus at several Mediterranean locations. Eur. J. Plant Pathol. 118, 73–83.
Torres, R., Teixidó, N., Usall, J., et al., 2011. Anti-oxidant activity of oranges after infection with the pathogen *Penicillium digitatum* or treatment with the biocontrol agent *Pantoea agglomerans* CPA-2. Biol. Control 57, 103–109.
Tu, Q., Chen, J., Guo, J., 2013. Screening and identification of antagonistic bacteria with potential for biological control of *Penicillium italicum* of citrus fruits. Sci. Horticulturae 150, 125–129.
Tuset, J.J., 1988. Podredumbres de los Frutos Cítricos. Conselleria d'Agricultura i Pesca, Generalitat, Valenciana, Valencia, Spain.
Valencia-Chamorro, S.A., Palou, L., del Río, M.A., Pérez-Gago, M.B., 2008. Inhibition of *Penicillium digitatum* and *Penicillium italicum* by hydroxypropyl methylcellulose-lipid edible composite films containing food additives with antifungal properties. J. Agric. Food Chem. 56, 11270–11278.
Valencia-Chamorro, S.A., Pérez-Gago, M.B., del Río, M.A., Palou, L., 2009a. Curative and preventive activity of hydroxypropyl methylcellulose-lipid edible composite coatings containing antifungal food additives to control citrus postharvest green and blue molds. J. Agric. Food Chem. 57, 2770–2777.
Valencia-Chamorro, S.A., Pérez-Gago, M.B., del Río, M.A., Palou, L., 2009b. Effect of antifungal hydroxypropyl methylcellulose (HPMC)-lipid edible composite coatings on postharvest decay development and quality attributes of cold-stored 'Valencia' oranges. Postharvest Biol. Technol. 54, 72–79.
Valencia-Chamorro, S.A., Pérez-Gago, M.B., del Río, M.A., Palou, L., 2010. Effect of antifungal hydroxypropyl methylcellulose-lipid edible composite coatings on Penicillium decay development and postharvest quality of cold-stored "Ortanique" mandarins. J. Food Sci. 75, S418–S426.
Valencia-Chamorro, S.A., Palou, L., del Río, M.A., Pérez-Gago, M.B., 2011a. Antimicrobial edible films and coatings for fresh and minimally processed fruits and vegetables: a review. Crit. Rev. Food Sci. Nutr. 51, 872–900.
Valencia-Chamorro, S.A., Palou, L., del Río, M.A., Pérez-Gago, M.B., 2011b. Performance of hydroxypropyl methylcellulose (HPMC)-lipid edible composite coatings containing food additives with antifungal properties during cold storage of 'Clemenules' mandarins. LWT - Food Sci. Technol. 44, 2342–2348.

Venditti, T., Molinu, M.G., Dore, A., Agabbio, M., D'hallewin, G., 2005. Sodium carbonate treatment induces scoparone accumulation, structural changes, and alkalinization in the albedo of wounded *Citrus* fruits. J. Agric. Food Chem. 53, 3510–3518.
Venditti, T., Dore, A., Molinu, M.G., Agabbio, M., D'hallewin, G., 2009. Combined effect of curing followed by acetic acid vapour treatments improves postharvest control of *Penicillium digitatum* on mandarins. Postharvest Biol. Technol. 54, 111–114.
Vilanova, L., Viñas, I., Torres, R., Usall, J., Jauset, A.M., Teixidó, N., 2012. Infection capacities in the orange-pathogen relationship: Compatible (*Penicillium digitatum*) and incompatible (*Penicillium expansum*) interactions. Food Microbiol. 29, 56–66.
Vilanova, L., Torres, R., Viñas, I., et al., 2013. Wound response in orange as a resistance mechanism against *Penicillium digitatum* (pathogen) and *P. expansum* (non-host pathogen). Postharvest Biol. Technol. 78, 113–122.
Wang, J.Y., Li, H.Y., 2008. *Agrobacterium tumefaciens*-mediated genetic transformation of the phytopathogenic fungus *Penicillium digitatum*. J. Zhejiang Univ. Sci. B 9, 823–828.
Wang, L.N., Liu, F., Jiang, Y.F., et al., 2011. Synergistic antimicrobial activities of natural essential oils with chitosan films. J. Agric. Food Chem. 59, 12411–12419.
Wang, H., Tao, N., Huang, S., Liu, Y., 2012a. Effect of Shatangju (*Citrus reticulata* Blanco) essential oil on spore germination and mycelium growth of *Penicillium digitatum* and *P. italicum*. J. Essent. Oil Bearing Plants 15, 715–723.
Wang, H.K., Yan, Y.H., Wang, J.M., Zhang, H.P., Qi, W., 2012b. Production and characterization of antifungal compounds produced by *Lactobacillus plantarum* IMAU10014. PLoS ONE 7, e29452.
Wardowski, W.F., Miller, W.M., Grierson, W., 2006. Degreening. In: Wardowski, W.F., Miller, W.M., Hall, D.J., Grierson, W. (Eds.), Fresh Citrus Fruits, second ed. Florida Science Source, Inc., Longboat Key, FL, USA, pp. 277–298.
Wehmer, C., 1894. Eine neue Sklerotien-bildende *Penicillium* species (*P. italicum*). Hedwigia 33, 211–214.
Whangchai, K., Saengnil, K., Singkamanee, C., Uthaibutra, J., 2010. Effect of electrolyzed oxidizing water and continuous ozone exposure on the control of *Penicillium digitatum* on tangerine cv. 'Sai Nam Pung' during storage. Crop Prot. 29, 386–389.
Yahyazadeh, M., Zare, R., Omidbaigi, R., Faghih-Nasiri, M., Abbasi, M., 2009. Control of *Penicillium* decay on citrus fruit using essential oil vapours of thyme or clove inside polyethylene and nano-clay polyethylene films. J. Horticultural Sci. Biotechnol. 84, 403–409.
Yildiz, F., Kinay, P., Yildiz, M., Karacali, I., 2005. Effects of preharvest applications of $CaCl_2$, 2,4-D and benomyl and postharvest hot water, yeast and fungicide treatments on development of decay on Satsuma mandarins. J. Phytopathol. 153, 94–98.
Youssef, K., Ligorio, A., Nigro, F., Ippolito, A., 2012a. Activity of salts incorporated in wax in controlling postharvest diseases of citrus fruit. Postharvest Biol. Technol. 65, 39–43.
Youssef, K., Ligorio, A., Sanzani, S.M., Nigro, F., Ippolito, A., 2012b. Control of storage diseases of citrus by pre- and postharvest application of salts. Postharvest Biol. Technol. 72, 57–63.
Yu, S., Oh, B., Lee, Y.H., 2012. Biocontrol of green and blue molds in postharvest satsuma mandarin using *Bacillus amyloliquefaciens* JBC36. Biocontrol Sci. Technol. 22, 1181–1197.
Zapata, P.J., Navarro, D., Guillén, F., et al., 2013. Characterisation of gels from different *Aloe* spp. as antifungal treatment: Potential crops for industrial applications. Ind. Crops Prod. 42, 223–230.
Zeng, R., Zhang, A., Chen, J., Fu, Y., 2013. Impact of carboxymethyl cellulose coating enriched with extract of *Impatiens balsamina* stems on preservation of 'Newhall' navel orange. Sci. Horticulturae 160, 44–48.
Zhang, J., Swingle, P.P., 2005. Effects of curing on green mold and stem-end rot of citrus fruit and its potential application under Florida packing system. Plant Dis. 89, 834–840.
Zhang, J., Timmer, L.W., 2007. Preharvest application of fungicides for postharvest disease control on early season tangerine hybrids in Florida. Crop Protect. 26, 886–893.

Zhang, Z., Zhu, Z., Ma, Z., Li, H., 2009. Molecular mechanism of azoxystrobin resistance in *Penicillium digitatum* UV mutants and a PCR-based assay for detection of azoxystrobin-resistant strains in packing- or store-house isolates. Int. J. Food Microbiol. 131, 157–161.
Zhang, T., Sun, X., Xu, Q., Zhu, C., Li, Q., Li, H., 2013. PdSNF1, a sucrose non-fermenting protein kinase gene, is required for *Penicillium digitatum* conidiation and virulence. Appl. Microbiol. Biotechnol. 97, 5433–5445.

CHAPTER 3

Crown Rot Disease of Bananas

Ludivine Lassois[1] and Luc de Lapeyre de Bellaire[2]
[1]Forest, Nature and Lanscape Department, Liège University, Gembloux Agro-Bio Tech, Gembloux, Belgium;
[2]CIRAD, Research Unit Banana, Plantain and Pineapple Cropping Systems, Montpellier, France

Contents

Postharvest Decay
http://dx.doi.org/10.1016/B978-0-12-411552-1.00003-X

OVERALL IMPORTANCE OF THE DISEASE

Crown rot disease is considered to be the main export banana postharvest disease (Reyes et al., 1998; Krauss and Johanson, 2000) and it affects export bananas in all banana-producing countries. Crown rot affects tissues of the so-called 'crown', which unites the peduncles (Fig. 3.1A). Since crowns are mainly infected at harvest time, future infection is unpredictable. The symptoms develop rapidly during shipping and ripening when the physiology of the fruit undergoes modifications that facilitate fungal development (Jones, 2000). The rot begins with mycelial development on the surface of the crown (Fig. 3.1B), followed by the peduncles and fruit (Fig. 3.1C–F). The bananas may detach from the peduncle in cases of severe infection. Crown rot results from the development in the crown of several common fungi, which constitute a fungal complex, and leads to softening and blackening of tissues at the site of the wound left when the cluster was cut from the bunch.

Crown rot affects the fruit quality because of the development of necrosis on the fruit, and it can also trigger early ripening of bananas during shipping (Peacock, 1973). Ripening may be caused by ethylene released by stressed and necrotic tissues (Simmonds, 1963; MacCracken and Swinburne, 1980; Dominguez and Vendrell, 1993) but also by ethylene produced by mycelia of fungi such as *Colletotrichum musae* (Peacock and Muirhead, 1974; Daundasekera et al., 2003). The result is a green life reduction and an accelerated maturation of the fruits. Overall, crown rot incidence involves a depreciation of the product inducing significant economic losses. Onset of the disease cannot be predicted and it spreads in an irregular pattern on all clusters in a shipping box. The same box may contain both healthy and infected clusters. The incidence of crown rot periodically increases during the rainy season, and losses of over 10% have been recorded in the UK in bananas coming from the Windward Islands that were harvested during this period (Krauss and Johanson, 2000). Losses of up to 86% have also been reported in non-chemically treated bananas from the Philippines (Alvindia et al., 2000).

Crown rot disease became a major problem in the banana industry during varietal reconversion initiatives in the 1960s. Up until 1960, only Gros-Michel subgroup cultivars were cropped for export, and for economic and practical reasons bananas were shipped in complete bunches consisting of double layers of fruits, called hands, which are arranged helicoidally around a central axis called the stalk. These cultivars were, however, susceptible to Panama disease (*Fusarium oxysporum* f.sp. *cubense*) and

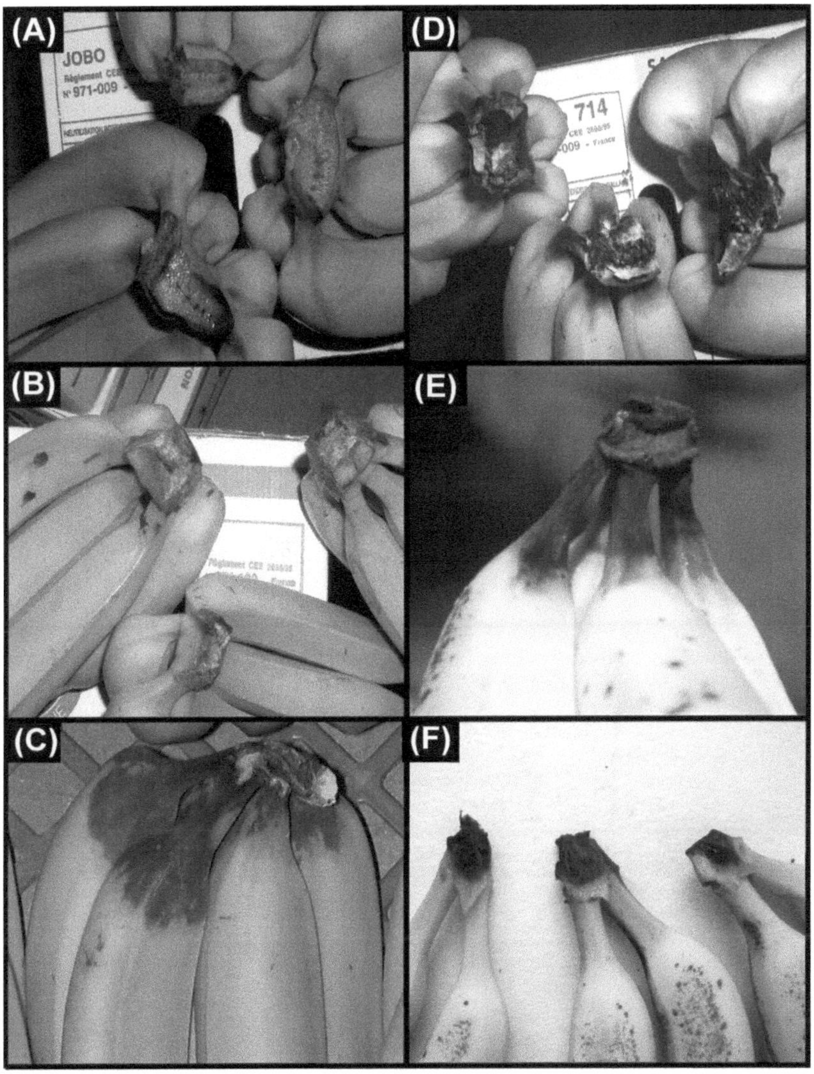

Figure 3.1 Crown rot symptoms. (A) Healthy crowns after maritime shipping before being placed in ripening rooms. (B) Crowns diseased with a superficial mycelium after maritime shipping. (C) Rot noted on bananas after maritime shipping. (D) Rot on the peduncles inducing banana ripening upon their arrival after maritime shipping. (E) External crown rot symptoms after fruit ripening. (F) Internal crown rot symptoms after fruit ripening (Lassois et al., 2010b).

thus were replaced by Cavendish subgroup cultivars because of their resistance to Panama disease (Stover, 1972). However, Cavendish bananas are more fragile during shipping and this triggered a revolution in the banana industry as bunch shipping was discontinued. Bananas were instead cut

Figure 3.2 Successive operations carried out from the field to the banana marketing stage. *Deflowering is done in the packing station when it has not been done in the field (Lassois et al., 2010b).

into clusters consisting of several banana fruits joined by the crown tissues and boxed for shipping (Eckert and Ogawa, 1985). The crown became a prime site for infection by different pathogens (Lukezic et al., 1967; Griffee and Burden, 1976; Jones, 2000).

The successive operations carried out from the field to the banana marketing stage are presented in Figure 3.2 and will help to a better understanding of the contamination process and the control methods.

TAXONOMY AND MORPHOLOGY

Crown rot is the result of the activity of a fungal complex. The role of bacteria in the complex has not been confirmed by detailed research (Lukezic et al., 1967). The microorganisms most commonly isolated in crown rot are: *Musicillium theobromae*, *Colletotrichum musae*, *Ceratocystis paradoxa*, *Lasiodoplodia theobromae*, *Nigrospora sphaerica*, *Cladosporium* spp., *Acremonium* spp., *Penicillium* spp. and *Aspergillus* spp., as well as many *Fusarium* species including *F. semitectum*, *F. verticillioides*, *F. sporotrichoides*, *F. oxysporum* and *F. solani* (Greene and Goos, 1963; Lukezic et al., 1967; Wallbridge and Pinegar, 1975; Griffee and Burden, 1976; Shillingford, 1976; Wallbridge, 1981; Ogundero, 1987; Mesturino and Ragazzi, 1988; Johanson and Blasquez, 1992; Marin et al., 1996; Anthony et al., 2004). Several organisms may be involved in disease development, however, disease severity and the nature of the complex may vary substantially, depending on the production area and season (Table 3.1). Lukezic and Kaiser (1966) showed that fungal populations may differ between banana plants and even between clusters originating from the same plant.

Fungi of the complex do not all have the same pathogenicity and variations have been reported between regions. Knight (1982) considered that *F. oxysporum*, *F. verticillioides* and *F. graminearum*, which have been isolated frequently from crowns of Windward Island banana production, are primary pathogens, whereas *L. theobromae*, *M. theobromae*, and *N. sphaerica* are considered to be relatively non-pathogenic species. According to the findings of Marin et al. (1996) in Costa-Rica, *F. verticillioides* and *F. semitectum* are the most pathogenic species, while Griffee (1976) considered *C. musae*, *L. theobromae*, *C. paradoxa*, *F. semitectum* and *F. graminearum* as major pathogens involved in this disease in the Caribbean. However, many authors agree on the strong pathogenicity of *C. musae*, which may trigger an infection from a very small amount of inoculum since other pathogens require larger amounts of inoculum to induce crown rot symptoms (Greene and Goos, 1963; Lukezic et al., 1967; Griffee, 1976; Shillingford, 1976; Finlay and Brown, 1993; Krauss, 1996; Krauss et al., 1998; Lassois et al., 2008).

The nature of the complex, the specific pathogenicity of the different microorganisms involved as well as interactions among them may alter the

Table 3.1 Bibliographic Data on the Fungal Species Isolated from Crown Rot in Different Banana-producing Regions.

Countries	References	*Colletotrichum musae*	*Fusarium semitectum*	*Fusarium verticillioides*	*Fusarium sporotrichoides*	*Fusarium oxysporum*	*Fusarium solani*	*Lasiodiplodia theobromae*	*Musicillium theobromae*	*Gliocladium roseum*	*Nigrospora sphaerica*	*Acremonium sp.*	*Penicillium sp.*
Honduras	1	3%	80%	14%				12%	81%			93%	
Central and South America	2	F	MF	X				R	F		R		
	3	0-10%	4-50%	0-28%						0-30%		0-8%	0-33%
Windward Islands (WI)	4	23-33%							5-7%	5%			
	5	36%	27%	6%				2%	8%		<1%		
	6	24%	18%	3%		3%	2%	2%	3%	2%	<1%	2%	<1%
	7	26-44%	7-23%	6-21%				3-9%	4-18%	0-10%	24-26%	0-10%	
	8	I		I								I	
	9	X	X	X				X					
	10	I	I										
	11	I	I	I		I		I	I		I		
	12	I	I	I		I		I	I		I		
Jamaica	13	11%	F	X		X		3%	13%		1%	3%	
Sri Lanka	14	X		X				X					
Nigeria	15	27%	3%	10%		4%	6%	26%	2%				
Somalia WI, Guatemala	16	38%	X	X	20%				2%			4%	

Percentage values correspond to isolation frequencies when given by the authors. "X" indicates that the pathogen was identified on the crown by the author(s), but without providing any isolation frequency data. F: frequently isolated; MF: most frequently isolated; R: rarely isolated. "I" indicates that authors evaluated fungal pathogenicity without information about the isolation frequencies. Information about the pathogenicity are shown when given by authors. [black]: Highly pathogenic; [dark grey]: Medium pathogenicity; [light grey]: Slightly or non pathogenic. 1: Lukezic et al., 1967; 2: Green and Goos, 1963; 3: Marin et al., 1996; 4: Wallbridge and Pinegar, 1975; 5: Griffee and Burden, 1976; 6: Wallbridge, 1981; 7: Johanson and Blasquez, 1992; 8: Lassois et al., 2008; 9: Finlay and Brown, 1993; 10: Knight et al., 1977; 11: Knight, 1982; 12: Griffee, 1976; 13: Shilingford, 1976; 14: Anthony et al., 2004; 15: Ogundero, 1987; 16: Mesturino and Rgazzi, 1988.

pathogenicity of the complex. *Colletotrichum musae* was more pathogenic when it was inoculated alone than when it was co-inoculated with other species (Lassois et al., 2008), however, Anthony et al. (2004) found that *L. theobromae*, *F. verticillioides*, and *C. musae* were more highly pathogenic when co-inoculated than when they were inoculated separately.

The respective roles of the different species that could belong to the fungal complex have been thoroughly studied, whereas little information is available on the antagonistic or synergistic relationships among these different species.

More specifically, *C. musae*, the most pathogenic fungi associated with crown rot disease is a specific pathogen of banana. The genus *Colletotrichum* comprises imperfect fungi (Deuteromycetes) belonging to the order of Melanconiales. *Colletotrichum musae* is a distinct lineage in the 'gloeosporiodes' species complex (Su et al., 2011). The mycelium of *C. musae* is aerial, white turning an orange color with age (de Lapeyre de Bellaire, 1999; Abd-Elsalam et al., 2010). Conidia of *C. musae* are hyaline, non-septate, and mostly ellipsoid, ranging from 12–17 μm × 4–8 μm in size (Lim et al., 2002; Abd-Elsalam et al., 2010) and are produced on septate conidiophores that are up to 30 μm long and 3.5 μm wide. Appressoria are highly melanized, dark brown and irregularly shaped ranging from 6–11 × 5–10 μm in size (de Lapeyre de Bellaire, 1999; Lim et al., 2002). Conidiophores are gathered within acervuli that are up to 400 μm (Jones and Slabaugh, 1994). *Colletotrichum musae* produces its conidia in a mucilage which protects them and makes it possible to ensure their survival within acervuli (Mondal and Parbery, 2005).

BIOLOGY OF CROWN ROT INFECTION PROCESS

The large number of pathogens involved in the complex and the fact that the characteristics of each species differ complicates studies on the epidemic cycle and very little information is available on the epidemiology of the crown rot complex. Fruit contamination might occur in the field on the bunch or in the washing tanks in the packing station but, although field infections cannot be excluded, infections mainly occur during harvest when clusters are trimmed from bunches.

Field Contamination

Conidial Dispersion

Spores of some species, such as *C. musae*, are mainly disseminated by rainwater (Jeffries et al., 1990; de Lapeyre de Bellaire and Mourichon, 1997; de Lapeyre de Bellaire et al., 2000). Particularly, de Lapeyre de Bellaire et al.

(2000) have shown that the streaming of rainwater is an essential factor for the transport of the conidia of *C. musae* on the surface of the fruit. Other banana pathogens are airborne and most species involved in the fungal complex are found in the air spora of a banana plantation (Meredith, 1962a), and it has also been shown that some *Fusarium* species can be wind disseminated (Lukezic and Kaiser, 1966).

Fruit Contamination in the Field

Fruit pollution and fruit contamination must be distinguished. Fruit pollution, which occurs in the field, concerns the quantity of conidia that reach the fruit surface. Fruit contamination is related to the quantity of conidia that infect the crown and leads to disease development. In banana plantations, fungal species sporulate abundantly on all putrescent organs such as leaves (Simmonds, 1941; Meredith, 1962a), floral parts (Agati, 1992; de Lapeyre de Bellaire and Mourichon, 1997), and bracts (Lukezic and Kaiser, 1966; de Lapeyre de Bellaire et al., 2000). *Colletotrichum musae* and some *Fusarium* species are primary colonizers of decomposing leaves (Meredith, 1962b) and floral parts (de Lapeyre de Bellaire et al., 2000). This fungus does not sporulate on the still green parts of the banana tree but only on the senescent organs. It develops as a saprophyte on all decomposition matter of the banana plantation. According to Meredith (1962b), *C. musae* is a primary colonizer of the decomposing leaves. This author explains that the infectious structures are formed on the leaves when they were still green, which confers a decisive advantage for the colonization of senescent organs. de Lapeyre de Bellaire and Mourichon (1997) have shown in a study carried out in Guadeloupe that the pollution of bananas by *C. musae* and various *Fusarium* species usually occurs during the first 40 days following bunch emergence and sharply decreases thereafter. For these authors, the floral parts and the last bunch bract are the two main effective sources of inoculum for the contamination of the fruits but, because they are closest to the fruit, the floral parts are the most effective inoculum source for fruit pollution.

Fruit Contamination in the Packing Station

The fungus *C. musae*, which is also the causal agent of anthracnose, establishes quiescent infections in the field during the first month following banana flowering (de Lapeyre de Bellaire et al., 2000). These quiescent infections could also contribute to the onset of crown rot if the pathogen has an opportunity to colonize the crown region (Griffee and Pinegar,

1974). However, although field infections cannot be excluded, infections of the crown mainly occur during harvest and when clusters are trimmed from bunches (Meredith, 1971). Contamination generally takes place when hands are cut with a contaminated knife (Greene and Goos, 1963; Stover, 1972; Wallbridge, 1981; Finlay et al., 1992) or when clusters are cleaned with contaminated washing water (Shillingford, 1977) (see Fig. 3.2). At harvest, the fruit and senescent floral organs bear high quantities of spores that could potentially contaminate the crowns. Some of these spores could be removed by washing and then accumulate in the washing water (Slabaugh and Grove, 1982). The washing tanks are the main source of inocula according to Shillingford (1977). After the banana clusters are dipped in the washing water, the spores can passively penetrate a few millimeters into the vascular vessels of the crown, and disease is then hard to control with a fungicide spray treatment (Eckert and Ogawa, 1985). Greene and Goos (1963) have shown that a suspension of *C. musae* spores could penetrate 5–7 mm into the crown tissues within only 3 min. Airborne spores may also contaminate the crowns before the fruit are placed in boxes in the packing station. The risk is especially high when the facilities are dirty or if debris (pistils, stalks, and fruits) is piled up nearby.

PREHARVEST FACTORS FAVORING CROWN ROT DISEASE

Most of the studies on crown rot focused on the etiology of the disease, especially on the identification of the most pathogenic fungal species involved, and on postharvest control methods. Unfortunately, these studies provided neither a good understanding of the conditions most conducive for disease expression nor adequate control of the disease at all spatio-temporal scales. There is now some evidence that fruit physiology at harvest influences crown rot development and that all interactions among the microorganisms implicated should be considered. Taking these into account, a new concept of 'banana quality potential at harvest' and a model of elaboration of this quality potential were proposed by Lassois et al. (2010b) (Fig. 3.3). In this model, the fruit quality potential at harvest is presented as a key factor in crown rot development. The quality potential develops during banana growth in the field and depends on two components: (1) a fruit physiological component, which determines the fruit susceptibility to the disease; and (2) a parasitic component that reflects a level of crown contamination by the fungal complex, as well as the pathogenicity of this complex. Both these components are major in

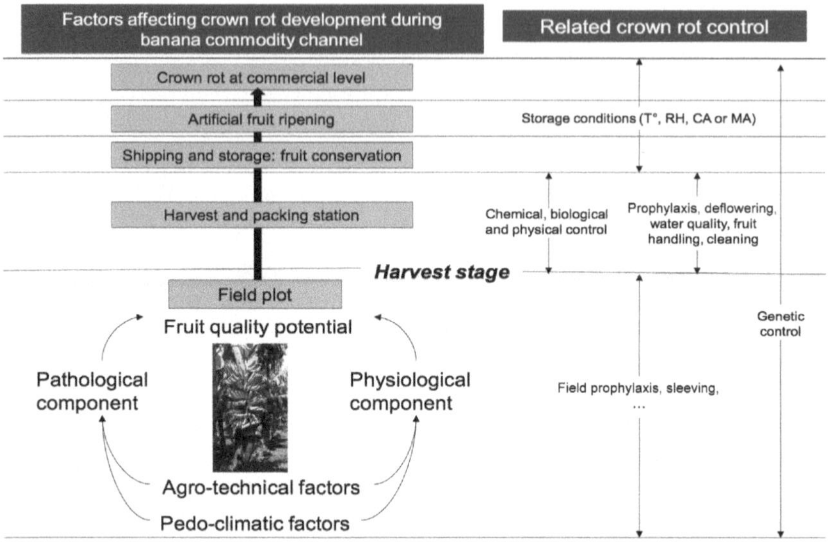

Figure 3.3 Diagram representing different key factors that arise throughout the commodity channel concerning crown rot development at commercial stage, and related control methods. T° = temperature; RH = relative humidity; CA = controlled atmosphere; MA = modified atmosphere (Lassois et al., 2010b).

the postharvest development of the crown rot disease and are greatly influenced by all environmental and agrotechnical preharvest factors. Only a few studies link them to either fruit susceptibility or to fruit contamination, and these are presented hereafter.

Influence of Environmental Factors on the Potential of Quality

Geographical and seasonal variations have been observed in the incidence of crown rot disease of bananas (Lukezic et al., 1967; Shillingford, 1978; Krauss and Johanson, 2000). Lukezic et al. (1967) have shown that the incidence of this disease varied throughout the year in Honduras. It is generally higher during the summer (March to September) and declines during the coldest period (October to February). Moreover, in Jamaica, a high disease incidence was found to be correlated with periods during the year when temperatures were highest (Shillingford, 1978), whereas in the Windward Islands, the incidence was reported to be highest during the rainy period (Krauss and Johanson, 2000). Therefore, it was important to disentangle whether these spatio-temporal fluctuations affected the parasitic and/or the physiological component of the fruit quality potential. Some experiments were carried out through controlled inoculations and demonstrated that the spatio-temporal factors have an impact on the physiological component of the fruit quality potential. A multilocal study carried out in Cameroon over a 1-year period showed less crown rot development at higher altitude (500 m) than at lower altitude (80 m) (Ewanéet al., 2013b). This study also suggested that fruits harvested in a low-altitude plantation during the rainy season might be more susceptible to crown rot than those harvested during the dry season. On the other hand, continuous variations of fruit susceptibility have been shown in the same banana plot through studies realized in Cameroon (Lassois, et al., 2010a) and Guadeloupe (Lassois et al., 2008).

Influence of Harvest Stage on Fruit Susceptibility

Harvest stage is expressed in accumulated degree day (dd) (Jullien et al., 2008) from flowering to harvest. It has been shown that harvest stage affects fruit susceptibility to crown rot: (1) an intra-bunch susceptibility gradient has been documented with a linear correlation (r = 0.95) between the hand position on the bunch and crown rot development (Lassois, et al., 2010a). Clusters growing on the first hands of a bunch are, on average, 70 dd more advanced than the hands initiated last (Jullien et al., 2001); (2) a linear relationship between fruit age (in dd) at harvest and the susceptibility of the fruit to crown rot was reported, with the oldest fruit being most susceptible to this disease (Forret, 2008).

Influence of Source/Sink Ratio on Fruit Susceptibility

Leaves are considered as 'source' tissues because they produce an excess of assimilates, while fruit are considered as 'sink' organs because they are not able to produce sufficient amounts of assimilate. It was demonstrated that the source/sink ratio during bunch growth affects the pomological characteristics of the fruits but also their susceptibility to crown rot (Lassois et al., 2010a). Fruits of bunches from which six hands were removed (high source/sink ratio, two hands remaining on the bunch) proved to be significantly less susceptible to crown rot than those from bunches with eight hands remaining on the bunch. It has also been shown that in the case of severe defoliations (low source-sink ratio), with only one or two leaves remaining on the tree, fruit susceptibility to crown rot disease was increased (Ewané, 2012). The authors considered that source/sink ratio modification leads, among other factors, to changes in the rate and characteristics of fruit filling which influences the formation of secondary metabolites involved in plant–pathogen interaction. A study was conducted to highlight underlying molecular mechanisms implied in the susceptibility variations observed when the source/sink ratio was modified (Lassois, et al., 2011a). Some genes involved in the susceptibility variation of bananas to crown rot disease were identified and results suggest that the catecholamine pathway could be implied in this phenomenon (Lassoiset al., 2011b). In the past, it had been suggested that preformed fungitoxic polyphenolic compounds could be involved in banana resistance to postharvest diseases (Mulvena et al., 1969; Muirhead and Deverall, 1984; Abdel-Saatar and Nawwar, 1986; Alam et al., 1993; Chillet et al., 2007), and it has also been shown that the modification of the source/sink ratio might induce modifications in the content of polyphenolics that might be implied in the modification of fruit susceptibility (Ewané, 2012).

Effect of Some Biotic Factors on Fruit Susceptibility

A study was carried out to evaluate the influence of some preharvest biotic factors on the fruits' susceptibility to postharvest crown rot development (Ewané et al., 2013a). It has been proven that the black leaf streak disease (*Mycosphaerella fijiensis*) severity at flowering and at harvest is positively correlated with the postharvest development of crown rot (Ewané et al., 2013a). This must be linked to the fact that black leaf streak disease reduces the source/sink ratio.

Influence of Sanitation Practices on Fruit Contamination

It is well known that most species involved in the fungal complex are saprophytes that occur on senescent banana organs, especially on decomposing leaves (Meredith, 1962b). Old leaves and other senescent organs present in the banana plantation may harbor inocula that could be responsible for severe contamination of bananas by this fungal complex (Simmonds and Mitchell, 1940).

POSTHARVEST FACTORS FAVORING CROWN ROT DISEASE

As can be seen in Figure 3.3, agronomic and environmental factors influence the fruit quality potential in the field and its propensity to develop crown rot. However, this figure also shows the importance of postharvest stages, such as harvesting, handling, transport and storage conditions in the development of crown rot. Fruit injuries enhance fruit ethylene production which has an impact on crown rot development. Furthermore, wounds on the fruits favor infection by pathogens. Fruit injuries must thus be avoided at all steps of the banana commodity channel. The inoculum pressure must be limited in all steps of the postharvest chain production to control efficiently crown rot development.

Organization of the Banana Commodity Channel

Due to the highly perishable nature of banana fruit, the commodity channel has to be well organized at all levels (Krauss and Johanson, 2000). This includes timely harvesting, a rapid cooling of fruit which is a prerequisite for effective crown rot control, an effective 'in' and 'out' quality control, an integrated efficient postharvest crown rot control, rapid loading into the boats and a continuous cooling chain from the beginning of the voyage to ripening centers and retailers in the importation countries. Shortcomings here can negate other efforts made to maintain healthy fruits.

It is difficult to control most of the preharvest factors in order to reduce crown rot impact. Inversely, most of the postharvest factors have to be controlled to reduce crown rot disease development. The impact of these different stages on disease development, as well as the methods applied to limit them, will be described in the following section concerning postharvest control methods.

CONTROL METHODS AND ALTERNATIVES TO CHEMICAL CONTROL

Conventional Chemical Control

Banana crown rot, like other storage diseases, has an especially detrimental impact on export produce. Routine postharvest fungicide treatment is still the most efficient and used crown rot control method (Table 3.2). This strategy was introduced in the late 1960s with the discovery of systemic benzimidazole derivative fungicides (thiabendazole and benomyl). They are classified as antimitotic compounds. Gradually, other fungicides that inhibit ergosterol biosynthesis, such as imazalil and bitertanol, were also introduced (Frossard et al., 1977; de Lapeyre de Bellaire and Nolin, 1994). The recent exclusion of bitertanol led to the use of QoI fungicides like azoxystrobin. The application methods vary markedly, including dipping, spraying and cascade treatment (see Fig. 3.2) but, in every case, the bananas must be thoroughly wetted to ensure fungicide treatment efficacy (de Lapeyre de Bellaire and Nolin, 1994). Alum is often combined with fungicide slurries to neutralize latex residue remaining on crowns when the bananas are removed

Table 3.2 Summary of Different Fungicides Used for Controlling Postharvest Diseases of Bananas in Commercial Banana Plantations.

Active Ingredient	Mode of Action	Recommended Rate (mg L^{-1})	Commercial Name	Observations
Benomyl	antimicotic	250	Benlate®	Banned
Thiabendazole	antimicotic	450	Tecto®, Metect®, Decco®	Efficient on crown rot and anthracnose
Imazalil	ergosterol biosynthesis inhibitor (DMI)	375	Fungaflor®, Sulima®, Neozil®	Effective on crown rot, not on anthranose
Bitertanol	ergosterol biosynthesis inhibitor (DMI)	200	Baycor®	Efficient on crown rot and anthracnose. Recently banned
Azoxystrobin	Quinone oustside inhibitor (QoI)	60	Ortiva®, Bankit®	Efficient on crown rot and anthracnose

from the washing bath (Muirhead and Jones, 2000), however, this mixture seems to have a negative impact on the performance of some active ingredients such as thiabendazole (Joas and Malisart, 2001). The time between crown trimming and fungicide application is critical. Crown rot severity seems to increase when fungicide application is delayed (Greene and Goos, 1963). Bananas are, nevertheless, usually treated in time in packing stations, i.e. just after they are removed from the washing bath. Problems concerning the development of resistant strains may arise with this practice. Postharvest fungicides generally have the same mode of action as those sprayed in the field to control Sigatoka disease. *Colletotrichum musae* strains resistant to thiabendazole, and generally to all antimitotic compounds, have been detected in many banana-producing countries (Hostachy et al., 1990; Johanson and Blasquez, 1992; de Lapeyre de Bellaire and Dubois, 1997; Krauss and Johanson, 2000). In Guadeloupe, it has been shown that thiabendazole-resistant *C. musae* strains developed after exclusive foliar applications of benomyl over a decade (1972–1982) for Sigatoka disease control (de Lapeyre de Bellaire and Dubois, 1997). Postharvest use of chemicals could ultimately be prohibited because the number of active substances registered for postharvest treatment has been reduced by current regulations, and legislation is becoming increasingly restrictive. Moreover, the efficacy of intensive fungicide treatments is not always adequate. In some areas where very few resistant strains are present, the efficacy of fungicides for postharvest disease control was found to vary substantially depending on season and production area (Chillet and de Lapeyre de Bellaire, 1996). Finally, consumers are now highly selective with respect to food safety concerns, and demand is increasing for produce that has not undergone any postharvest chemical treatments. Alternative non-fungicide solutions are therefore being sought.

Non-Conventional Control

Because the systematic fungicide control approach is not satisfactory and is increasingly criticized, several biological control approaches have been developed in recent years to find alternatives. We present here a set of methods that should be implemented in an integrated way to achieve efficient crown rot control (see Fig. 3.3) as an alternative to chemical control.

Field Control Methods

Bunch Sanitation

- Leaf removal: since senescent leaves hanging on the bunch might enhance fruit contamination (Meredith, 1962b; de Lapeyre de Bellaire et al., 2000), these leaves must be removed to reduce fruit contamination.
- Bunch sleeving (see Fig. 3.2): no accurate studies have been carried out to assess the impact of sleeving on crown rot development rate. It is, nevertheless, known that bunch sleeving with perforated plastic film protects bunches from fungal contamination and can reduce contamination by *C. musae* by over 80% (de Lapeyre de Bellaire et al., 2000). These results suggest that sleeving directly reduces crown contamination in the field, or that sleeved fruits release fewer spores in washing water during packing operations. Furthermore, this practice protects fruits from injuries.
- Deflowering (see Fig. 3.2): the main inoculum sources for *C. musae* and several *Fusarium* species are the floral remnants of the fruits and their early removal, generally when the young hands assume a horizontal position, is an important step further to limit bunch contamination (de Lapeyre de Bellaire et al., 2000).

Controlling the Banana Harvest Stage

Bananas are climacteric fruits, and the period between harvest and climacteric, the preclimacteric life, is commonly called the green life.

Bananas should be harvested at an age that will ensure a sufficient preclimacteric life during husbandry, because they should reach the ripening rooms still unripe (green). Furthermore, bananas are exported to Europe when they have grown to a commercial grade specified by European regulations (Commission Regulation (EC) No 2257/94 of 16 September 1994, setting quality standards for bananas) and market requirements. Thus, both age and grade of bananas are closely monitored during bunch development to determine the optimum date for harvest. Generally, once a week, newly emerged bunches are tagged with a colored plastic ribbon; different colors correspond to different weeks. Based on this colored system, the age of the banana is either estimated (1) in days or (2) in physiological age of the fruits. This age is then expressed as a sum of accumulated temperatures from flowering to harvest rather than in days (Jullien et al., 2008). It has been shown that physiological age is the best estimator of the green life. It has also been shown that younger fruits are less susceptible to crown rot than older fruits of the same grade (Forret, 2008), so good harvest stage management is an essential step to reduce harvesting of very susceptible fruits.

Black Leaf Streak Control

Good control of black leaf streak (Ewané et al., 2013a) or other foliar disease is important, since these disease promote fruit crown rot susceptibility by reducing the source/sink ratio. This requires proper chemical strategies as well as other integrated practices (de Lapeyre de Bellaire et al., 2010).

Genetic Control

The range of varieties grown for dessert banana export is very narrow as all clones belong to the Cavendish subgroup (Bakry et al., 1997). Moreover, breeding is complicated because triploid banana varieties are generally sterile. In the past, banana genetic improvement programs were mainly focused on obtaining varieties resistant to Sigatoka and Panama diseases. Resistance to crown rot of improved varieties has not been considered in any breeding programs to date. According to Marin et al. (1996), FHIA-01 and FHIA-02, two hybrids produced by the FHIA breeding program, are partially resistant to crown rot. Conversely, Perez-Vicente and Hernandez (2002) consider that these two varieties are more susceptible to crown rot caused by *F. semitectum* and *C. musae* compared with Cavendish varieties. According to these authors, only FHIA-23 is more resistant to crown rot than Grande-Naine. Finally, the first FHIA hybrids were introduced in the late 1980s, but their characteristics differ from those of Cavendish bananas and consumers have shown little interest in them, thus limiting their distribution.

Postharvest Control Methods

Deflowering and Sanitization

Deflowering involves the removal of all floral parts that could potentially harbor inocula before the dehanding operation (see Fig. 3.2). When the floral parts have not been removed in the field, deflowering is done in the packing station before bunch trimming, thus reducing the risk of contamination of the washing baths (Shillingford, 1976; Krauss and Johanson, 2000).

Packing Station Water Quality

Contamination of crowns when the bananas are processed in the dehanding and washing tanks is a key step in crown rot development because the wounds resulting from the clustering of banana hands are the main portal for infection by crown rot pathogens. Spore accumulation in the water can be reduced by regularly changing the washing water in the washing and delatexing baths (Slabaugh and Grove, 1982). These baths can contain a high quantity of spores, mainly *Fusarium* spp. and *Verticillum* spp., that detach from the peel surface (Shillingford, 1977), whereas *C. musae* conidia are less often found (Shillingford, 1976). Care should be taken to avoid contaminating the baths with plant debris (pistils, leaves, trimming waste, etc.) and the water should be regularly refreshed.

It is also recommended that the bath water be treated with active chlorine at the rate of 1–2 mg L^{-1} (Arneson, 1971; Slabaugh and Grove, 1982; Eckert and Ogawa, 1985) or quaternary ammonium disinfectants in order to hamper contamination (Shillingford, 1977), but the efficacy of this disinfection procedure is controversial (Slabaugh and Grove, 1982; Eckert and Ogawa, 1985). The chlorine concentration should be regularly adjusted to offset the high observed losses by volatilization or through redox reactions with latex or other organic matter in the washing baths (Shillingford, 1977).

Packing Station Sanitation

Protective measures implemented in the packing station are aimed at keeping the crowns of freshly trimmed bananas away from all inoculum sources. To ensure efficient crown rot control, it is thus essential to keep the packing station and the adjoining facilities clean. Waste (stalks, low-grade fruit, etc.) located in the vicinity of the banana packing area, which could contaminate the air in the station, must be eliminated (Finlay et al., 1992; Krauss and

Johanson, 2000). It has been shown that trimming clusters in a clean environment rather than in the field can reduce crown rot incidence by 50%.

Banana and Crown Trimming

The fruits should also be cleaned before being trimmed from the bunch. A contaminated trimming knife could spread inocula from the peel into the crown tissues (Lukezic et al., 1967). This may be avoided by washing the bunches in chlorinated water before they are trimmed into clusters (Krauss and Johanson, 2000). Bananas should be trimmed with a clean stainless steel blade. It was shown that roughly trimming the crowns, or ripping them off the hands, significantly increased the level of crown rot disease. Tissue fragments on the surface of the crowns dry out and quickly become senescent, thus providing an ideal site for rot development. Moreover, banana trimming knife tips are rounded to avoid banana fruit wounds. It is also important to cut wide crown sections containing as much crown tissue as possible, a technique that seems to enhance crown resistance to rot and seldom leads to the spread of rot into the fruit pedicels (Muirhead and Jones, 2000).

Packing

Although most pathogens require a high relative humidity for their *in vitro* development (Badger, 1965), bananas are less susceptible to crown rot under these conditions. A high relative humidity seems to hinder transpiratory water loss from the fruit, which is essential to ensure a long green life. Indeed, this ripening stage is markedly reduced under low relative humidity atmospheric conditions (30–40%) as a result of ethylene production from the fruit peel (Peacock, 1973). The senescence of banana crown tissues can then be hampered by maintaining their turgidity in plastic bag packaging (Muirhead and Jones, 2000).

Shipping

Bananas must be cooled as soon as possible to 13°C after packing and delivered promptly to a ripening facility (see Fig. 3.2). One way to slow down banana metabolism, and thus to delay crown rot development, is to refrigerate the fruit in boxes at the lowest possible temperature without inducing physiological disorders. Cooling should be continuous throughout the shipping phase which, in turn, should be as short as possible. Containers designed for maritime shipping of bananas are climate controlled at 13–14°C, because temperatures below 12°C provoke peel browning due to oxidization (Muirhead and Jones, 2000). Fungal growth is slowed down at 13°C (Finlay and Brown, 1993) which is a good trade-off with respect to preserving the bananas and hampering fungal colonization of the crowns. However, this is much lower than the optimal temperatures for germination and growth of fungi responsible for crown rot, but it is not low enough to slow down the activity of these fungi, which can take place within a broad temperature range, i.e. 8–36°C (Simmonds and Mitchell, 1940). As a result, crown rot incidence and severity are higher in bananas exposed to temperatures over 16°C (Slabaugh and Grove, 1982).

Controlled and Modified Atmospheres

The composition of the atmosphere around bananas during shipping can be manipulated to slow down metabolic activity. Moreover, high CO_2 (>15%) and low O_2 (<1%) levels are toxic to many fungi (Goos and Tschirsch, 1962; Al Zaemey et al., 1994). Unfortunately, bananas cannot be stored under these conditions because major physiologic alterations generally occur during ripening when the atmospheric CO_2 level is above 7–12% or when the O_2 level is below 1–2% (Daun et al., 1973; Wilson, 1976; Thompson, 1998). Then, ideally, bananas should be stored in an atmosphere of 5–7 % CO_2 and 2% O_2 (Daun et al., 1973; Thompson, 1998). Modified atmospheres (MA) can be achieved by packing bananas in sealed plastic bags (polybags), and controlled atmospheres (CA) are obtained by injecting pure nitrogen into the storage rooms. It was shown that crown rot can be partially controlled (20%) by packing bananas in MA (1–2% O_2 and 12–14% CO_2) (Bastiaanse et al., 2010). For MA, the balance of O_2 and CO_2 contents depends on the extent of fruit respiration, bag permeability, and the composition of the air outside the bag (Thompson, 1998). The O_2 and CO_2 contents generally range from 1 to 10% and 2 to 14%, respectively, depending on the quality and thickness of the plastic packaging (Marriott and Palmer, 1980; Thompson, 1998). These plastic bags must also have a high mechanical strength because even small punctures will upset the MA (Chillet and de Lapeyre de Bellaire, 1996).

Hot Water Treatments (HWT)

To date, there has been no commercial adaptation of this technique to bananas grown for exportation (Cavendish), like in other tropical fruits such as papaya (Couey et al., 1984; Nishijima et al., 1992) and mango (Spalding and Reeder, 1978; Coates et al., 1993). However, some recent work has been done on local cultivars grown in Asian countries. Hot water treatment (20 min at 50°C) is ineffective in killing dormant spores but causes a delay in mycelium growth and conidial germination on cv 'Bungulan'(AAA) grown in the Philippines (Alvindia, 2012). De Costa and Erabadupitiya (2005) have shown that the optimal temperature and exposure time for controlling crown rot on 'Embon' (AAA) and 'Kolikuttu' (AAB) varieties from Sri Lanka was 50°C for 3 min. Higher temperatures lead to pale fruit, and exposure times of over 5 min reduce Brix values (De Costa and Erabadupitiya, 2005) and caused unacceptable fruit browning on cv 'Kluai-Hom Thong' (AAA) from Thailand (Win et al., 2007). In Brazil, a 20 min longer treatment at a temperature under 45°C was effective for crown rot control in cv 'Santa CatarinaPrata' (AAB) and 'Williams' (Cavendish, AAA) with the percentage of infected bananas decreasing from 100% to less than 15% (Reyes et al., 1998). However, commercial tests on naturally infected fruit have not achieved crown rot control, and ripening delays have also been noted.

UV and Gamma Radiation Treatment

Ultraviolet light (UV) treatments have considerable potential for fruit and vegetable postharvest disease control. Stevens et al. (2005) discussed the possibility of inducing apple resistance to *C. gloeosporioides* through UV-C treatments. In bananas, however, the peel seems to be too sensitive to UV for the use of this technique for postharvest control of some pathogens of the complex of *C. musae* (Joas, 1997).

Kanapathipillai et al. (1987) pointed out that gamma ray treatment (38 min at 4 kGy) inhibited spore germination, the formation of *C. musae* appressoria, and all fungal development on the surface of fruit pieces. However, these authors did not investigate the effects of gamma rays on whole fruit, or on their complex components. Although dosages of around 0.5 kGy can extend banana green life (Marriott and Palmer, 1980; Wills et al., 1998), Marriott and Palmer (1980) noted that irradiation can alter the banana peel; the maximum dose tolerated by the fruits is likely around 0.5 kGy (Wills et al., 1998). Finally, the development of postharvest fruit irradiation has been hampered especially by the high cost of such treatments and their low consumer acceptance.

Biological Control

Currently, there is a continuation of a study on developing a biocontrol formulation to treat crown rot of banana and to replace the systematic fungicide treatment. The following examples are far from exhaustive. Postharvest biological control is very promising because the crown rot infection site on the fruit is limited, the environmental conditions during storage are clearly defined and stable, and bananas have a high added value (Jijakli and Lepoivre, 1999). The results of many studies have suggested that the use of microorganisms such as fungi, bacteria, and yeasts could provide partial crown rot control (Postmaster et al., 1997; De Costa and Subasinghe, 1998; East and Kenyon, 1998; Krauss et al., 1998; Gunasinghe et al., 2004; De Costa and Erabadupitiya, 2005; Alvindia and Natsuaki, 2008; Lassois et al., 2008; Williamson et al., 2008; Sangeetha et al., 2010a). It has also been shown that biological control efficacy increases with the antagonistic agent concentration and with the incubation time between the application of the antagonist and crown contamination by the fungal complex (Postmaster et al., 1997; Lassois et al., 2008). Antagonists can be used to reduce significantly lesions induced by the fungal complex that causes crown rot, but the control efficacy is limited and variable (Lassois et al., 2008). This type of control, when used alone, cannot provide total crown rot control, and so should be combined with other control tactics such as inorganic salts and MA packaging to improve the effectiveness (Bastiaanse et al., 2010; De Costa and Gunawardhana, 2012; Alvindia, 2013).

Natural Compounds

The impact of several natural substances or non-synthetic fungicides, such as preparations of calcium, plant extracts or organic acids, on crown rot development has also been evaluated. Garlic (*Allium sativum*) extracts (Krauss and Johanson, 2000) and essential oils of cinnamon (*Cinnamomum zeylanicum*), clove (*Syzygium aromaticum*) (Ranasinghe et al., 2002, 2005), citronella (*Cymbopogon nardus*), and basil (*Ocimum basilicum*) (Anthony et al., 2004; Sangeetha et al., 2010b) have also been found to have fungicidal activity. Win et al. (2007) showed that cinnamon extracts (5 g L^{-1}) partially reduced crown rot severity (25%), increased green life, and had no negative effects on postharvest banana quality. These plant extracts have fungistatic and fungitoxic activity and can inhibit completely conidial germination and mycelial growth of *C. musae*, *Fusarium* spp., and *L. theobromae*. However, cases of phytotoxicity have been reported and the level of control provided by these natural substances is not sufficient to meet market requirements. The use of antioxidants (Khan et al., 2001) and organic salts (Alvindia et al., 2004), sometimes combined with surfactants (Alvindia and Natsuaki, 2007), can also enable partial crown rot control (up to 60%). Finally, experiments showed that some treatments were found to provide the same level of crown rot control as fungicide treatment (Alvindia, 2013; Sangeetha et al., 2013).

This is the case with Biocto 6® (seed extract from citrus) (Productos Biogenicos S.A., San Jose, Costa Rica) combined with a wax-based additive (Verdiol, Productos Biogenicos S.A., San Jose, Costa Rica) (Demerutis et al., 2008). In the same way, Sangeetha et al. (2013) showed that the dipping of banana in zimmu leaf extract at 25% concentration exhibited 100% inhibition of crown rot disease in cold storage (14°C) up to 35 days and increased the shelf-life to 64 days without altering organoleptic properties. These seem to be a promising potential ecofriendly technology substituting fungicides in the control of the disease but further studies would be needed prior to recommending them in commercial packing houses, including validation of the results in real full-scale commercial exportation conditions, identification of the principle compounds responsible for the control and also the performance of safety tests.

CONCLUDING REMARKS

Crown rot studies and control are especially complex because of the observed diversity and variability in the composition and pathogenicity of the complex involved in the development of this disease. The broad range of possible situations complicates studies on both the parasitic and physiologic components of fruit quality. Further studies are thus required to gain better insight into this disease, especially because little documentation is available on certain aspects such as the epidemiology of crown rot. It is also very important to enhance the overall understanding of banana physiological mechanisms involved in the induction of fruit resistance to crown rot pathogens with the aim of improving control.

There is considerable growing interest in alternative methods to reduce or even completely eliminate fungicide treatments. To be efficient, these alternative control methods should not be too specific considering the broad spectrum of pathogens involved in the complex. Crown rot research should enhance the overall understanding of this disease and thus lead to the development of an effective integrated control strategy.

REFERENCES

Abd-Elsalam, K.A., Roshdy, S., Amin, O.E., Rabani, M., 2010. First morphogenetic identification of the fungal pathogen *Colletotrichum musae* (Phyllachoraceae) from imported bananas in Saudi Arabia. Genet. Mol. Res. 9, 2335–2342.

Abdel-Saatar, M., Nawwar, M., 1986. 3,4-dimethoxybenzaldehyde, a fungistatic substance in peel of green banana fruits in relation to resistance at different degrees of maturity. Horticultural Science 21, 812.

Agati, J.A., 1992. Banana stem and fruit rot. Philippine Agric. 10, 411–422.
Al Zaemey, A.B., Magan, N., Thompson, A.K., 1994. In vitro studies of the effect of environmental conditions on the anthracnose pathogen of bananas, *Colletotrichum musae*. Int. Biodeterioration Biodegradation 33, 369–381.
Alam, M.S., Alam, S., Islam, S., Alam, N., 1993. Biochemical changes in banana fruits in response to crown rot pathogens. Bangladesh J. Bot. 22, 143–148.
Alvindia, D.G., 2012. Revisiting hot water treatments in controlling crown rot of banana cv. Buñgulan. Crop Prot. 33, 59–64.
Alvindia, D.G., 2013. Enhancing the bioefficacy of *Bacillus amyloliquefaciens* DGA14 with inorganic salts for the control of banana crown rot. Crop Prot. 51, 1–6.
Alvindia, D.G., Kobayashi, T., Natsuaki Keiko, T., Tanda, S., 2004. Inhibitory influence of inorganic salts on banana postharvest pathogens and preliminary application to control crown rot. J. Gen. Plant Pathol. 70, 61–65.
Alvindia, D.G., Kobayashi, T., Yaguchi, Y., Natsuaki, K.T., 2000. Symptoms and the associated fungi of postharvest diseases on non-chemical bananas imported from the Philippines. Japanese J. Trop. Agric. 44, 87–93.
Alvindia, D.G., Natsuaki, K.T., 2007. Control of crown rot-causing fungal pathogens of banana by inorganic salts and a surfactant. Crop Prot. 26, 1667–1673.
Alvindia, D.G., Natsuaki, K.T., 2008. Evaluation of fungal epiphytes isolated from banana fruit surfaces for biocontrol of banana crown rot disease. Crop Prot. 27, 1200–1207.
Anthony, S., Abeywickrama, K., Dayananda, R., Wijeratnam Shanthi, W., Arambewela, L., 2004. Fungal pathogens associated with banana fruit in Sri Lanka, and their treatment with essential oils. Mycopathologia 157, 91–97.
Arneson, P.A., 1971. Sensitivity of postharvet rot fungi of bananas to chlorine. Phytopathology 61, 344–345.
Badger, A.M., 1965. Influence of relative humidity on fungi causing crown rot of boxed bananas. Phytopathology 55, 688–692.
Bakry, F., Carreel, F., Caruana, M.L., Côte, F.X., Jenny, C., Tezenas du Montcel, H., 1997. Les bananiers. In: Charrier, A., Hamon, S., Jacquot, M., Nicolas, D. (Eds.), L'amélioration des plantes tropicales. CIRAD ORSTOM, Montpellier, pp. 109–139.
Bastiaanse, H., de Lapeyre de Bellaire, L., Lassois, L., Misson, C., Jijakli, M.H., 2010. Integrated control of crown rot of banana with *Candida oleophila* strain O, calcium chloride and modified atmosphere packaging. Biol. Control 53, 100–107.
Chillet, M., de Lapeyre de Bellaire, L., 1996. Conditionnement en polybag pour le contrôle de l'anthracnose de blessure des bananes. Fruits 51, 163–172.
Chillet, M., Hubert, O., de Lapeyre de Bellaire, L., 2007. Relationship between physiological age, ripening and susceptibility of banana to wound anthracnose. Crop Prot. 26, 1078–1082.
Coates, L.M., Johnson, G.I., Cooke, A.W., 1993. Postharvest disease control in mangoes using high humidity hot air and fungicide treatments. Ann. Appl. Biol. 123, 441–448.
Couey, H.M., Alvarez, A.M., Nelson, M.G., 1984. Comparison of hot-water spray and immersion treatments for control of pstharvest decay of papaya. Plant Dis. 68, 436–437.
Daun, H., Gilbert, S.G., Ashkenazi, Y., Henig, Y., 1973. Storage quality of bananas packaged in selected permeability films. J. Food Sci. 38, 1247–1250.
Daundasekera, M., Joyce, D.C., Aked, J., Adikaram, N.K.B., 2003. Ethylene production by *Colletotrichum musae* in vitro. Physiol. Mol. Plant Pathol. 62, 21–28.
De Costa, D.M., Erabadupitiya, H.R.U.T., 2005. An integrated method to control postharvest diseases of banana using a member of the *Burkholderia cepacia* complex. Postharvest Biol. Technol. 36, 31–39.
De Costa, D.M., Gunawardhana, H.M.D.M., 2012. Effects of sodium bicarbonate on pathogenicity of *Colletotrichum musae* and potential for controlling postharvest diseases of banana. Postharvest Biol. Technol. 68, 54–63.

De Costa, D.M., Subasinghe, S.S.N.S., 1998. Antagonistic bacteria associated with the fruit skin of banana in controlling its post-harvest diseases. Trop. Sci. 38, 206–212.
de Lapeyre de Bellaire, L., 1999. Bio-écologie de *Colletotrichum musae* (Berk. & Curt.) Arx, agent de l'anthracnose des bananes, dans les conditions tropicales humides de la Guadeloupe. Université de Paris-Sud, Orsay. p. 100.
de Lapeyre de Bellaire, L., Chillet, M., Dubois, C., Mourichon, X., 2000. Importance of different sources of inoculum and dispersal methods of conidia of *Colletotrichum musae*, the causal agent of banana anthracnose, for fruit contamination. Plant Pathol. 49, 782–790.
de Lapeyre de Bellaire, L., Dubois, C., 1997. Distribution of thiabendazole-resistant *Colletotrichum* isolates from Guadeloupe banana plantations. Plant Dis. 81, 1378–1383.
de Lapeyre de Bellaire, L., Fouré, E., Abadie, C., Carlier, J., 2010. Black leaf streak disease is challenging the banana industry. Fruits 65, 327–342.
de Lapeyre de Bellaire, L., Mourichon, X., 1997. The pattern of fungal contamination of the banana bunch during its development and potential influence on incidence of crown-rot and anthracnose diseases. Plant Pathol. 46, 481–489.
de Lapeyre de Bellaire, L., Nolin, J., 1994. Amélioration du contrôle du chancre sur les bananes d'exportation et traitements post-récolte. Fruits 49, 179–185.
Demerutis, C., Quirós, L., Martinuz, A., Alvarado, E., Williams, R.N., Ellis, M.A., 2008. Evaluation of an organic treatment for post-harvest control of crown rot of banana. Ecol. Eng. 34, 324–327.
Dominguez, M., Vendrell, M., 1993. Wound ethylene biosynthesis in preclimacteric banana slices. Acta Horticulturae 343, 270–274.
East, L., Kenyon, L., 1998. Development of biological control methods for post-harvest rots of banana. In: Council, B.C.P. (Ed.), Proceedings of the Brighton Crop Protection Conference: Pests and Diseases. United Kingdom edn. Farnham (United Kingdom), Brighton, UK, pp. 549–554. 16–19, November 1998.
Eckert, J.W., Ogawa, J.M., 1985. The chemical control of post-harvest diseases: subtropical and tropical fruits. Ann. Rev. Phytopathol. 23, 421–454.
Ewané, C.A., 2012. Etude de la composante physiologique impliquée dans le développement des pourritures de couronne de bananes et rôle des composés phénoliques dans les mécanismes de variation de sensibilité. Ph D thesis. Plant pathology Unit, Gembloux Agro-Bio Tech.
Ewané, C.A., Chillet, M., Castelan, F., et al., 2013a. Impact of the extension of black leaf streak disease on banana susceptibility to post-harvest diseases. Fruits 68, 351–365.
Ewané, C.A., Lassois, L., Brostaux, Y., Lepoivre, P., de Lapeyre de Bellaire, L., 2013b. The susceptibility of bananas to crown rot disease is influenced by geographical and seasonal effects. Can. J. Plant Pathol. 35, 27–36.
Finlay, A.R., Brown, A.E., 1993. The relative importance of *Colletotrichum musae* as a crown-rot pathogen on Windward Island bananas. Plant Pathol. 42, 67–74.
Finlay, A.R., Lubin, C., Brown, A.E., 1992. The banana stalk as a source of inoculum of fungal pathogens which cause crown rot. Trop. Sci. 32, 343–352.
Forret, M., 2008. Etude de la variation de sensibilité des bananes d'exportation aux pourritures de couronne en fonction du stade de récolte. Travail de fin d'étude, Faculté des Sciences Agronomiques de Gembloux.
Frossard, P., Laville, E., Plaud, G., 1977. Etude des traitements fongicides appliqués aux bananes après récolte. III/. Action de l'imazalil. Fruits 32, 673–676.
Goos, R.D., Tschirsch, M., 1962. Effect of environmental factors on spore germination, spore survival, and growth of *Gloeosporium musarum*. Mycologia 54, 353–366.
Greene, G.L., Goos, R.D., 1963. Fungi associated with crown rot of boxed bananas. Phytopathology 53, 271–275.
Griffee, P.J., 1976. Pathogenicity of some fungi isolated from diseased crowns of banana hands. J. Phytopathol. 85, 206–216.

Griffee, P.J., Burden, O.J., 1976. Fungi associated with crown rot of boxed bananas in the Windward Islands. J. Phytopathol. 85, 149–158.
Griffee, P.J., Pinegar, J.A., 1974. Fungicides for control of the crown rot complex: in vivo and in vitro studies. Trop. Sci. 16, 107–120.
Gunasinghe, R.N., Ikiriwatte, C.J., Karunaratne, A.M., 2004. The use of *Pantoea agglomerans* and *Flavobacterium* sp. to control banana pathogens. J. Horticultural Sci. Biotechnol. 79, 1002–1006.
Hostachy, B., Vegh, I., Leroux, P., Jacquemot, E., Foucher, S., Pigou, R., 1990. Bananes de la Martinique. Incidence des problèmes fongiques sur la qualité. Phytoma, 37–44.
Jeffries, P., Dodd, J.C., Jeger, M.J., Plumbey, R.A., 1990. The biology and control of *Colletotrichum* species on tropical fruit crops. Plant Pathol. 39, 343–366.
Jijakli, M., Lepoivre, P.C.G., 1999. Yeast species for biocontrol of apple postharvest diseases: an encouraging case of study for practical use. In: Upadhyay, R., Mukerij, K. (Eds.), Biotechnological approaches in biocontrol of plant pathogens. Kluwer Academic/Plenum publishers, New York, pp. 31–49.
Joas, J., 1997. Programme qualité de la banane d'exportation. Rapport d'essai sur l'emploi des UV en post récolte. Cirad-Flhor, Fort de France.
Joas, J., Malisart, S., 2001. Incidence des conditions d' application sur l' efficacite des fongicides utilises en post-recolte pour la banane [Effect of the application techniques on the effectiveness of the fungicides used in the banana postharvest]. Fruits 56, 383–394.
Johanson, A., Blasquez, B., 1992. Fungi associated with banana crown-rot on field-packed fruit from the Windward Islands and assessment of their sensitivity to the fungicides thiabendazole, prochloraz, and imazalil. Crop Prot. 11, 79–83.
Jones, D.R., 2000. Diseases of Banana. Abaca and Enset, Wallingford, UK.
Jones, D.R., Slabaugh, W.R., 1994. Banana diseases caused by fungi. Anthracnose. In: Ploetz, R., Zentmyer, G.A., Nishijima, W., Rohrbach, K.G., Ohr, H.D. (Eds.), Compendium of Tropical Fruit Diseases. APS Press, pp. 4–9.
Jullien, A., Chillet, M., Malezieux, E., 2008. Pre-harvest growth and development, measured as accumulated degree days, determine the post-harvest green life of banana fruit. J. Horticultural Sci. Biotechnol. 83, 506–512.
Jullien, A., Malezieux, E., Michaux-Ferriere, N., Chillet, M., Ney, B., 2001. Within-bunch variability in banana fruit weight: importance of developmental lag between fruits. Ann. Bot. 87, 101–108.
Kanapathipillai, V.S., Ahmad, R., Mahamad, M.I., 1987. The effect of sterile filtrates of *Trichoderna* spp and *Penicillium* spp and 4 kGy irradiation on the spore germination of *Colletotrichum musae*. In: Singh, K.G., Manalo, P.L., Sastrontomo, S.S., et al. (Eds.), Movements of Pests and Control Strategies, pp. 283–292. Kuala Lumpur, Malaysia, ASEAN Plant Quarantine Centre and Training Institute, Selangor, Malaysia.
Khan, S.H., Aked, J., Magan, N., 2001. Control of the anthracnose pathogen of banana (*Colletotrichum musae*) using antioxidants alone and in combination with thiabendazole or imazalil. Plant Pathol. 50, 601–608.
Knight, C., 1982. Pathogenicity of some fungi associated with crown rot of bananas. Phytopathologische Zeitschrift 104, 13–18.
Krauss, U., 1996. Establishment of a bioassay for testing control measures against crown rot of banana. Crop Prot. 15, 269–274.
Krauss, U., Bidwell, R., Ince, J., 1998. Isolation and preliminary evaluation of mycoparasites as biocontrol agents of crown rot of banana. Biol. Control 13, 111–119.
Krauss, U., Johanson, A., 2000. Recent advances in the control of crown rot of banana in the Windward Islands. Crop Prot. 19, 151–160.
Lassois, L., Bastiaanse, H., Chillet, M., Jullien, A., Jijakli, M.H., De Lapeyre de Bellaire, L., 2010a. Hand position on the bunch and source–sink ratio influence the banana fruit susceptibility to crown rot disease. Ann. Appl. Biol. 156, 221–229.

Lassois, L., Jijakli, H., Chillet, M., de Lapeyre de Bellaire, L., 2010b. Crown rot of bananas: Preharvest factors involved in postharvest disease development and integrated control methods. Plant Dis. 94, 648–658.
Lassois, L., Frettinger, P., de Lapeyre de Bellaire, L., Lepoivre, P., Jijakli, H., 2011a. Identification of genes involved in the response of banana to crown rot disease. Mol. Plant Microbe Interact. 24, 143–153.
Lassois, L., De Clerck, C., Frettinger, P., De Lapeyre De Bellaire, L., Lepoivre, P., Haissam Jijakli, M., 2011b. Catecholamine biosynthesis pathway potentially involved in banana defense mechanisms to crown rot disease. Commun. Agric. Appl. Biol. Sci. 76, 591–601.
Lassois, L., de Lapeyre de Bellaire, L., Jijakli, M.H., 2008. Biological control of crown rot of bananas with *Pichia anomala* strain K and *Candida oleophila* strain O. Biol. Control 45, 410–418.
Lim, J., Lim, T.H., Cha, B., 2002. Isolation and identification of colletotrichum musae from imported bananas. Plant Pathol. J. 18, 161–164.
Lukezic, F.L., Kaiser, W.J., 1966. Aerobiology of *Fusarium roseum* 'Gibbosum' associated with crown rot of boxed bananas. Phytopathology 56, 545–548.
Lukezic, F.L., Kaiser, W.J., Martinez, M.M., 1967. The incidence of crown rot of boxed bananas in relation to microbial populations of the crown tissus. Can. J. Bot. 45, 413–421.
MacCracken, A.R., Swinburne, T.R., 1980. Effect of bacteria isolated from surface of banana fruits on germination of *Colletotrichum musae* conidia. Trans. Br. Mycol. Soc. 74, 212–214.
Marin, D.H., Sutton, T.B., Blankenship, S.M., Swallow, W.H., 1996. Pathogenicity of fungi associated with crown rot of bananas in Latin America on Grande Naine and disease-resistant hybrid bananas. Plant Dis. 80, 525–528.
Marriott, J., Palmer, J.K., 1980. Bananas – Physiology and biochemistry of storage and ripening for optimum quality. CRC Crit. Rev. Food Sci. Nutr. 13, 41–88.
Meredith, D.S., 1962a. Some components of the air-spora in Jamaican banana plantations. Ann. Appl. Biol. 50, 577–594.
Meredith, D.S., 1962b. Some fungi on decaying banana leaves in Jamaica. Trans. Br. Mycol. Soc. 45, 335–347.
Meredith, D.S., 1971. Transport and storage diseases of bananas: biology and control. Trop. Agric. 48, 35–50.
Mesturino, L., Ragazzi, A., 1988. Microorganismi fungini associati ai frutti di *Musa*, spp. presenti sul mercato italiano. Rivista di Agricoltura Subtropicale e Tropicale 82, 503–515.
Mondal, A.H., Parbery, D.G., 2005. The protective role of the spore matrix of *Colletotrichum musae* during rehydratation and exposure to extreme temperature and UV radiation. Aust. Plant Pathol. 34, 229–235.
Muirhead, I.F., Deverall, B.J., 1984. Evaluation of 3,4-dihydrobenzaldehyde, dopamine and its oxydation products as inhibitors of *Colletotrichum musae* (Berk. et Curt.) Arx in green banana fruits. Aust. J. Bot. 32, 575–582.
Muirhead, I.F., Jones, D.R., 2000. Fungal diseases of banana fruit. Postharvest diseases. In: Jones, D.R. (Ed.), Diseases of Banana. Abaca and Enset, CABI, Wallingford, pp. 190–206.
Mulvena, D., Webb, E.C., Zerner, B., 1969. 3,4-Dihydrobenzaldehyde, a fungistatic substance from green cavendish bananas. Phytochemistry 8, 393–395.
Nishijima, K.A., Miura, C.K., Armstrong, J.W., Brown, S.A., Hu, B.K.S., 1992. Effect of forced, hot-air treatment of papaya fruit on fruit quality and incidence of postharvest diseases. Plant Dis. 76, 723–727.
Ogundero, V.W., 1987. Crown rot fungi of Nigerian bananas cv. Robusta and the effects of benomyl on their exo-enzymes. J. Basic Microbiol. 27, 43–47.
Peacock, B.C., 1973. Effect of *Colletotrichum musae* infections on the preclimacteric life of bananas. Queensland J. Agric. Animal Sci. 30, 239–246.
Peacock, B.C., Muirhead, I.F., 1974. Ethylene production by *Colletotrichum musae*. Queensland J. Agric. Animal Sci. 31, 249–252.

Perez Vicente, L., Hernandez, A., 2002. Reaction of Black Sigatoka resistant FHIA hybrids to *Fusarium pallidoroseum* and *Colletotrichum musae* causal agents of crown rot and anthracnose diseases of bananas. Phytopathology 92, S136.

Postmaster, A., Kuo, J., Sivasithamparam, K., Turner, D.W., 1997. Interaction between *Colletotrichum musae* and antagonistic microorganisms on the surface of banana leaf discs. Sci. Horticulturae 71, 113–125.

Ranasinghe, L., Jayawardena, B., Abeywickrama, K., 2002. Fungicidal activity of essential oils of *Cinnamomum zeylanicum* (L.) and *Syzygium aromaticum* (L.) Merr et L.M. Perry against crown rot and anthracnose pathogens isolated from banana. Lett. Appl. Microbiol. 35, 208–211.

Ranasinghe, L., Jayawardena, B., Abeywickrama, K., 2005. An intergrated strategy to control post-harvest decay of embul banana by combining essential oils with modified atmosphere packaging. Int. J. Food Sci. Technol. 40, 97–103.

Reyes, M.E.Q., Nishijima, W., Paull, R.E., 1998. Control of crown rot in 'Santa Catarina Prata' and 'Williams' banana with hot water treatments. Postharvest Biol. Technol. 14, 71–75.

Sangeetha, G., Thangavelu, R., Rani, S.U., Muthukumar, A., Udayakumar, R., 2010a. Induction of systemic resistance by mixtures of antagonist bacteria for the management of crown rot complex on banana. Acta Physiol. Plantarum 32, 1177–1187.

Sangeetha, G., Thangavelu, R., Rani, S.U., 2010b. Evaluation of plant oils for suppression of crown rot disease and improvement of shelf life of banana (*Musa* spp. AAA subgroup, cv. Robusta). Int. J. Food Sci. Technol. 45, 1024–1032.

Sangeetha, G., Thangavelu, R., Usha Rani, S., Muthukumar, A., 2013. Antimicrobial activity of medicinal plants and induction of defense related compounds in banana fruits cv. Robusta against crown rot pathogens. Biological Control, 64, 16–25.

Shillingford, C.A., 1976. Occurence of banana fruit-rot fungi in Jamaican boxing plants. Plant Dis. Rep. 60, 788–793.

Shillingford, C.A., 1977. Control of banana fruit rots and of fungi that contaminate washing water. Trop. Sci. 19, 197–203.

Shillingford, C.A., 1978. Climatic factors affecting post-harvest decay of Jamaican bananas. J. Agric. Univ. Puerto Rico, 45–49.

Simmonds, J.H., 1941. Latent infection in tropical fruits discussed in relation to the part played by species of *Gloeosporium* and *Colletotrichum*. Proc. Royal Soc. Qld. 52, 92–120.

Simmonds, J.H., 1963. Studies in the latent phase of *Colletotrichum* species causing ripe rots of tropical fruits. Qld. J. Agric. Animal Sci. 20, 373–424.

Simmonds, J.H., Mitchell, R.S., 1940. Black end and anthracnose of the banana with special reference to Gloeosporium musarum Cke. and Mass. Bulletin of the. Council for Scientific and Industrial Research of Australia, Melbourne.

Slabaugh, W.R., Grove, M.D., 1982. Postharvest diseases of bananas and their control. Plant Dis. 66, 746–750.

Spalding, D.H., Reeder, W.F., 1978. Controlling market diseases of mangoes with heated benomyl. Proc. Florida State Horticultural Soc. 91, 186–187.

Stevens, C., Khan, V.A., Wilson, C.L., Lu, J.Y., Chalutz, E., Droby, S., 2005. The effect of fruit orientation of postharvest commodities following low dose ultraviolet light-C treatment on host induced resistance to decay. Crop Prot. 24, 756–759.

Stover, R.H., 1972. Banana, Plantain and Abaca Diseases. . Kew, Surrey, UK.

Su, Y.Y., Noireung, P., Liu, F., et al., 2011. Epitypification of Colletotrichum musae, the causative agent of 319 banana anthracnose. Mycoscience 52, 376–382.

Thompson, A.K., 1998. Controlled Atmosphere Storage of Fruits and Vegetables. Wallingford CABI.

Wallbridge, A., 1981. Fungi associated with crown-rot disease of boxed bananas from the Windward Islands during a two year survey. Trans. Br. Mycol. Soc. 77, 567–577.

Wallbridge, A., Pinegar, J.A., 1975. Fungi associated with crown-rot disease of bananas from St Lucia in the Windward Islands. Trans. Br. Mycol. Soc. 64, 247–254.

Williamson, S.M., Guzmán, M., Marin, D.H., Anas, O., Jin, X., Sutton, T.B., 2008. Evaluation of Pseudomonas syringae strain ESC-11 for biocontrol of crown rot and anthracnose of banana. Biol. Control 46, 279–286.
Wills, R., McGlasson, W.B., Graham, D., Joyce, D.C., 1998. Postharvest. An Introduction to the Physiology and Handling of Fruit. Vegetables and Ornamentals, Wallingford.
Wilson, L.G., 1976. Handling of postharvest tropical fruit. Horticultural Sci. 11, 120–121.
Win, N.K.K., Jitareerat, P., Kanlayanarat, S., Sangchote, S., 2007. Effects of cinnamon extract, chitosan coating, hot water treatment and their combinations on crown rot disease and quality of banana fruit. Postharvest Biol. Technol. 45, 333–340.

CHAPTER 4

Botrytis cinerea (Gray Mold)

Gianfranco Romanazzi and Erica Feliziani
Marche Polytechnic University, Department of Agricultural, Food, and Environmental Sciences, Ancona, Italy

Contents

OVERALL IMPORTANCE OF THE FUNGUS

Botrytis cinerea is a fungus that has a wide range of hosts that make it very easy to find as a pathogen in many crops both pre- or postharvest. According to a recent review, the fungus *B. cinerea* is included at second place into the word top 10 pathogen list based on scientific and economic importance (Dean et al., 2012). More than 200 dicotyledonous crop species worldwide are estimated to be hosts for *B. cinerea* (Williamson et al., 2007). Altogether, the global expense of gray mold control (cultural measures, botryticides, broad-spectrum fungicides, biocontrol) easily surmounts €1 billion/annum. The impact of product loss occurring despite disease control, and the quality loss during the retail chain, are likely to be far higher (Dean et al., 2012).

A recent study from the FAO estimated that with respect to the total amounts of fruits and vegetables produced globally, somewhere between 15% and 50% are lost at the postharvest stage, mainly because of pathogen decay (FAO, 2011). The highest losses were recorded in the developing

Postharvest Decay
http://dx.doi.org/10.1016/B978-0-12-411552-1.00004-1

countries of Africa and Asia, which lack the necessary technologies to prolong the storage life of fresh produce. Among the postharvest pathogens, *B. cinerea* is regarded as one of the most common and important. Vegetables and small fruit crops that are susceptible to gray mold after harvest include artichoke, apple, asparagus, bean, beet, blackberry, blueberry, broccoli, cabbage, carrot, cauliflower, celery, cucumber, currant, eggplant, endive, grape, kale, kaki, kiwi, leek, lentil, lemon, lettuce, onion, pea, peanut, pepper, pear, peach, plum, pomegranate, potato, pumpkin, raspberry, strawberry, sweet cherry, tomato and many others (Droby and Lichter, 2007; Romanazzi, 2013) (Fig. 4.1). Infections by *B. cinerea* can occur before harvest, at the field stage, and they can remain latent until storage, when the pathogen takes advantage for disease development from higher relative humidity and low temperatures, that slow down host defenses. At the beginning of the infections, a darker circular area is visible where the fruit tissues are softer than the other fruit parts, subsequently abundant sporification, whose color ranges from white to gray, depending on the quantity of light, can develop from the site of infection. Infection starts from natural openings or

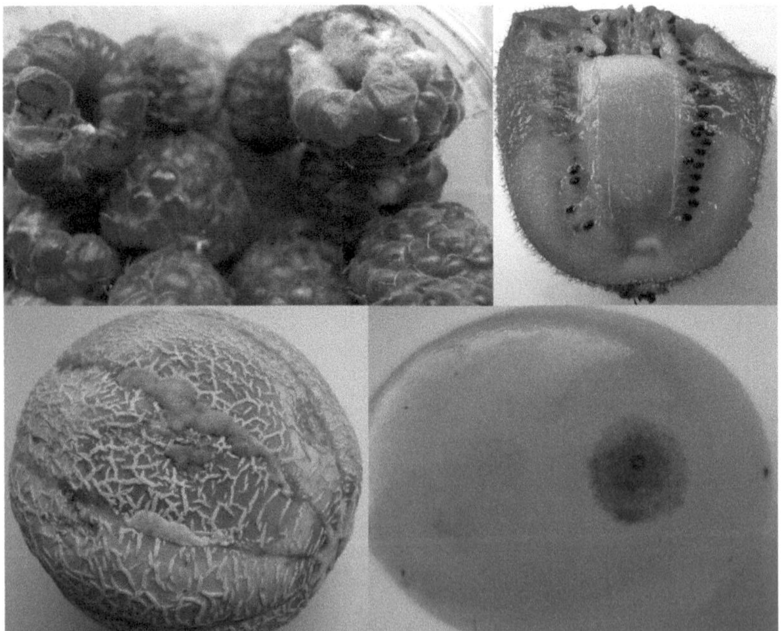

Figure 4.1 Gray mold infections on stored raspberries (top left), kiwi fruit (top right), melon (bottom left), and initial spot on table grapeberry (bottom right).

mechanical wounds that occur on fruit. Often, *B. cinerea* can develop from rotted fruit next to healthy fruit, causing extensive breakdown of the commodity, and sometimes spoiling entire lots. Indeed, the pathogen is also able to progress in infections at storage temperatures (usually 0–5°C), when fruit resistance is decreased. However, in some niche sectors, the presence of the fungus is considered beneficial. The grapes affected by noble mold are appreciated for wine production since infected grapes have a beneficial role in the production of rare dessert wines, thus another name for gray mold is 'noble rot'.

TAXONOMY

B. cinerea is the asexual stage (anomorph) of the teleomorph *Botryotinia fuckeliana*. Table 4.1 shows the classification of *B. cinerea*. The cosmopolitan genus *Botrytis* contains 22 recognized species and one hybrid (Staats et al., 2005), all of which are plant pathogens. *B. cinerea* is the only species in the genus with a broad host range, whereas all other *Botrytis* species are considered to be specialized on a single plant species (Staats et al., 2005).

The common classification of the genus *Botrytis* is largely based on morphological characters and, to a minor extent, on physiology and host range, while recently, a classification of the genus was constructed based on DNA sequence data of three nuclear protein-coding genes, [such as glyceraldehyde-3-phosphate dehydrogenase (G3PDH), heat-shock protein 60 (HSP60), and DNA-dependent RNA polymerase subunit II (RPB2)] (Staats et al., 2005). The genome of *B. cinerea* was sequenced using Sanger technology at low coverage (Amselem et al., 2011) and then updated (Staats and van Kan, 2012).

Table 4.1 *Botryotinia fuckeliana* Classification

Superkingdom	**Eukaryota**
Kingdom	Fungi
Subkingdom	Dikarya
Phylum	Ascomycota
Subphylum	Pezizomycotina
Class	Leotiomycetes
Order	Helotiales
Family	Sclerotiniaceae
Genus	Botryotinia

(Source: from Genbank http://www.ncbi.nlm.nih.gov/Taxonomy/Browser/wwwtax.cgi?mode=Info&id=40559&lvl=3&keep=1&srchmode=1&unlock&lin=s)

MORPHOLOGY

The mycelium of *B. cinerea* is branched, septate, hyaline to brown. Conidiophores, which arise directly from the mycelia or from the sclerotia, are tall, slender, irregularly branched in the terminal portion, with apical cells enlarged or rounded, bearing clusters of conidia simultaneously on short denticles. Conidia are smooth, hyaline or gray, egg-shaped, with mean length of 10 μm and mean width of 5 μm (Barnett and Hunter, 1998) (Fig. 4.2A). Sclerotia, which are survival structures, are often present. Figure 4.3 shows the life cycle of the fungus, both at the sexual and asexual stage. Sclerotia are formed from mycelial branches that fuse together to form a globular mass, which is at first hyaline, but later turns brown or black because of the deposition of melanic pigments in the outer rind. The melanized coat and β-glucans encasing the internal mycelium protect sclerotia from desiccation, UV radiation and microbial attack over long periods (Williamson et al., 2007). When the environmental conditions are favorable for the fungus, sclerotia germinate through the emission of conidiophores or, after a sexual process, the elongation of an apothecium. However, the most frequent mode of germination is conidiophore production (Jarvis, 1977). Chlamydospores are short-term survival structures which may help the fungus to overcome short unfavorable periods encountered on plant surfaces. They are formed

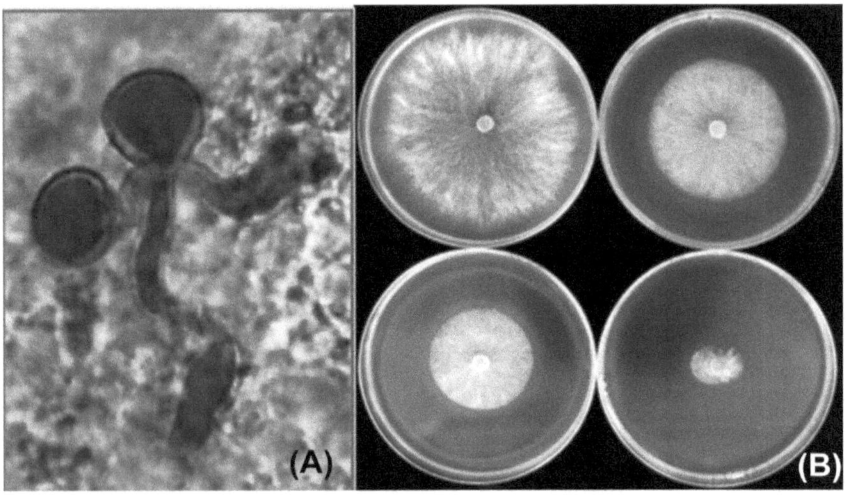

Figure 4.2 Germinating conidia of *B. cinerea* (kindly provided by Joe Smilanick). (A) Effects of chitosan amended PDA on radial growth of *Botrytis cinerea*. (B) Control (top left), 0.12% chitosan (top right), 0.25% chitosan (bottom left), and 0.50% chitosan (bottom right).

by transformation of vegetative mycelium parts and are released by hyphal disintegration (Holz et al., 2007).

In the sexual reproductive cycle of *B. cinerea*, large numbers of microconidia are observed in all the crosses that form sexual bodies (Fukumori et al., 2004). Microconidia provide an alternative microscopic propagule for these fungi when subjected to adverse conditions. In general, they are found in aging cultures of the fungus or those which are contaminated by other organisms, and in association with sclerotia. Microconidia develop from germ tubes produced by macroconidia, or endogenously, inside empty cells of mature hyphae, and from appressoria and sclerotia (Fukumori et al., 2004; Holz et al., 2007). Fukumori et al. (2004) demonstrated that the microconidia are able to function as spermatia during the sexual reproduction of *B. cinerea*. Indeed, in their experimentation, sexual bodies formed in the mating of sclerotium of one isolate with the microconidia harvested from another isolate. The apothecia that were formed on the sclerotium were at first deeply concave or funnel-shaped; as the heads expanded they became flat and ultimately convex. The size ranged from 10 to 15 mm in height and from 5 to 7 mm in cap diameter. Asci arose in a fascicle from the base of the caps of the apothecium. They were cylindrical, hyaline, and evanescent at maturity, and contained eight ascospores.

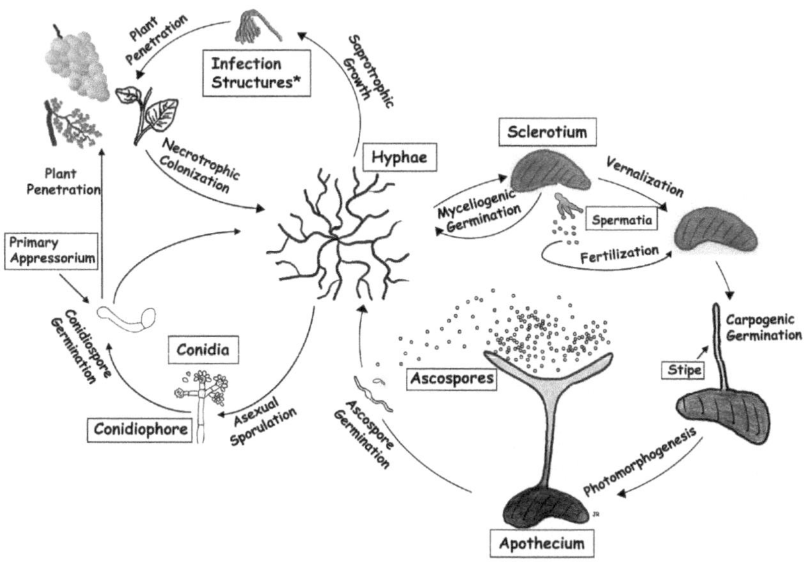

Figure 4.3 Life cycle of *Botrytis cinerea*, with different stages of sexual and asexual development. (Modified from Amselem et al., 2011.)

The ascospores were hyaline, one-celled, and ovoid or ellipsoid in shape (Fukumori et al., 2004).

Most of *B. cinerea* isolates collected in Italy were heterothallic (self-sterile and able to produce ascospore progeny when crossed with other strains) and only 16% of the field isolates and 6% of the ascospore were homothallic (self-fertile). These data suggested that sexual reproduction and meiotic recombination might be an important source of genetic variation in this pathogenic fungus (Faretra et al., 1988).

BIOLOGY OF *BOTRYTIS CINEREA* INFECTION PROCESS

The fungus exists in different habitats as mycelia, micro- and macroconidia, chlamydospores, sclerotia, apothecia and ascospores and these are dispersed by diverse means (Jarvis, 1977), even if *B. cinerea* releases its macroconidia mainly in air-dry currents (Holz et al., 2007). The life cycle of *Botrytis* comprises a somatic or vegetative or anamorph stage in which the mycelium produces asexual conidia (macroconidia) (see Fig. 4.2A) and in difficult environmental conditions sclerotia (Beever and Weeds, 2007). The pathogen also forms microconidia from phialides abundantly in aging cultures, which function primarily as spermatia (Williamson et al. 2007). Sclerotia usually germinate to produce mycelium or conidia, but after appropriate preconditions and fertilization, they may germinate to produce aphotecia (teleomorph stage), containing ascospores resulting from meiosis (Beever and Weeds, 2007). Sexual reproduction of *B. cinerea* was reported for the first time by Groves and Drayton (1939) and then studied *in vitro* (Faretra and Antonacci, 1987; Faretra et al., 1988).

In temperate climates, the sclerotia germinate in spring serving as a primary source of inoculum within a crop (Williamson et al., 2007). Besides sclerotia during winter or periods that are not favorable for the fungus, mycelium of *B. cinerea* also survive within infected dead host tissues left as crop debris and inside some seeds to serve as primary inoculum. In perennial crops, the dead leaves, flowers and mummified fruits contain masses of mycelium that can often be ideally situated within a crop canopy to produce conidia and initiate infections.

B. cinerea is a typical example of a necrotrophic fungus: it first kills host plant cells and then colonizes the dead tissue (Amselem et al., 2011). The minimum temperature for growth is 0°C, the optimum is 20°C, and the maximum is 30°C. Therefore, *B. cinerea* develops well during cold storage of fruit and vegetables.

AGRONOMIC AND ENVIRONMENTAL FACTORS INFLUENCING *BOTRYTIS CINEREA* INFECTION

According to classical paradigm in plant pathology, the existence of a disease caused by a biotic agent absolutely requires the interaction of a susceptible host, a virulent pathogen, and an environment favorable for disease development (Francl, 2001). The aim of disease management strategies is to eliminate one of these three conditions. Concerning the pathogen, human actions are not very effective in interfering with the ability of *Botrytis* species to infect living host plants. This may result from a combination of different factors such as the possession of pathogenicity factors (e.g. toxins and cell-wall degrading enzymes) that confer the ability to kill and invade plant tissues, the ability to avoid or counteract plant resistance mechanisms (e.g. toxicity of phytoalexins), the ability to survive outside host-plant tissues under less favorable conditions, and the ability to reproduce and disperse (Staats et al., 2005).

Concerning hosts, some management can be done, if agronomic choices and environmental conditions are the main aspects that can be artificially modified. In the field, treatments with resistance inducers could activate the plant defenses and prevent the onset of infections. This strategy could be a new methodology to act on the disease triangle on the side of the host. Having a fortified plant could be helpful in reducing the possibility of disease development. The choice of plant varieties that are suited to the growing area, and growing areas that are not suitable for the gray mold development is fundamental. Regions characterized by warm and dry spring weather or areas where wind is prevalent at some point during the day are to be preferred when the cultivation area is chosen. Concerning cultural practices, the most important means of inhibiting *B. cinerea* activity is to avoid high humidity and stagnant air, especially in the fruit zone, since these are ideal conditions for fungal development. Appropriate leaf removal and canopy management could minimize high humidity conditions. Microirrigation should be preferred to canopy irrigation. For strawberry production, outdoor planting should be planned to provide good air circulation patterns. Using plastic mulches can prevent berry–soil contact, and it is important to be sure that water puddles do not form under the fruit on the plastic. In greenhouse cultivation, heating and ventilation can avoid high humidity conditions. This may require extra venting early in the day when moisture has condensed and before sunlight has warmed the air. Sanitation is another fundamental point. The practice of removing dead or dying tissues from the plants and from the soil surface either in the field or in greenhouse or removing all ripe fruits during harvest as well as any fruit with signs of decay or rain damage and not to throw debris under benches or on walks are all necessary to eventually prevent gray mold infections. In strawberries, for example, the pathogen colonizes senescent leaves and overwinters here, sporulating in the next season when flowering occurs (Powelson, 1960).

POSTHARVEST FACTORS FAVORING *BOTRYTIS CINEREA* INFECTION

Overripe fruit are more prone to gray mold infection, therefore the correct choice of fruit maturity at harvest is as important as the other cultural practices. During storage, the rate of postharvest deterioration depends on several external factors, including temperature, relative humidity, air speed, atmospheric composition (concentrations of oxygen, carbon dioxide, and ethylene), and sanitation procedures (Kader, 2005). Careful handling and the prevention of mechanical damage or pathophysiological injuries due to excessive chilling or lighting or to anomalies in the atmospheric gaseous composition are fundamental for decay control. In many instances, these causes are interrelated; i.e. mechanical injury can be associated with postharvest decay from many causes (Mari et al., 2009). In the food industry during storage, the use of technologies (e.g. exposure to modified or controlled atmosphere) that delay fruit respiration then ripening and senescence is common to prolong the shelf-life of fruit.

CONTROL

Traditional Control Methods

Fungicides

Infection by *B. cinerea* that causes postharvest decay usually occurs before the harvest at the field stage, and can remain latent until storage. The traditional control of gray mold infection consists of field application of synthetic fungicides during the crop growing cycle. Because of the lack of effective methods to predict the risk of this disease, preventive applications, before the appearance of disease symptoms, are recommended. For example, for table grapes, four fungicide applications at the end of flowering, bunch closure, veraison and 3 weeks before harvest are usually carried out (Feliziani et al., 2013b). For strawberries, fungicides are applied around flowering, and are repeated up to harvest, depending on weather conditions and the preharvest interval of the formulations (Feliziani and Romanazzi, 2013). For kiwi fruit, some postharvest treatments with fungicides, as well as field applications, are carried out before commercialization, i.e. the fungicide fluodioxonil is approved in Italy as a postharvest treatment to be applied 30 days before consumption.

Several families of synthetic fungicides able to control gray mold infections are available (Table 4.2). They can be classified into categories according to their biochemical modes of action: fungicides affecting fungal respiration, microtubule function, osmoregulation, methionine biosynthesis or sterol biosynthesis (http://www.frac.info). The mode of action of anilinopyrimidine, which includes cyprodinil, mepanipyrim, pyrimethanil, consists of inhibition of methionine biosynthesis and secretion of hydrolytic enzymes. Boscalid, which belong to the pyridine-carboxamides group, prevents energy production, since it is a succinate dehydrogenase inhibitor. It blocks the synthesis of essential cell components and, hence, disrupts fungal growth with deleterious effects on spore germination, germ tube elongation, mycelial growth, and sporulation. Similarly, the mode of action of the active principal of fluopyram, which is a pyridinyl-ethyl-benzamide fungicide, mainly consists of inhibition of succinate dehydrogenase, and thus the fungal respiratory chain complex II. Pyraclostrobin acts through the inhibition of mitochondrial respiration by binding at the ubiquinol oxidation center (Qo site) of cytochrome bc1in Complex III. Fenhexamid, which is in the class of hydroxyanilides, inhibits the C3-keto-reductase step in ergosterol biosynthesis. Like fenhexamid, fenpyrazamine, belonging to the same class, inhibits sterol biosynthesis, preventing fungal germ tube and mycelium elongation. Besides fenhexamid and fenpyrazamine, the sterol biosynthesis inhibitors are a wide class of fungicides that are commonly used in conventional agriculture; these include imidazoles, triazoles, and morpholin. Iprodione, belonging to the class of dicarboximides, prevents the germination of spores and growth of mycelium by inhibiting DNA and RNA synthesis and cell division in fungi. The mode of action of fluazinam, which belongs to the chemical group of 2,6-dinitro-anilines, is based on the ability of this active ingredient to uncouple oxidative phosphorylation, thus to inhibit ATP synthesis. The proposed mode of action of fludioxonil and the other fungicides of the phenylpyrrole group is the interference with the transport process of sugars and amino acids in the plasmatic membrane of fungi.

B. cinerea is a classical 'high-risk' pathogen, and development of resistance to several classes of fungicides has been frequently reported worldwide. In many countries, the use of some of them (i.e. benzimidazole) has been revoked or is restricted in anti-resistance strategies. Mutants of *B. cinerea*

Table 4.2 List of some of the most Common Fungicides used to Control Gray Mold Infection Registered for at Least one Fruit Species

Active Ingredient (%)	Trade Name	Company	Chemical group	FRAC Code
Boscalid (70)	Cantus	Basf	Pyridine-carboxamides	7
Cyprodinil (37.5) + fluodioxonil (25)	Switch	Syngenta	Anilinopyrimi dines + phenyl-pyrroles	9 + 12
Cyprodini (75)	Vangard	Syngenta	Anilinopyrimi dines	9
Fenhexamid (50)	Elevate/ Teldor	ArystaLife Science/ Bayer Crop Science	Hydroxyanilides	17
Fenpyrazamine (50)	Prolectus	Sumitomo Chemical Company	Amino-pyrazolinone	17
Fluazinam (40)	Omega	Syngenta	2,6-Dinitro-anilines	29
Fluodioxonil (20.4)	Scholar	Syngenta	Phenylpyrroles	12
Fluodioxonil (50)	Geoxe	Syngenta	Phenylpyrroles	12
Fluopyram (50)	Luna Privilege	Bayer Crop Science	Pyridinyl-ethyl-benzamides	7
Iprodione (50)	Rovral	Basf	Dicarboximides	2
Pyraclostrobin (12.8) + boscalid (25.2)	Pristine	Basf	Qolfungicides + pyridine-carboxamides	7 + 11
Pyraclostrobin (6.7) + boscalid (26.7)	Signum	Basf	QoI fungicides + Pyridine-carboxamides	7 + 11
Pyrimethanil (54.6)	Scala	Bayer Crop Science	Anilinopyrimi dines	9

resistant to fungicides were isolated in laboratory studies and in the field for fenexamide (De Guido et al., 2007; Fillinger et al., 2008), boscalid (Bardas et al., 2010; De Miccolis Angelini et al., 2010), pyraclostrobin (Markoglou et al., 2006; Bardas et al., 2010), carbendazim, pyrimethanil (Moyano et al.,

2004; Myresiotis et al., 2007), cyprodinil, iprodione (Myresiotis et al., 2007), procymidone, diethofencarb (Moyano et al., 2004) and fluazinam (Korolev et al., 2011). The fact that *B. cinerea* strains resistant to fungicides are very common emphasizes the need for an accurate adoption of Fungicide Resistance Action Committee (FRAC) guidelines suggesting a limited number of applications per year. Furthermore, the traditional control of postharvest infection in cold-stored table grapes can be achieved by repeated fumigation with sulfur dioxide or by the use of sulfur dioxide generating pads closed in polyethylene-lined grape containers.

Alternatives to Synthetic Fungicides

Although field application of synthetic fungicides remains the most common method to control postharvest rot of fruit and vegetables, their application has several negative aspects that need to be taken in account. The fungicides decline in their residues on fruit after harvest and storage decay can develop. For some commodities, the application of postharvest fungicides is not permitted, due to the several normative restrictions and, in general, the use of fungicide is banned in organic agriculture. In addition, increasing public concern towards fungicide residues has contributed to the interest in the development of alternative methods for controlling postharvest decay of fruit and vegetables, which need to be integrated into, if not totally replace, the use of synthetic fungicides. Also, the appearance of pathogen isolates resistant to one or more fungicides has discouraged their repeated use. Research efforts have led to the development of novel control tools as alternatives to synthetic fungicide treatments. For schematic reasons, these can be grouped into four main categories: (1) biological control agents (BCAs); (2) natural compounds; (3) compounds generally recognized as safe (GRAS); and (4) physical methods alone or the combination of all four groups (Mari et al., 2009; Romanazzi et al., 2012).

BCAs are mainly bacteria and yeasts that are 'antagonistic' to pathogens that can cause postharvest fruit spoilage. These can act through five mechanisms: competition for nutrients and space; antibiosis; parasitism; induction of resistance in the host tissue; and production of volatile metabolites (Jamalizadeh et al., 2011). Several products based on microorganisms are registered against *B. cinerea*. Some examples of commercially available biofungicides are: *Pseudomonas syringae* (BioSave, JET Harvest Solutions, Longwood, FL, US), *Cryptococcus albidus* (YieldPlus, Lallemand, Montreal, Canada), *Bacillus subtilis* (Serenade, Bayer, Leverkusen, Germany), *Candida sake* (Candifruit, IRTA, Lleida, Spain), *Pantoea agglomerans* (Pantovital,

Domca, Granada, Spain), *Aurobasidiumpullulans* (Boni Protector, Bio-Ferm, Tulln, Austria; Botector, Manica, Rovereto, Italy), *Candida oleophila* (Nexy, BioNext, Paris, France), *Bacillus amyloliquefaciens* (Amylo-X, Biogard CBC, Grassobbio, Italy), and *Metschnikowia fructicola* (Shemer, Bayer, Leverkusen, Germany) (Feliziani and Romanazzi, 2013; Spadaro et al., 2013; Teixidó et al., 2013).

Natural or GRAS compounds are substances that are known not to be harmful to the environment and to human health, and these are used for their antimicrobial properties or their induction of plant defenses. Among the natural compounds, plant extracts and essential oils have been reported to control postharvest diseases, both *in vitro* and *in vivo*, and to prolong the overall quality and storage life of fresh commodities (Antunes and Cavaco, 2010; Feliziani et al., 2013a). Inorganic salts have been shown to be active antimicrobial agents against a range of phytopathogenic fungi, and among these agents, bicarbonates have been proposed as safe and effective alternative means to control postharvest rot of fruit and vegetables. Also, as well as these salts being non-toxic and having minor environmental impact at effective concentrations, they are inexpensive (Sanzani et al., 2009). Several sanitizers classified as GRAS have been applied to extend the postharvest storage of various produce, including acetic acid, electrolyzed oxidizing water, and ethanol (Romanazzi et al., 2012). Resistance inducers are plant or pathogen constituents, or their analogs, that act as plant elicitors, as they can activate plant defense mechanisms, and thus simulate the presence of pathogens. Among the resistance inducers, the natural biopolymer chitosan (see Fig. 4.2B) and the synthetic elicitor benzothiadiazole have been reported to activate systemic acquired resistance in horticultural produce (Terry and Joyce, 2004; Romanazzi et al., 2013; Feliziani et al., 2013b).

Physical means include heat treatment, UV-C light, hypobaric and hyperbaric treatments, and exposure to modified or controlled atmospheres or to ozone, among other measures (Romanazzi et al., 2012). Physical treatments can have dual effects on the fruit, as these are active against the pathogen and, at the same time, they can induce host defense responses (Wilson et al., 1994).

Depending on the characteristics of the commodity and on the specific situation, one strategy of controlling postharvest decay of fruit might be more appropriate than another. To overcome the drawbacks that can arise with the use of a unique strategy, the integration of methods might provide additive or synergistic effects for disease control (Mari et al., 2007). An ideal alternative means of controlling postharvest decay should improve on current practices, should be affordable and easy to implement, and should not have any negative influence on the fruit to which it is applied or on the environment or human health (Romanazzi et al., 2012).

CONCLUDING REMARKS

Botrytis cinerea has a wide host range and is in second place into the word top 10 pathogens list based on scientific and economic importance (Dean et al., 2012). Considering this figure, it is clear that its management is crucial for many fruit and vegetables grown worldwide. The current management practice based on synthetic fungicides, even if effective in some cases, needs to be complemented with alternatives considering the high risk of fungal resistances. The new trend of the European Union policies with Directive 128/2009 has made integrated pest management mandatory from January 1, 2014. This Directive states that the European Member States need to take all the necessary measures to promote the use of products with low risk to human health and the environment among those available to control the same pathogen. In addition, this Directive considers that where there is the risk of the development of pathogen resistance against a plant protection measure, the available anti-resistance strategies should be applied to maintain the effectiveness of the products. This can include the use of multiple pesticides with different modes of action, or the integration of several means that might even be non-chemical. Depending on the characteristics of the commodity and on the specific situation, one strategy of controlling gray mold disease on fruits and vegetables might be more appropriate than another. To overcome the drawbacks that can arise with the use of a unique strategy, the integration of methods might provide additive or synergistic effects for disease control (Mari et al., 2007). An ideal alternative of controlling postharvest decay should improve on the current practice, should be affordable and easy to implement, and should not have any negative influence on the fruit to which it is applied or on the environment or human health (Romanazzi et al., 2012).

The recent availability of the sequence of the entire *B. cinerea* genome could open new insight into the mechanisms of infection of the fungus. Furthermore, the knowledge at molecular level on the interaction of host–plant–chemicals applied will be helpful to understand better changes that occur in host plants following treatments or the effects of treatments on the pathogen, and then optimize application to get the highest effects with the minimum number/time/concentration of applications.

ACKNOWLEDGMENT

The financial support within the projects 'Induced resistance to control grape diseases', was granted by Marche Polytechnic University and 'EUBerry Project' (EU FP7 KBBE 2010-4, Grant Agreement No. 265942).

REFERENCES

Amselem, J., Cuomo, C.A., van Kan, J.A.L., et al., 2011. Genomic analysis of the necrotrophic fungal pathogens *Sclerotinia sclerotiorum* and *Botrytis cinerea*. PLoS Genet. 7, 1–27.

Antunes, M.D., Cavaco, A.M., 2010. The use of essential oils for postharvest decay control. A review. Flavour Fragrance J. 25, 351–366.

Bardas, G.A., Veloukas, T., Koutita, O., Karaoglanidis, G.S., 2010. Multiple resistance of *Botrytis cinerea* from kiwifruit to SDHIs, QoIs and fungicides of other chemical groups. Pest Manag. Sci. 66, 967–973.

Barnett, H.L., Hunter, B.B., 1998. Illustrated genera of imperfect fungi. APS Press, St Paul, Minnesota.

Beever, R.E., Weeds, P.L., 2007. Taxonomy and genetic variation of *Botritys* and *Botryotynia*. In: Elad, Y., Williamson, B., Tudzynski, P., Delen, N. (Eds.), Botrytis: Biology, Pathology and Control. Springer, Heidelberg, pp. 29–52.

Dean, R., van Kan, J.A.L., Pretorius, Z.A., et al., 2012. The top 10 fungal pathogens in molecular plant pathology. Mol. Plant Pathol. 13, 414–430.

De Guido, M.A., De Miccolis Angelini, R.M., Pollastro, S., Santomauro, A., Faretra, F., 2007. Selection and genetic analysis of laboratory mutants of *Botryotinia fuckeliana* resistant to fenhexamid. J. Plant Pathol. 89, 203–210.

De Miccolis Angelini, R., Habib, W., Rotolo, C., Pollastro, S., Faretra, F., 2010. Selection, characterization and genetic analysis of laboratory mutants of *Botryotinia fuckeliana* (*Botrytis cinerea*) resistant to the fungicide boscalid. Eur. J. Plant Pathol. 128, 185–199.

Droby, S., Lichter, A., 2007. Postharvest Botritys. In: Elad, Y., Williamson, B., Tudzynski, P., Delen, N. (Eds.), Botrytis: Biology, Pathology and Control. Springer, Heidelberg, pp. 349–367.

FAO, 2011. Global food losses and food waste. Study conducted for the International Congress "SAVE FOOD!" at Interpack 2011 Düsseldorf, Germany.

Faretra, F., Antonacci, E., 1987. Production of apothecia of *Botryotinia fuckeliana* (de Bary) Whetz. Under controlled environmental conditions. Phytopathol. Mediterranea 26, 29–35.

Faretra, F., Antonacci, E., Pollastro, S., 1988. Sexual behavior and mating system of *Botryotinia fuckeliana*, teleomorph of *Botrytis cinerea*. J. Gen. Microbiol. 134, 2543–2550.

Feliziani, E., Romanazzi, G., 2013. Preharvest application of synthetic fungicides and alternative treatments to control postharvest decay of fruit. Stewart Postharvest Rev. 9, (in press).

Feliziani, E., Santini, M., Landi, L., Romanazzi, G., 2013a. Pre and postharvest treatment with alternatives to synthetic fungicides to control postharvest decay of sweet cherry. Postharvest Biol. Technol. 78, 133–138.

Feliziani, E., Smilanick, J.L., Margosan, D.A., et al., 2013b. Preharvest fungicide, potassium sorbate, or chitosan use on quality and storage decay of table grapes. Plant Dis. 97, 307–314.

Fillinger, S., Leroux, P., Auclair, C., Barreau, C., Al Hajj, C., Debieu, D., 2008. Genetic analysis of fenhexamid-resistant field isolates of the phytopathogenic fungus *Botrytis cinerea*. Antimicrob. Agents Chemother. 52, 3933–3940.

Francl, L.J., 2001. The Disease Triangle: A plant pathological paradigm revisited. Plant Health Instructor. http://dx.doi.org/10.1094/PHI-T-2001-0517-01.

Fukumori, Y., Nakajima, M., Akutsu, K., 2004. Microconidia act the role as spermatia in the sexual reproduction of *Botrytis cinerea*. J. Gen. Plant Pathol. 70, 256–260.

Groves, J.W., Drayton, F.L., 1939. The perfect stage of *Botrytis cinerea*. Mycologia 31, 485–489.

Holz, G., Coertze, S., Williamson, B., 2007. The ecology of Botrytis on plant surface. In: Elad, Y., Williamson, B., Tudzynski, P., Delen, N. (Eds.), Botrytis: Biology, Pathology and Control. Springer, Heidelberg, pp. 9–27.

Jamalizadeh, M., Etebarian, H.R., Aminian, H., Alizadeh, A., 2011. A review of mechanisms of action of biological control organisms against postharvest fruit spoilage. EPPO Bullettin 41, 65–71.

Jarvis, W.R., 1977. Botryotinia and Botrytis Species: Taxonomy, Physiology, and Pathogenicity. Hignell Printing Limited, Winnipeg, Canada.

Kader, A.A., 2005. Increasing food availability by reducing postharvest losses of fresh produce. Acta Horticulturae 682, 2169–2175.

Korolev, N., Mamiev, M., Zahavi, T., Elad, Y., 2011. Screening of *Botrytis cinerea* isolates from vineyards in Israel for resistance to fungicides. Eur. J. Plant Pathol. 129, 591–608.

Mari, M., Neri, F., Bertolini, P., 2007. Novel approach to prevent and control postharvest diseases of fruits. Stewart Postharvest Rev. 3, 1–7.

Mari, M., Neri, F., Bertolini, P., 2009. Management of important diseases in Mediterranean high value crops. Stewart Postharvest Rev. 5, 1–10.

Markoglou, A.N., Malandrakis, A.A., Vitoratos, A.G., Ziogas, B.N., 2006. Characterization of laboratory mutants of *Botrytis cinerea* resistant to QoI fungicides. Eur. J. Plant Pathol. 115, 149–162.

Moyano, C., Gómez, V., Melgarejo, P., 2004. Resistance to pyrimethanil and other fungicides in *Botrytis cinerea* populations collected on vegetable crops in Spain. J. Phytopathol. 152, 484–490.

Myresiotis, C.K., Karaoglanidis, G.S., Tzavella-Klonari, K., 2007. Resistance of *Botrytis cinerea* isolates from vegetable crops to anilinopyrimidine, phenylpyrrole, hydroxyanilide, benzimidazole, and dicarboximide fungicides. Plant Dis. 91, 407–413.

Powelson, R.L., 1960. Initiation of strawberry fruit rot caused by *Botrytis cinerea*. Phytopathology 50, 491–494.

Romanazzi, G., 2013. Innovative control strategies for *Botrytis cinerea* in different postharvest fruit systems. Proceedings of XVI International Botrytis Symposium, Locorotondo, Italy. June 23–28, 2013, p. 72.

Romanazzi, G., Lichter, A., Mlikota Gabler, F., Smilanick, J.L., 2012. Recent advances on the use of natural and safe alternatives to conventional methods to control postharvest gray mold of table grapes. Postharvest Biol. Technol. 63, 141–147.

Romanazzi, G., Feliziani, E., Santini, M., Landi, L., 2013. Effectiveness of postharvest treatment with chitosan and other resistance inducers in the control of storage decay of strawberry. Postharvest Biol. Technol. 75, 24–27.

Sanzani, S.M., Nigro, F., Mari, M., Ippolito, A., 2009. Innovations in the control of postharvest diseases of fresh fruits and vegetables. Arab J. Plant Protect. 27, 240–244.

Spadaro, D., Garibaldi, A., Gullino, M.L., 2013. Discovery, development and technology transfer of biocontrol agents for postharvest disease control. In Proceedings of II International Symposium on Discovery and Development of Innovative Strategies for Postharvest Disease Management, Kusadasi, Turkey. April 28–May 2, p. 12.

Staats, M., Van Baarlen, P., vanKan, J.A.L., 2005. Molecular phylogeny of the plant pathogenic genus Botrytis and the evolution of host specificity. Mol. Biol. Evol. 22, 333–346.

Staats, M., van Kan, J.A.L., 2012. Genome update of *Botrytis cinerea* strains B05.10 and T4. Eukaryot. Cell 11, 1413–1414.

Teixidó, N., Torres, R., Viñas, I., Abadias, M., Usall, J., 2013. Biological and natural materials for postharvest disease management: success and failure. In Proceedings of II International Symposium on Discovery and Development of Innovative Strategies for Postharvest Disease Management, Kusadasi, Turkey. April 28–May 2, p.11.
Terry, L.A., Joyce, D.C., 2004. Elicitors of induced disease resistance in postharvest horticulture crops: a brief review. Postharvest Biol. Technol. 32, 1–13.
Williamson, B., Tudzynski, B., Tudzynski, P., van Kan, J.A.L., 2007. *Botrytis cinerea*: the cause of grey mould disease. Mol. Plant Pathol. 8, 561–580.
Wilson, C.L., El Ghaouth, A., Chalutz, E., et al., 1994. Potential of induced resistance to control postharvest diseases of fruits and vegetables. Plant Dis. 78, 837–844.

CHAPTER 5

Alternaria alternata (Black Rot, Black Spot)

Rosalba Troncoso-Rojas and Martín Ernesto Tiznado-Hernández
Coordinación de Tecnología de Alimentos de Origen Vegetal, Centro de Investigación en Alimentación y Desarrollo, Hermosillo, Sonora, México

Contents

OVERALL IMPORTANCE OF THE FUNGUS

Alternaria alternata is a fungus that has been related to food poisoning due to the production of mycotoxins which include alternariol, altenuene, alternariol monomethyl ether, altertoxins and L-tenuazonic acid (Scott, 2001). Some of these are dangerous and, indeed, alternariol and alternariol methyl ester can increase the rate of breaks in the DNA of human carcinoma cells by inhibiting DNA relaxation and stimulating the cleavage of DNA by

Postharvest Decay
http://dx.doi.org/10.1016/B978-0-12-411552-1.00005-3

topoisomerases I, IIα and IIβ (Fehr et al., 2009). This species is also known to cause fungal keratitis in humans (Xu et al., 2013).

There are several causes of fruit and vegetables postharvest losses; however, one of the most important is fungal attack (Tripathi and Dubey, 2004). This is of paramount importance because fungi can sporulate and resist the postharvest treatments aimed at eliminating the microorganisms present in the produce. Therefore, they can start infecting the fruit while on its way to the international markets.

Of the fungi which can infect fruits and vegetables, *A. alternata* can attack a wide range of agricultural products in many areas of the world and, indeed, its presence had been reported in Pakistan (Fatima et al., 2009), China (Gao et al., 2013), Australia (Harteveld et al., 2013), Bangladesh (Bashar et al., 2012), Mexico (Espinoza-Verduzco et al., 2012), Korea (Kwon et al., 2011), Greece (Elena, 2006), Oman (Al-Sadi et al., 2011), Argentina (Pose et al., 2010), Japan (Taba et al., 2009), India (Hubballi et al., 2010), Bulgaria (Mirkova and Konstantinova, 2003), the USA, Colombia, Turkey, South Africa and Israel (Peever et al., 2002), Spain (Vicent et al., 2000), Brazil (Peres et al., 2003), Peru (Marín et al., 2006) and Iran (Golmohammadi et al., 2006).

From this, although there is a lack of written reports, we can conclude that this fungus is most likely spread all over the world and able to proliferate in many different environments. As mentioned above, besides *A. alternata*'s presence in many regions of the world, this fungus has been found to be responsible for different diseases during the postharvest shelf-life of many different horticultural products including stem-end rot of mangoes (Amin et al., 2011), black rot in cherry tomatoes (Wang et al., 2010), core browning and moldy core of fuji apples (Gao et al., 2013), fruit spot on apples (Harteveld et al., 2013), moldy core and core rot in apples (Shtienberg, 2012), black rot of mandarins (Nemsa et al., 2012), black rot of kiwi fruit (Kwon et al., 2011), fruit rot in immature cucumbers (Al-Sadi et al., 2011), black spot in melons (Wang et al., 2010), moldy heart on peaches (Pose et al., 2010), brown spot disease on tangerines as well as in its related hybrids (Reis et al., 2006), fruit rot of *Capsicum annuum* (Anand et al., 2009), brown spot of mandarins (Fourie et al., 2009), *Alternaria* rot on figs (Doster and Michaillides, 2007), pericarp browning on litchi fruits (Sivakumar et al., 2007), *Alternaria* rot of pingguoli pear (*Pyrus bretschneideri* Rehd. cv. 'Pingguoli') which is characterized for black spots (Li et al., 2007), *Alternaria* brown spot on citrus fruit (Timmer et al., 2000), black spot on persimmon fruit (Prusky et al., 1981), side and stem-end rot of mango (Kobiler et al., 2001), brown apical necrosis on persian walnuts (Belisario et al., 2002), *Alternaria* late blight of pistachios (Pryor and

Michailides, 2002), brown spot on the hybrids of tangerine × grapefruit as well as on grapefruit (Peever et al., 2002).

From the above, it can be clearly seen that *A. alternata* is a fungus that can infect many fruits, however, although it is not within the scope of this chapter, this fungus has been reported to infect other parts of plants like seeds (Szopinska et al., 2012), leaves (Harteveld et al., 2013), stems (Choi et al., 2010) and flowers (Espinoza-Verduzco et al., 2012) reducing the agricultural production either directly by infecting the fruit or indirectly by impairing the physiology of plant photosynthesis (Lagopodi and Thanassoulopoulos, 1998; Qiang et al., 2010).

In summary, besides the possible negative effects on human health, *A. alternata* is one of the most important fungi inducing postharvest decay because of its ability to infect both fruit and vegetative tissues in a wide range of hosts in most of the world's natural environments.

TAXONOMY AND MORPHOLOGY

Alternaria is a dictyosporic genus of the family Dematiaceae, order Hyphomycetes, Fungi Imperfecti. The genus was established in 1817 by Nees, with *A. alternata* (originally *A. tenuis*) as the type species. Then the dispute about its taxonomy started and, finally, in 1912, it was named *A. alternata* (Fr.) Keissl (Rotem, 1998).

Several attempts have been made to identify and differentiate this species using morphology, physiology, metabolic profile, DNA sequences of coding regions, DNA molecular markers and a combination of two or more of the mentioned approaches. Morphological studies had shown that typical colonies of *A. alternata* are lettuce-green to olive green in color and usually have a prominent (2–5 mm) white margin when growing on potato dextrose agar. Isolates typically produce colonies over 70 mm in diameter after 7–10 days. Based upon the sporulation habit of single-spored colonies, *A. alternata* is distinguished by the formation of conidial chains six to 14 conidia in length and the development of numerous secondary, and occasionally tertiary, chains two to eight conidia in length. Chain branching occurs in a sympodial manner through the elongation of secondary conidiophores from distal terminal conidial cells and subsequent conidium formation. Small conidia (20–50 μm long) are an important characteristic of this species. Conida are ovate in shape, divided by transverse and vertical walls, with minimum development of apical extensions (Fig. 5.1). The hyphae and conidiophores are light brown and septate (Simmons and Roberts, 1993; Simmons, 1999; Pryor and Michailides, 2002).

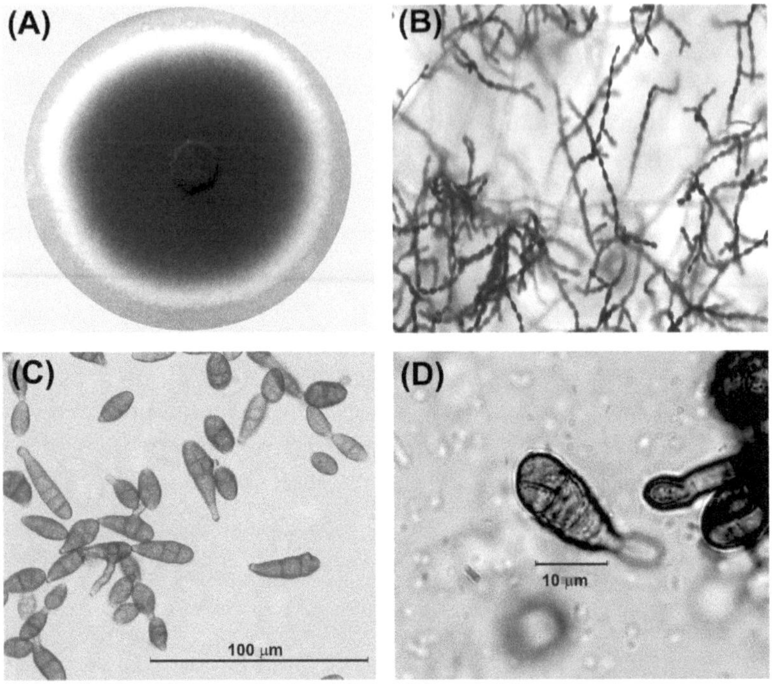

Figure 5.1 *Alternaria alternata* morphology. (A) Colonies of *A. alternata* growing on potato dextrose agar. (B) Conidial chains and chain branching of *A. alternata*. (C) Conidia at ×40. (D) Conidia at ×100. (*Photographs (B) and (C) are a kind gift from Professor Barry Pryor, Plant Pathology, University of Arizona, Tucson, Arizona, USA.*)

Unambiguous identification and classification of *A. alternata* has been always difficult and that is why some authors had concluded that all plant pathogenic *Alternaria* species are in fact *A. alternata* and further suggested differentiating them from each other by using the term pathotype based on their host specificity (Masunaka et al., 2005). For instance, an *Alternaria* species which infects apple leaves was previously called *A. mali*, however, now it is classified as *A. alternata* apple pathotype and the protocol developed for its identification is based on the presence or absence of the AMT gene encoding a cyclic peptide synthetase enzyme which plays a crucial role in the infection of plant tissues (Johnson et al., 2000).

Several studies had concluded that, indeed, there is a large variability within the *A. alternata* species which, most likely, is one of the most important factors making identification difficult. An analysis of the DNA intergenic space region of 322 *A. alternata* isolates was carried out by restriction fragment length polymorphism using a DNA region containing two entire copies of the nuclear rDNA as a probe. The isolates of *A. alternata* were obtained from 13 areas of Japan. The analysis found only eight different genotypes with different frequencies depending upon the sampled area. However, by using a DNA region of low repetitive DNA as a probe, the presence of 46 genotypes within 104 isolates from one region was recorded. Furthermore, analysis of 68 isolates of one of the groups created by using the ribosomal DNA was able to find 24 different genotypes (Adachi et al., 1993). Analysis of the variability of 56 *A. alternata* strains isolated from pistachio, obtained from four different regions of California, was carried out using the same approach as the experiment described before. Furthermore, the probe utilized was the one including two entire copies of the nuclear rDNA. In this case, the presence of 34 genotypes was recorded, suggesting a much larger variability (Aradhya et al., 2001). Analysis of *A. alternata* strains isolated from tomato using random amplified length polymorphism with 29 primers was carried out. In this experiment, it was possible to differentiate 65 genotypes within 69 different isolates analyzed, suggesting again that there is a large variation within the *A. alternata* species (Morris et al., 2000). In agreement with this work, analysis with random amplified length polymorphism using five primers was able to find 30 different genotypes within 30 isolates of *A. alternata* (Weir et al., 1998).

In another experiment, 112 strains of *A. alternata* isolated from mature pines (*Pinus tabulaeformis*) were analyzed by random amplified microsatellites including the next two triplets at the 3′ end: CCA and CGA. The phylogenetic analysis clearly showed 105 different genotypes within the 112 strains suggesting that there is a large variation within the species (Guo et al., 2004). This study suggests that perhaps there is a large variation within the species *A. alternata*; however, the identification was done using morphological characteristics which can easily lead to mistakes in classification.

Microsatellite regions have also been utilized to analyze the variability within the species. Using five microsatellites isolated from an *A. alternata* genomic library, it was possible to find between eight and 14 alleles per locus and, using this tool, the authors were able to find 62 different genotypes within 64 isolates (Tran-Dinh and Hocking, 2006). The analysis of the inter microsatellite region in the genome of *A. alternata* using four primers was able to find five genotypes in five isolates from tomato and five genotypes in five isolates from cabbage (Troncoso-Rojas et al., 2013).

The use of morphology involves the analysis of the spore width, spore length, number of transverse septa and presence and length of beaks

(Pusz, 2009). Indeed, specific protocols were developed to induce conidia production in *A. alternata* isolated from citrus to help in the fungal identification (Carvalho et al., 2008). Further, it had been found that the use of standardized protocols to analyze metabolite profiles, the growth rates at different temperatures and colony morphology can be useful in the differentiation of *Alternaria* spp. (Andersen et al., 2005). In agreement with this study, the use of morphology, metabolite profile and the physiological characteristics of the colony growing in the media dichloran rose bengal yeast extract sucrose agar, allowed differentiation of *A. infectoria* from *A. alternata* (Andersen and Thrane, 1996). However, it had been found in other studies that the use of only the physiological behavior of the colony growing in the media dichloran rose bengal yeast extract sucrose agar was not enough to differentiate between these species (de Hoog and Horré, 2002). Also, *A. infectoria* was originally identified based on colony morphology including the three-dimensional structure of the conidia, however, it was found that these characteristics can be similar with other *Alternaria* species which render this characteristic rather useless in *Alternaria* taxonomy (Simmons and Roberts, 1993).

In agreement, growth conditions can alter the colony characteristics which make identification by using conidia and conidiophore morphology rather difficult (Misaghi et al., 1977). Indeed, it was observed that physiological alterations in *A. alternata* growing in malt extract such as excessive branching, cell wall swelling and hyphal size reduction occur when treated with 500 μg $mL^{-1}$of chitosan prepared by deacetylacion of chitin with alkaline treatment (de Oliveira et al., 2012). Also, the use of a defined medium to grow different species of *Alternaria* was able to group together several isolates of *A. arborescens* and several isolates of *A. alternata* and clearly separate the two groups. However, RAPD analysis of the isolates from each group, based on three primers, showed genotypic differences between them (Roberts et al., 2000).

An analysis based on colony morphology, conidial physiological characteristics and metabolite profile did not differentiate between *A. alternata* and *A. tenuissima* (Serdani et al., 2002). In contrast, using the same variables, 39 isolates from India, Taiwan, Japan, Korea, the USA, and Zimbabwe (formerly Rhodesia) were studied. Cluster and principal component analysis of the physiological characteristics of colony growth in nine media at five different temperatures and 38 metabolites was able to differentiate between *A. alternata*, *A. longipes* and *A. gaisen* (Andersen et al., 2001).

The use of the sequences of gene coding proteins had been shown to be inefficient because most of the genes that had been chosen for the taxonomical identification of fungi are invariant in these taxa (Peever et al., 2004). Therefore, several attempts had been carried out based on DNA markers either alone or in combination with genes encoding proteins.

An analysis using random amplified length polymorphism was unable to differentiate *A. alternata* isolated from tomato or potato, although it did discriminate between *A. solani* and *A. alternata* (Weir et al., 1998). Also, random amplified length polymorphism analysis failed to differentiate between *A. alternata* and *A. infectoria* (Roberts et al., 2000).

The intergenic space region of DNA has been tested to help in the identification of the taxon *A. alternata*. In this context, the utilization of a restriction map developed by using 11 enzymes of the intergenic spacer region of the nuclear genes encoding the ribosomal RNA did not provide enough polymorphism to distinguish between *A. tenuissima*, *A. alternata*, and *A. arborescens* (Hong et al., 2005). However, the use of the ITS1 region sequence was enough to differentiate *A. infectoria* from *A. alternata* due to the presence of a 26 bp insert in the former species. However, using this approach it was not possible to distinguish clearly *A. longipes*, *A. mali*, *A. tenuissima* from *A. alternata* (de Hoog and Horré, 2002). In partial agreement, analysis using the ITS1 and ITS2 did not differentiate *A. alternata* from *A. arborescens*, or *A. tenuissima* from *A. infectoria* (Serdani et al., 2002). In agreement with these results, a cluster analysis based on the ITS4–ITS5 sequences did not differentiate between *A. alternata*, *A. gaisen*, *A. yaliinficiens*, *A. arborescens*, *A. tenuissima* and *A. brassicicola* (Roberts et al., 2012).

The use of the 5.8s ribosomal DNA region sequence along with the sequences of the ITS1 and ITS2 region failed to differentiate between *A. alternata* and *A. brassicae* (Kai et al., 2001). In contrast, it was reported that the use of the same three regions was able to differentiate between *A. alternata* and *A. brassicae*, however, in this study, the analysis did not show phylogenetic differences between *A. alternata* and *A. tenuissima* species. However, by using the sequence of the small mitochondrial subunit, the authors were able to differentiate *A. alternata* from other *Alternaria* species (Pryor and Gilbertson, 2000).

From the above mentioned, clearly there is controversy about the use of the sequence of the ITS regions, which suggests the need to find another DNA genomic region to differentiate between different strains of *Alternaria*.

A differential display study was carried out using the protocol of selective substractive hybridization in *A. gaisen* growing in total darkness and total darkness after scarification using the two conditions as a driver and tester of growth. Out of all the expressed sequence tags found, a 445 bp partial sequence of the clone L152 was utilized to analyze the genomics of different *Alternaria* species. Cluster analysis using this sequence was able to differentiate between *A. alternata*, *A. gaisen*, *A. yaliinficiens*, *A. arborescens*, *A. tenuissima* and *A. brassicicola* suggesting that this sequence is variable enough within the *Alternaria* species and it can be used to differentiate among several species of this genus (Roberts et al., 2012).

Phylogenetic analysis based on sequence characterized amplified region markers along with the gene encoding the endopolygalacturonase enzyme failed to differentiate in some cases the *A. alternata* from *A. arborescens* using isolates from Italy, South Africa, India, France, Japan, Turkey, the UK and several states of the USA. Although the results seem to support the presence of an *A. arborescens group*, all the other isolates must be considered *A. alternata* until the utilization of other novel characteristics to help in identification (Andrew et al., 2009). In another study, *Alternaria* species were isolated from pistachio trees growing in orchards in California and morphologically classified as *A. alternata*, *A. arborescens*, *A. tenuissima* and *A. infectoria*. The molecular classification was carried out using random amplified polymorphism DNA, polymerase chain reaction restriction fragment length polymorphism of the nuclear intergenic spacer ribosomal DNA as well as by sequence analysis of nuclear internal transcribed spacer of the gene encoding the ribosomal RNA. However, the use of these molecular tools together did not give enough variation to differentiate *A. alternata*, *A. arborescens* and *A. tenuissima* (Pryor and Michailides, 2002).

In one study, 14 strains of *Alternaria* found in several hosts in Washington and California were studied using several analyses. The physiological analysis of growth morphology was carried out in several media, including potato dextrose agar, dichloran rose bengal yeast extract sucrose agar and weak potato dextrose agar. The amplification and sequences of the rDNA ITS region, a region of the gene encoding the glyceraldehyde-3-phosphate dehydrogenase and a region of the gene encoding the plasma membrane ATPase was carried out. Using the ITS, ATPase and the glyceraldehyde-3-phosphate dehydrogenase sequences, the same result was found, that is, it was not possible to differentiate *A. tenuissima* and *A. alternata* (Lawrence et al., 2013).

From the above, it can be clearly seen that it is not easy to identify unambiguously the species *A. alternata* although it can be suggested that a multiple approach using the physiological characteristics of colony growth in different media, cluster analysis based on the use of the partial sequence of the clone L152 and the metabolite profile can be a good option for now until future research develop an easier, inexpensive and sensitive protocol.

BIOLOGY OF *A. ALTERNATA* INFECTION PROCESS

Initial events in the infection process caused by fungi in general, as well as by *A. alternata,* are spore adhesion to the cuticle and directed growth of the germ tube on the plant surface. Germ tubes and hyphae elongate by apical deposition of wall glycoproteins and polysaccharides such as chitin and glucans. During extension of the fungal apex, these components are assembled into microfibrils as a result of hydrogen bonding and cross-linking of adjacent polysaccharide chains. Another minor component of *Alternaria*'s hyphae is melanin (Mendgen et al., 1996). This initial, not metabolically demanding (passive) adhesion is followed by a second stage involving secretion of a film unsheathing the germ tube and parts of the cuticle in the vicinity of the hyphae (Jones, 1994). These fungal sheaths, which are associated with the germ tube of many fungi, are assumed to mediate adhesion and preparation of the infection court. Once conidia germinate on the surface of the host tissue, a germ tube and an appressorium are formed. The germ tube is a specialized structure distinct from the fungal mycelium, often growing only a very short distance before it differentiates into an appressorium. From the appressorium, a specialized narrow hyphal strand, called the penetration peg, arises and advances into and through the cuticle and cell wall. Penetration of the plant takes place only if melanin (dark pigment) accumulates in the appressorial cell wall. It appears that melanin produces a rigid structural layer and, by trapping solutes inside the appressorium, causes water to be absorbed. This increases the turgor pressure in the appressorium and, in this way, it induces the physical penetration of the plant by the penetration pegs (Rotem, 1998).

Alternaria alternata is generally considered as a weak and opportunistic pathogen that follows different routes for penetrating plant tissue, like wounds (Pearson and Hall, 1975), natural openings such as lenticels, stem ends, and pedicels (Prusky, 1996), and by direct breaching of the host cuticle (Mersha et al., 2012), which enables the pathogen to enter the unripe tissue and remain quiescent for weeks until the fruit ripens. Typically, tissues weakened by different stresses, senescence, or wounding are more susceptible to *Alternaria* infections than healthy tissues (Mmbaga et al., 2011).

Although it is known that there is a quiescent phase in the life cycle of *Alternaria*, it is not entirely clear whether the ungerminated or the germinated appressorium represents the quiescent stage. Experimental evidences indicate that appressoria germinate to produce infection hyphae prior to the onset of quiescence. In any case, the fungus ceases growth soon after appressorium formation and remains in a quiescent state until fruit ripens. Prusky et al. (2002) reported that *A. alternata* can remain quiescent on cuticles of unripe mango and avocado fruits although it can colonize the mesocarp of the same peeled fruits, suggesting that specific changes during fruit ripening affect the transition into an active infection. It is considered that activation of the quiescent stage is the result of (1) induced accessibility of disassembled cell wall substrates during fruit softening and ethylene induction; (2) a reduction in the amount of preformed antifungal compounds, such as polyphenols, phytoalexins, and other fungitoxic substances; (3) the decline of inducible host defense responses, such as pathogenesis related proteins; and (4) changes in the pH conditions in the host (Prusky et al., 2013). It has been reported that the activation of quiescent infection in *A. alternata* is facilitated by large families of genes encoding cell-wall degrading enzymes, such as glucanases (Eshel et al., 2002).

The most common path for fruit infection used by *A. alternaria* is through wounds produced by mechanical damage, sunburn or chilling injury, before, during or after harvest, as observed on tomatoes (Grogan et al., 1975; Pearson and Hall, 1975), blueberries (Ceponis and Capellini, 1978), apples (Shtienberg, 2012), persimmon (Prusky and Ben-Arie, 1981), and mango (Prusky, 1996). In addition, *Alternaria* infection via natural openings has been observed in table grapes, mandarin, and tangelo, among others. In table grapes, it was observed that *Alternaria* hyphae penetrated the berries through stomata, lenticels and microcracks in the epidermis (Swart et al., 1995). The formation of appressoria at the tips of germ tubes, hyphal tips, and on short side branches was observed. According to the authors, the adhesion of appressoria was very prominent, and some appressoria were completely embedded in cuticular wax. The fungus did not form an extensive network of intercellular hyphae within the living tissue after entry via natural openings. Rather, hyphae remained localized in the substomatal cavities or lenticels and, although the fungus was able to grow inside epidermal cells adjacent to microcracks in the epidermis, or those surrounding wounds, it would appear that this process is a slow one. The authors suggested that the formation of thick-walled hyphae underneath the wax layer, and colonization of stomata and lenticels, might help *A. alternata* to survive for long periods under unfavorable conditions. In another study carried out in 'Minneola' tangelo leaves, Solel and Kimchi (1998) reported that after inoculation with conidia of *A. alternata* pv. *citri*, germ tubes grew randomly over the leaf surface. Twenty-four hours after inoculation, appressoria appeared over the epidermal cells; they formed infection hyphae that penetrated into the leaf, and developed intercellular branching hyphae giving rise to minute necrotic lesions.

Direct penetration of *Alternaria* through the fruit cuticle has not been reported as a common mode of infection, but it was observed in persimmon fruit by scanning electron microscopy. According to Prusky and Ben-Arie (1981), *A. alternata* was seen to penetrate directly through the persimmon cuticle following inoculation of fruit. During the first stages of pathogenesis, fungal development in the tissue was rather superficial (three to four cells in depth) and cellular collapse was observed.

During the infection process, *A. alternata* secretes substances like enzymes, polysaccharides and toxins that seem to be involved in disease development. The germinated appressoria breach the fruit cuticle and penetrate through the fruit wax, fruit cuticle, and become quiescent close to the host cell wall. *A. alternata* is known to produce a wide spectrum of cell wall-degrading enzymes. Eshel et al. (2002) demonstrated that the production of endoglucanases in *A. alternata* was triggered by a pathogen-induced pH increase in the host. In another study, the infection behavior of *A. alternata* Japanese pear pathotype during the interaction between Japanese pear leaves and the pathogen was examined (Suzuki et al., 2003). The authors observed that cutinase released from penetration peg functions as an aggression factor during the first step of fungal invasion of pear tissues. They also observed that pectin layers were degraded by pectinases in the vicinity of the hyphae, suggesting that *A. alternata* utilizes the metabolites derived from the degradation of pectins as a carbon source. In general, plant pathogenic enzymes disintegrate the structural components of host cells, break down substances in the cell, or affect components of its membranes and the protoplast directly, thereby interfering with the vital function of the cell.

In addition to enzymes, *A. alternata* produces a group of mycotoxins, host-specific toxins (HST), which have been shown to be critical determinants of pathogenicity or virulence in several plant–pathogen interactions. These chemically diverse secondary metabolites are employed by the fungi during pathogenesis and are not required for normal growth or reproduction (Eshel et al., 2002). These toxins induce susceptibility in the host to the pathogen by suppressing the host defense mechanism. The wild *A. alternata* is a saprophyte or a weak pathogen, and the transition from a non-specific and non-pathogenic type to a pathogenic and host-specific type involves the production of host-specific toxins (Gilchrist and Grogan, 1976; Rotem, 1998). The virulence and host specificity arise from the production of a distinctive HST by specific cultivars. In the tomato pathotype of *A. alternata*–tomato interactions, a major factor in pathogenicity is the production of host-specific AAL-toxins that are capable of inducing cell death only in susceptible cultivars (Yamagishi et al., 2006). Other HSTs produced by different pathotypes of this fungus, are: AM-toxin, from *A. alternata* on apple; AC-toxin, from citrus; AK-toxin, from pear; AF-toxin, from strawberry; and AT-toxin, from tobacco (Nishimura and Kohmoto, 1983). In the case of the *A. alternata* tangerine pathotype, the most important toxin produced was identified as ACT 1b, which plays an important role in the infection process affecting the plasma-membrane integrity in susceptible genotypes, leading to an electrolyte leakage and necrosis (Kohmoto et al., 1993). From these studies, it is possible that the toxins could be involved in any or all stages of infection, from initial penetration of the tissue to establishment, colonization, and death of the tissue.

Upon pathogen attack, infected plant cells generate signaling molecules to activate the defense mechanisms in surrounding cells to limit pathogen spread (Kunkel and Brooks, 2002), which is considered to be dependent on specific plant–pathogen interactions. The signaling molecules such as jasmonic acid, salicylic acid, the polypeptide systemin, xylanase and ethylene, among others, play key roles in mediating disease resistance against pathogens (Sheard et al., 2010). Recent studies reported that the challenge of tomato plants by *Alternaria* AAL toxins initiates a cascade of signaling events that involve the synthesis and subsequent activation of the ethylene, jasmonic acid, and salicylic acid metabolic pathways. The ethylene response and jasmonic acid signaling enhance the susceptibility of tomato plants to AAL toxin, while the salicylic acid pathway is involved in resistance against AAL (Jia et al., 2013). In apple leaves, plant growth regulators such as indole-3-acetic acid, zeatin riboside, gibberellin A_3, abscissic acid, and the polyamines, putrescine, spermidine and spermine, were observed to interact with each other to modulate signaling and metabolic networks during infection by *A. alternata* apple pathotype (Chen et al., 2012).

The defensive signal-transduction cascades culminate in the accumulation of constitutive antifungal chemicals (e.g. phenolics), the synthesis of inducible antifungal proteins (e.g. pathogenesis-related proteins), and the laying down of physical barriers to pathogen invasion (e.g. callose) (García-Brugger et al., 2006), that limit fungal growth, penetration and development. It was reported that natural antifungal compounds present in fruit tissue may be involved in regulating the quiescence of fungal infections. Previous studies have shown an induction of the phenylpropanoid pathway in lemon seedlings as a response to the presence of *A. alternata* (Ortega et al., 2002). In tomatoes, an increase in the synthesis of phenolic acids was observed as a response to the infection caused by the fungus (Ruelas et al., 2006). As mentioned before, the synthesis of inducible antifungal proteins such as chitinase and β-1,3-glucanase has been related to the plant defense response against pathogens. In tomatoes, one study was conducted to evaluate the chitinase and β-1,3-glucanase activity during ripening of three varieties of tomato fruit inoculated with *A. alternata*. The tomato varieties used in this study were susceptible or resistant to fungal infection. The authors observed high levels of activity in both enzymes as a response to infection caused by the pathogen. Chitinase activity was higher in the mature green stage and also in the resistant tomato variety 'Jeronimo' inoculated with *A. alternata*. They concluded that chitinase and β-1,3-glucanase induction are part of the tomato fruit defense mechanism against *A. alternata* infection which shows a different behavior depending upon the stage of development and tomato variety (Cota et al., 2007).

The natural resistance of fruit and vegetables to disease declines with the increase in the storage time and when fruits begin to ripen. Weak pathogens that normally require a wound in order to infect can become a problem in produce that has been stored for long periods of time (Prusky, 1996; Prusky et al., 2013). Treatments that help to maintain the natural 'vitality' of fruit and vegetables aid in delaying the onset of disease in stored produce. But when control is not adequate, postharvest losses to disease can be substantial.

AGRONOMIC AND ENVIRONMENTAL FACTORS FAVORING *A. ALTERNATA* INFECTION

Postharvest diseases vary each year and a wide range of preharvest factors influences their development. These include the weather (rainfall, temperature, humidity, etc.), production area, cultivar, cultural practices (pesticide application, fertilization, irrigation, planting density, pruning, mulching, fruit bagging, etc.) and planting material. These factors may have a direct influence on the development of the disease by reducing inoculum sources or by discouraging infection. Alternatively, they may affect the physiology of the produce in a way that impacts on disease development after harvest.

Examples are given below that demonstrate how these preharvest factors lead to disease in specific crop.

Weather

Alternaria species is a hardy fungus and can live in extreme conditions. *A. alternata* can pass the winter in the soil, seed, infected crop debris or perennial host tissue, such as bark, nodes, and scaly leaves, as mycelia and/or conidia. Some strains can produce survival structures to resist the unfavorable conditions. While some *Alternaria* species need stimuli to initiate conidiophore formation and sporulation, *A. alternata* can sporulate easily without triggering. The spores of *A. alternata* can be propelled into the air by a shift from wetness to dryness, a rapid increase in humidity or exposure to red light (Rotem, 1998).

The germination of *Alternaria* species is improved in the presence of free moisture. At high relative humidity, germination may be facilitated by transient condensation of water vapor caused by fluctuating temperature. For example, Stevenson and Pennypacker (1988) observed that germination of *A. solani* spores in 96% relative humidity was associated with microscopic condensation of water. In another study, spores of *A. alternata* in containers inside temperature-controlled baths (20 ± 0.01°C) germinated only when wetted occasionally at 100% relative humidity, but never at 98% relative humidity (Rotem, 1998).

Dehpour et al. (2007) published that conidia of *A. alternata* germinate quickly if moisture is present and begin to produce toxin even before they penetrate the tissue. *Alternaria* brown spot infection is most associated with environmental conditions (Canihos et al., 1999) and leaf or fruit maturity of tangerines (Whiteside, 1976). In addition, it has been reported that, in these fruits, the fungus sporulates most prolifically on necrotic lesions of mature or senescent leaves when the substrate is lightly moistened or at high relative humidity. Conidia are also released after rainfall and/or sudden changes in humidity and dispersed by wind (Timmer et al., 2000).

The minimum wetting period needed for the establishment of various *Alternaria* species in host tissue ranges from 3 to 72 h. The requirements for a long wetting period may result from the need to weaken invaded tissue by means of toxic or enzymatic agents, and it is possible that this requirement is more common in infections by weakly pathogenic species. However, a long wetting period may also be required for infection of organs with high resistance by virulent species.

Sawamura (1972) reported that the toxin-producing *A. alternata* f. sp *mali* infects apple leaves after a wetting of 3 h but needs a wetting of 48 h for infecting apple fruits. In another study, the effect of eight wetting periods (from 2 to 48 h) and nine temperatures (from 4 to 36°C) on the infection of 'Delicious' apple seedlings by *A. alternata* f. sp. *mali* was tested. The authors observed that at all temperatures infection increased with the duration of wetness and were most intensive between 12 and 28°C, 23.5°C being the optimum. At this temperature, 5.1 and 12.7 h of wetness were required for light and severe infection, respectively (Filajdič and Sutton, 1991).

In addition, the effect of temperature on the germination of *Alternaria* species has been reported. In almost all *Alternaria* species, the optimum temperature for germination is around 25°C, and the maximum temperature is around 35°C, although different values are given in different reports (Rotem, 1998). Malathrakis (1983) reported that the optimum temperature for germination of *A. alternata* in tomato is around 29°C, and the maximum temperature is 40°C while, in apple, the optimum temperature is 28°C (Filajdič and Sutton, 1991). In 'Minneola' tangelo leaves, the optimum temperature for infection by *A. alternata* is 27°C. As temperature declines, longer wetting periods are needed for infection to occur. In citrus, the optimum conditions for *Alternaria* infections are temperatures from 23 to 27°C and 8–12 h of continuous leaf wetness in the form of dew or rainfall. From these studies, it is clear that *A. alternata* is able to germinate at a wide range of temperatures (Timmer et al., 2000).

Under controlled conditions, infection occurs over a wide range of temperatures, from less than 10°C up to 35°C or above. Various researchers have sometimes arrived at different conclusions. Thus, the optimum temperature for infection of tobacco by *A. alternata* is 20°C according to Stavely and Main (1970), 25°C according to Ramm and Lucas (1962), and 30°C according to Riley (1949). In some cases, the optimum temperature for infection is similar to the optimum for germination, which may reflect the general requirement of the fungus. In other cases, these temperatures can be different, which may possibly reflect a specific reaction of the host, succumbing to infection under conditions not necessarily optimal for spore germination or penetration. For instance, the optimum temperature for infection of tomato by *A. alternata* is 25°C (Malathrakis, 1983) while, in apple, it is 28–31°C (Filajdič and Sutton, 1991).

Cultural Practices

As reviewed by Rotem (1998), cultural practices can be related with certain cases of association between vigor and susceptibility of plants to infection by *Alternaria* pathogens. Among these cultural practices, soil moisture is a major factor affecting the growth conditions of plants. But the predisposition to *Alternaria* diseases is affected by a deficient as well as by an abundant supply of water. The effect of low moisture is reflected in enhanced development of some *Alternaria* diseases in dry rather than in rainy years. However, the effect of soil moisture on *Alternaria* rot development is not clear.

Another cultural factor that affects the growth conditions of plants and also has an effect on the development of *Alternaria* diseases is soil nutrition. Reports on the effects of fertilization on *Alternaria* diseases have focused on the amounts of balanced fertilizers, the action of nitrogen (N), and the effect of potassium. Various reports describe the action of one fertilizer or another and often contradict each other. Well-fertilized and poorly fertilized plants have been observed to have different levels of susceptibility in pot experiments, but the results have not always been confirmed by field trials. For instance, specific doses of N, P, and K fertilizers led to a very low level of tomato early blight in sand culture but had little if any effect in the field. This was particularly evident in highly fertile soil, which supported growth by itself and yielded large plants. Previous reports indicate the effect of nitrogen on *Alternaria* diseases, and it is different from the susceptibility-increasing influence of the same element in other diseases. For instance, tomatoes abundantly fertilized with nitrogen were found to be more resistant to *A. solani*, although they produced a lower yield than plants supplied with lower doses of N (Thomas, 1948). These responses were attributed to the effect of N to delay the senescence of plants (Rotem, 1998). Another aspect of N application was demonstrated with *A. alternata* on tobacco. Nitrogen delayed maturation in this crop too, but the effect of the delay was different from that described previously. Since the vigorous plants were left in the field for a longer period, they were exposed to inocula longer and developed more disease (Cole et al., 1961). Also, in soft white spring wheat, the incidence of black point disease caused by *A. alternata* increased with the N supply from 0 to 120 kg ha^{-1}. However, further additions of N were shown to reduce disease slightly (Conner et al., 1992).

The effect of other minerals related to the development of *Alternaria* diseases has been reported in a number of cases, too limited for an evaluation of their effect. For instance, in the *A. alternata* f. sp. *mali*-apple system, the susceptibility of leaves was not correlated with the N, P, K, Mg, or sodium content. In contrast, the calcium content was higher in resistant than susceptible cultivars. It was observed that foliar application of Ca compounds inhibited the disease in apple (Yoon et al., 1989).

POSTHARVEST FACTORS FAVORING *A. ALTERNATA* INFECTION

Fruit fungal infection may occur, in addition to the growing season, during harvesting, handling, transport, postharvest storage and marketing conditions, or after purchase by the consumer. Fruits contain high levels of sugars and nutrient elements and their low pH values make them particularly susceptible to fungal decay.

Packing Process

As mentioned before, *A. alternata* is generally considered a weak pathogen that gains entry into the fruit via wounds or natural openings, and remains quiescent until the fruit ripens. Thus, it is important to avoid fruit injuries caused during their handling in the packing house. The machinery used during all packing processes, such as harvesting, cleaning, selection, and packing, should be designed and operated according to the commodities. In addition, care should be taken with the height of drop, presence of sharp corners or projections on boxes and machinery and constantly renewing washing water (Snowdon, 1990). A data compilation from the USDA inspection carried out with fruits arriving in the New York (USA) market illustrated problems encountered in marketing fresh produce. According to these inspections, the fruits showed different kinds of disorders, mechanical damage being one of them. The kinds of mechanical injuries found in different commodities such as avocado, apricots, cucumber, mango, papaya, pineapple, tomato, squash, watermelon, among others, were bruising, scarring, cuts/punctures, cracking, transit rubs, and others (Ceponis et al., 1987; Cappellini et al., 1988). In addition, it was observed that these injuries were the entry point for fungal contamination including *A. alternata*.

Cold Storage

The storage environment markedly influences the rates of fungal growth. Low temperatures used during storage of fruits and vegetables induce a low respiration rate, and a reduced moisture loss in many commodities, as well as inhibiting the growth of decay pathogens (Sommer, 1985). Various types of physiological injury such as chilling and heat injury can predispose commodities to infection by postharvest pathogens. For example, the incidence of *Alternaria* rot in pawpaw, apple and various vegetable crops is increased by exposure to excessive cold.

In addition, at higher temperatures, a saturated atmosphere (high relative humidity) is conducive to mold growth. Therefore, the minimum temperature for fungal growth varies according to species. In the case of *A. alternata*, it can grow at temperatures between 0 and 10°C, and it is considered the major storage decay agent of some harvested commodities. It has been reported in tomatoes that *Alternaria* fruit rot increases in proportion to the magnitude and duration of exposure to chilling temperatures (0–10°C). In the study carried out by McColloch and Worthington (1952), it was observed that *A. alternata* was unable to provoke active fruit rot on healthy, mature green tomatoes. However, inoculation of ripe fruits frequently resulted in large, sunken lesions, whereas inoculations of green fruit resulted in quiescent lesions that enlarged after the fruit ripened.

In another study carried out in peppers and reported by Halfon-Meiri and Rylski (1983), the black mold indicative of *Alternaria* fruit rot appeared most often on mature red peppers before or after harvest. In addition, it was observed that red chili peppers harvested during wet periods and improperly stored before processing quickly develop *Alternaria* rots. According to the studies carried out by Wall and Biles (1993) green chili peppers packaged, stored, and shipped for the fresh market can also develop *A. alternata* infections. In sweet peppers, it was reported that this commodity is highly susceptible to *Alternaria* rot when it is stored at temperatures less than 7°C (Meir et al., 1995).

Some studies reported that *A. alternata* causes bunch rot of exported table grapes (Swart et al., 1995). It has been reported that the disease develops in cold-stored fruit and its sporadic occurrence in some consignments is a serious challenge to the prolonged storage of table grapes at low temperatures. The superficial growth of *A. alternata* occurred predominantly on flaccid or dry rachis and pedicels of exported grapes. Also, those lesions were first noticeable during storage at 10°C. According to the authors, *A. alternata* colonizes berries, pedicels and rachises during the entire period of bunch development, but the stress factors during cold storage might predispose table grape bunches to *A. alternata* decay.

Alternaria rot often develops by mid-season during cold storage of apples showing injuries such as delayed sunscald, bruising, or chemical injury. In apples held late in storage, the rot may develop at skin checks, as in the 'York Imperial' variety, or even at enlarged, ruptured lenticels over bruises.

Alternaria infections frequently follow scald, soft scald, or Jonathan spot. The lesions following these physiological disorders are usually not sunken and often are without definite margins. Warmer temperatures favor the development of the rot, therefore keeping apples at temperatures of 0 to 4.4°C when moving them from storage to the consumer helps to prevent spoilage (Chen et al., 2012; Shtienberg, 2012).

High temperatures in a saturated atmosphere (high relative humidity) can also increase susceptibility of harvested produce to disease. For example, hot water dipping of mangoes for excessive times and/or temperatures can result in increased levels of stem end rot. So, appropriate storage can minimize moisture loss, slow down respiration rate and inhibit the development of pathogens.

Fruit Ripening

The natural resistance of fruit and vegetables to disease declines with storage duration and ripeness. Natural antifungal compounds present in fruit tissue may be involved in reducing the fungal infections. During ripening, levels of these compounds decline to sub-fungitoxic levels which coincide with the development of fungal symptoms on fruit (Prusky, 1996). Pearson and Hall (1975) carried out a study to determine the effect of incipient infections on green fruit to subsequent development of *Alternaria* rot of ripe tomato fruit. The authors observed that the infections of green fruit by *A. alternata* produced tiny quiescent lesions that were of minimal importance in the development of severe rot in ripe fruit. They also found the presence of large severe lesions developed from infections that occurred after fruit ripening that were responsible for most of the losses incurred in processing tomatoes. In apples, Valiuškaitė et al. (2006) reported that the incidence of *Alternaria* rot was closely connected to fruit maturity. They found a significantly lower degree of *Alternaria* rot in fruits picked at optimum maturity as compared with those picked later. Weak pathogens that normally require a wound in order to infect can become a problem in commodities that have been stored for long periods of time. Treatments that help to maintain the natural freshness of fruit and vegetables aid in delaying the onset of disease in stored produce.

CONTROL

Alternaria alternata is a fungus that can infect fruits mainly via wounds, or natural openings, therefore control of *Alternaria* rot depends on careful handling during picking, washing, and packing to prevent physiological diseases and injuries that open the way for infection. Warmer temperatures favor the development of the rot, furthermore, prompt storage and cooling of the fruits is essential. However, these measures are not enough to control this disease and there are several methods employed for this purpose; some of them are described below.

Conventional Alternatives

Synthetic Fungicides

Fungicides are used extensively for postharvest disease control in fruit and vegetables (Eckert and Ogawa, 1988; González and Valenzuela, 2007). Several fungicides are used before and after harvest as treatment to prevent or control the development of *A. alternata*. Since 1980, chlorothalonil, captafol and imazalil had been labeled for control of *Alternaria* rot of tomatoes. However, Spalding (1980) reported that captafol can cause contact dermatitis in susceptible individuals and is, therefore, only used for tomatoes that are to be harvested mechanically. In a similar study, it was observed that imazalil controlled *Alternaria* rot of tomatoes and peppers. Spalding and King (1980) reported that dipping tomatoes and bell peppers for 10 seconds in an aqueous solution of 50–250 μg of the imazalil inhibited development of rot caused by *A. alternata*. In another study, imazalil controlled *Alternaria* rot on wound-inoculated apples and naturally inoculated pears during 0°C storage for 6 months (Prusky and Ben-Arie, 1981). In persimmons, a postharvest dip ($1\,g\,L^{-1}$) of prochloraz or imazalil inhibited the spread of black spot lesions on the fruit surface, but it did not prevent the establishment of an infection if the treatment was delayed several days after superficial inoculation with spores (Prusky et al., 1981). In blueberries, postharvest infections of *Alternaria* were significantly reduced by a one-minute dip in a chlorinated solution of ($0.5\,g\,L^{-1}$) sec-butylamine (Ceponis and Cappellini, 1978). Black pit caused by *A. alternata* in potatoes was controlled by spraying the tubers with $1\,g\,L^{-1}$ iprodione before storage (Droby et al., 1984). A mixture of benomyl ($500\,mg\,mL^{-1}$) and prochloraz ($1000\,mg\,L^{-1)}$ gave good control of *Alternaria* on pears stored at −0.5°C for 6–7 months (Sitton and Pierson, 1983). Several years later, Singh and Singh (2006) tested the efficacy of seven fungicides such as chlorothalonil, copper oxychloride, azoxystrobin, propineb, copper hydroxide, mancozeb at 2500, 2000, 1000, 500 and 250 ppm and hexaconazole at 1000, 500, 200, 100 and 50 ppm against *A. alternata* growth *in vitro*. The authors showed that all the fungicides tested significantly reduced the radial growth of the fungus. Further, hexaconazole was the most effective showing 100% fungus growth inhibition.

Alternaria brown spot, caused by *A. alternata* pv. *citri*, is a severe fungal disease of mandarins. Although cultural practices can greatly reduce disease severity in the orchard, fungicide applications are essential to produce quality fresh fruit for the market. Thus, copper products, mancozeb, dicarboximide, and triazole fungicides are recommended to control *Alternaria* brown spot. Vincent et al. (2007) evaluated the residual activity of several fungicides (mancozeb, difenoconazole, iprodione, pyraclostrobin, famoxadone, copper oxychloride, copper hydroxide, cuprous oxide, Bordeaux mixture) used for the control of *Alternaria* brown spot of citrus. Residual activity was measured in fruits 24 h after fungicide application. The authors reported that most of the fungicides tested showed a significant reduction of disease in 'Fortune' mandarin fruit in the orchard after natural rain. Copper products showed longer residual activity on fruit and higher rain fastness than did mancozeb, difenoconazole, iprodione, famoxadone and pyraclostrobin.

Recently, the effect of five different fungicides such as prochloraz, deconil, carbendazim, thiabendazol, and mancozeb were evaluated as dip treatments on black spot decay caused by *A. alternata* in mango during storage at 20°C. The results showed that mancozeb and prochloraz proved to be the most effective fungicides to control *Alternaria* rot, observing a decrease in lesion diameter of 67.43% and 64.25%, respectively (Mohsan et al., 2011).

Non-Conventional Alternatives

There are some other alternatives to the use of synthetic chemical fungicides to preserve fruit and vegetables during storage and shelf-life such as biological control, application of natural compounds including chitosan, essential oils, isothiocyanates, elicitors, shortwave radiation, ozone, etc. According to Romanazzi et al. (2012), the ideal alternative treatment for controlling postharvest diseases should be affordable and easy to implement, and should not have any negative influence on the fruit, the environment or human health, and should be in accordance with food safety. Some alternative treatments are described below.

Biological Control

The use of yeast or bacterial strains to control postharvest decay of several fruits using antagonistic microorganisms isolated from plant tissues has been extensively studied, and several examples of successful disease control have been reported (Sharma et al., 2009; Begum et al., 2010; Rathod, 2012).

Several microorganism antagonists have been identified and used for controlling postharvest diseases of different fruits and vegetables. Microbial antagonists have been identified and artificially introduced on a variety of harvested commodities including citrus, pomes, stone fruits, and vegetables for control of postharvest diseases. For example, effective control of *A. alternata* of citrus, litchi and muskmelon was observed with bacterial antagonists such as *Bacillus subtilis* (Jiang et al., 2001; Yang et al., 2006; Wang et al., 2010), fruit jujube was controlled by the antagonistic yeasts *Cryptococcus laurentii* and *Rhodotorula glutinis* (Tian et al., 2005), cherry with *Enterobacter aerogenes* or *Trichosporon pullulans* (Qin et al., 2004), tomato with the yeast *Pichia guilliermondii* (Zhao et al., 2011) or the bacterium *Trichoderma harzianum* (El-Katatny and Emam, 2012) (Table 5.1). Microbial cultures are applied either as postharvest sprays or as dips in an antagonist's solution. For example, postharvest application of *Trichoderma harzianum*, *T. viride*, *Gliocladium roseum* and *Paecilomyces variotii* resulted in better control of *Alternaria* rot in lemons than preharvest applications (Pratella and Mari, 1993). A significant reduction in storage decay was achieved by bringing several yeast species in direct contact with wounds in the peel of harvested fruits. For instance, direct contact of microbial antagonist and infested fruit peel has been quite useful for the suppression of *A. alternata* in tomatoes (Zhao et al., 2011). However, not all the pathogens react in a similar fashion to a given antagonist.

Heat Treatment

The use of these treatments had been widely available to control several fungal diseases, but they are not commonly used to control postharvest decay caused by *A. alternata*. Some studies have been conducted to demonstrate the effectiveness of heat treatment on the control of *A. alternata* and some other *Alternaria* species. For example, the use of water dips with a temperature between 38 and 60°C for 2 to 60 min has been reported to control *in vivo* and *in vitro* spore germination and decay development of *A. tenuis* in tomatoes (Barkai-Golan, 1973). In another study, the effectiveness of a prestorage dry heat treatment and hot water dip in reducing storage rots of *Capsicum* bell peppers and tomatoes caused by *A. alternata* was assessed. Treatment with hot water at 50–53°C for 2–3 min resulted in reduction (73.4%) of the pathogenicity and development of infection by these pathogens in inoculated peppers (Fallik et al., 1996). Prusky et al. (1998) reported that hot water spray and a fruit brushing treatment reduce the incidence of postharvest disease caused by *A. alternata* by almost 60% and keep the quality of mango fruit for a longer time. Meath (2000) observed that mature green hard mango fruits dipped for 3 minutes in heated (50°C) aqueous solutions of mancozeb (3 g L^{-1}) and iprodione

Table 5.1 Microbial Antagonists used to Control Postharvest Diseases in Horticultural Commodities Caused by *Alternaria Alternata*

Horticultural Commodities	Antagonist	Antagonist Concentration	Percentage Inhibition	References
Capsicum annuum	*Bacillus licheniformis*	8×10^6 CFU mL^{-1}	67	Sid et al., 2003
Chili	*T. harzianum*	5×10^5 spores mL^{-1}	72.27	Begum et al., 2010
Jujube fruit	*Cryptococcus laurentii*	10^8 CFU mL^{-1}	77	Qin and Tian, 2004
Jujube fruit	*Cryptococcus laurentii* *Rhodotonela glutinis*	5×10^7 CFU mL^{-1}	73-76	Tian et al., 2005
Jujube fruit	*Rhodoisporidium paludigenum*	10^8 cell mL^{-1}	80	Wang et al., 2009
Melon fruit	*Bacillus subtilis*	10^9 cell mL^{-1}	77.2	Wang et al., 2010
Litchi fruit	*Bacillus subtilis*	10^8 spores mL^{-1}	64.38	Jiang et al., 2001
Tomato (cherry)	*Trichoderma harzianum*	10^8 CFU mL^{-1}	100	El-Katatny and Emam, 2012
Tomato	*Pichia guilliermondii*	10^8 CFU mL^{-1}	100	Zhao et al., 2011
Sweet cherries	*Trichoderma pullulans* *Cryptococcus laurentii* *Rhodotorula glutinis* *Pichia membranefa ciens*	10^4	73	Qin et al., 2004

(0.5 and 0.75 g L^{-1}) gave excellent control of stem-end rot infections. Fallik et al. (2000) found that dipping cantaloupe fruits inoculated with *A. alternata* and others pathogens in hot water at 59°C for 15 sec inhibited decay (48%) caused by these fungi. Similar results were reported by Tohamy et al. (2004) who found that dipping tomato fruits in hot water at 55°C for 7 min, or keeping in hot air for 72 h at 38°C prevented decay development in fruits artificially inoculated with *A. alternata* for up to 15 days when stored at 20°C. Esguerra et al. (2006) developed a long time hot water treatment for disease control of mangoes, which consisted of a 10 min dip at 52–55°C followed by 10 min hydrocooling, and 30 min drying prior to packing. This treatment is considered a bottleneck in packing house operations, resulting in slow adoption or misapplication of hot water treatment.

In contrast to hot water dips, an extended hot air treatment can be given at a lower temperature (Lurie and Klein, 1991). Generally, fruit are held at 38°C for 3 or 4 days at a relative humidity of >85%. In previous experiments, apples, pomegranates, avocados, and tomatoes were held under those conditions, showing a water loss <5%, with no external signs of shriveling. Tomato fruits heated for 4 days at 38°C were not damaged, and preliminary data showed that resistance to pathogens was enhanced by the treatment. In this study, it was observed that decay due to *A. alternata* in tomato was decreased or delayed in fruit that had been heated. *A. alternata* caused decay (around 30– 50%) in non-heated tomatoes after storage at 2°C for 21 days, while <8% decay was found in tomatoes that were heated before storage.

Despite the evident success of hot water dips to reduce populations of microorganisms, the high temperature required by the short treatment time (up to 60 min) can easily damage fruit tissue if the recommended exposure time is exceeded. The low tolerance for variation in treatment time thus limits the value of such heat treatments (Teitel et al., 1991). In addition, the stage of fruit maturity may limit the period during which such treatments can be applied safely (Couey and Hayes, 1986).

Modified Atmosphere Packaging (MAP)

Modified atmosphere packaging has been used in combination with refrigeration since the last century to enhance fruit quality during prolonged storage. Ben-Arie et al. (1991) reported that persimmon fruits packed in low-density polyethylene bags resulted in a significant delay of black rot disease development. The authors reported that the atmosphere within the packaging bag showed increases in natural volatiles produced by the fruit, such as CO_2, acetaldehyde and ethanol. It has been reported that high levels of CO_2 suppress the development of most pathogens by inhibiting various metabolic functions (Sommer, 1985). The concentration of CO_2 required to inhibit mycelial growth varies with fungal species. The growth of mycelium of *A. alternata* decreased linearly with increasing CO_2 concentrations from 10 to 45%, in which total inhibition can be achieved (Wells and Uota,

1970). At 20% CO_2, mycelium development of *A. alternata* was inhibited by 50%. In a similar study, Ben-Arie and Zutkhi (1992) evaluated modified atmosphere packaging using low-density polyethylene bags 0.06 or 0.08 mm thick, to extend postharvest life of persimmon fruits. The authors observed that fruit packed showed much less decay (about 90% reduction) caused by *A. alternata*, and peel browning, especially in the 0.08 mm bags, even after 18 weeks of storage at 0°C.

Natural Compounds and Plant Extracts

Natural compounds have shown very promising results in the control of plant pathogens. The antifungal effect depends on their chemical characteristics, fungi species, host nature and storage conditions of fruits and vegetables (Troncoso-Rojas and Tiznado-Hernández, 2007; Phillips et al., 2012; Mahdavi et al., 2013). Some natural compounds that had been tested to control *A. alternata* diseases during postharvest of fruits and vegetables are described below.

Essential Oils

The essential oils produced by plants have long been known to have fungicidal properties. Most of them have been reported to inhibit postharvest fungi attack mainly under *in vitro* conditions (Singh and Tripathi, 1999; Hidalgo et al., 2002; Zaker and Mosallanejad, 2010). However, some studies had been carried out to prove the antifungal activity of essential oils in some fruits and vegetables, and among them, very few had been designed to test their effect against *A. alternata.*

Previous studies were conducted to test the antimicrobial activities of essential oils of fennel, peppermint, caraway, eucalyptus, geranium and lemongrass against some plant pathogens including *A. alternata* (Abo-El-Seoud et al., 2005). The results of this study confirmed the antimicrobial activity of the essential oils against the fungus. Feng and Zheng (2007) evaluated the antifungal activity of essential oils of five plants, such as thyme (*Thymus vulgaris* L), sage (*Salvia officinalis*), nutmeg (*Myristica fragans*), eucalyptus (*Eucalyptus* spp.), and cassia (*Cinnamomum* spp.), against *A. alternata* at different concentrations (100–500 ppm) in *in vitro* and *in vivo* conditions. The authors found that the cassia oil completely inhibited the growth of *A. alternata* at 300–500 ppm, while the thyme oil exhibited a lower degree of inhibition, 62% at 500 ppm. Hadizadeh et al. (2009) studied the antifungal effect of essential oils from some medicinal plants of Iran, such as nettle (*Urtica dioica*), thyme (*Thymus vulgaris*), eucalyptus (*Eucalyptus* spp.), rute (*Ruta graveolens*) and common yarrow (*Archillea millefolium*) on *A. alternata* infecting potato tubers. In this study, it was found that both nettle and the thyme oils showed antifungal activity against the pathogen. In another study, the antifungal activity of the essential oils of ajowan (*Trachypermum ammi* L), fennel (*Foeniculum vulgare*) and caraway (*Carumcarvi*) (2-caren-10-al) was evaluated *in vitro* and *in vivo* conditions. In this study, mature red tomato fruits were inoculated with *A. alternata* and were sprayed with different concentrations (0, 250 and 500 μL L^{-1}) of essential oils. The fruits were placed in plastic boxes and kept in cold storage (13°C) for 20 days. The authors found that ajowan (thymol) and fennel oils (trans-anethole) exhibited the highest antifungal activity, reducing around 95% of postharvest decays on tomatoes for short period during storage (Abdolahi et al., 2010). On the other hand, Phillips et al. (2012) observed that citrus essential oils (Citri-V) in vapor phase reduced spore germination of *A. alternata* under *in vitro* conditions, but this treatment was not effective in reducing the growth of the fungus on tomato fruits.

Another natural substance that has shown antimicrobial activity is propolis. This is a wax-like resinous substance collected by bees from tree buds and used as cement to seal cracks or open spaces in the hive. Ozcan (1999) tested the effect of propolis on *A. alternata*, *Aspergillus niger*, *Aspergillus parasiticus*, *Botrytis cinerea*, *Fusarium oxysporum* f. sp. *melonis* and *Penicillium digitatum* and found that *A. alternata* was the most sensitive to propolis. In a similar study, Ojeda-Contreras et al. (2008) evaluated the effect of caffeic acid phenethyl ester (CAPE), a component of propolis, to control *A. alternata* infecting tomato fruit. The commercial fungicide Captan was included in the study as a control. The authors concluded that CAPE controls *A. alternata* infection better than a commercial fungicide without negative effects on tomato fruit ripening and fruit quality.

Recently, the essential oils of three medicinal plants – basil (*Ocimum basilicum*), pennyroyal (*Mentha pulegium*) and sen (*Cassia angustifolia*) – were tested (200–60 ppm) for their antifungal activity against fungal pathogens of citrus fruit including *A. alternata*, and the inoculated fruits were stored at 6 and 25°C (Mahdavi et al., 2013). Among the essential oils tested, *C. angustifolia* decreased the infection percentage of tested fungi most efficiently. In addition, it was observed that the percentage of infection was similar at 6°C for all concentrations of the essential oils. However, at 25°C, *C. angustifolia* essential oils at 200 ppm caused a large reduction (about 92%) in the infection percentage.

Although some studies have reported on the antifungal activity of essential oils, the mechanism(s) of action of such oils is poorly understood. However, some researchers reported that there is a relationship between the chemical structure of the most abundant compounds in the essential oils and antimicrobial activity. According to Faid et al. (1996), antimicrobial activity of major oil compounds is in the order: phenols (highest activity) > alcohols > aldehydes > ketones > ethers > hydrocarbons.

Isothiocyanates

Isothiocyanates (ITC) are compounds produced by several plants belonging to the families *Brassicaceae*, *Capparaceae* and *Caricaceae* as a system of defense against pathogen attack, and they arise from the hydrolysis of glucosinolates by the enzyme myrosinase. It had been widely reported that the isothiocyanates have various biological effects, such as antifungal activity. The fungicidal effects of isothiocyanates have been reported since the last century (Tiznado-Hernández and Troncoso-Rojas, 2006; Troncoso-Rojas and Tiznado-Hernández, 2007; Troncoso-Rojas et al., 2013). However, there are few published studies concerning the postharvest effect of the isothiocyanates on the control of fungal diseases in fruits and vegetables, and even fewer published studies about use of isothiocyanates on the control of diseases caused by this fungus.

In previous studies, the use of benzyl isothiocyanate (BITC) was evaluated in the control of black rot in tomato, and the effect of this compound on postharvest physiology and quality. Tomatoes were inoculated with the pathogen and two concentrations of benzyl isothiocyanate were tested, 0.28 and 0.56 mg mL^{-1}. The commercial fungicide Captan®, was used as a control. The authors reported that both concentrations of benzyl isothiocyanate significantly reduced (45% and 80%, respectively) the growth and development of *A. alternata* on tomatoes. Furthermore, a significant disease control (60%) was observed in benzyl isothiocyanate treatment, when compared to the commercial fungicide (Captan®). No significant differences in ethylene production were observed between concentrations and exposure times of benzyl isothiocyanate. However, fruits exposed to benzyl isothiocyanate for a longer time (36 h) showed a low ethylene production, when compared with tomatoes exposed for 18 h (Troncoso-Rojas et al., 2005b). A similar study was carried out in bell peppers. A mixture of isothiocyanates (MTIC) formulated in base to the relative amounts of isothiocyanates detected in cabbage leaves (*Brassica oleracea* var. *Capitata*, 2-propenyl, benzyl, 2-phenylethyl and phenyl isothiocyanate in a ratio of 1:3.5:5.3:9.6, respectively) was evaluated to control *A. alternata* infection of bell pepper fruits, and on its postharvest quality. The experiment was performed with the same conditions as tomato, using the MTIC concentrations at 0.28 and 0.56 mg mL^{-1}, during 18 h of exposure. The authors observed a complete inhibition of *Alternaria* growth in pepper fruits treated with the MTIC at a concentration of 0.56 mg mL^{-1} during 10 days of storage. This treatment was more effective for controlling *Alternaria* rot as compared to the commercial fungicide Captan®. In addition, the postharvest quality of the peppers was not negatively affected by the application of isothiocyanates (Troncoso et al., 2005a). Similarly, another study was conducted in muskmelon. 2-Propenyl isothiocyanate, benzyl isothiocyanate and the MTIC described previously were tested to control the disease caused by this fungus in muskmelon. In addition, the effect of the isothiocyanates on postharvest fruit quality was evaluated. Two concentrations of the isothiocyanates were tested, 0.25 and 0.5 mg mL^{-1}, during 18 h of exposure. The authors found that isothiocyanates significantly reduced (83–88%) the development of *Alternaria* rot in muskmelon. The MTIC had the same effect on the disease control compared with the 2-propenyl and benzyl isothiocyanate. It was also observed that the antifungal effect of isothiocyanates was higher with respect to the commercial fungicide (Captan®). Like the bell peppers, the postharvest quality of the melons was not affected by the application of the treatments (Troncoso-Rojas et al., 2009). Recently, Wang and Chen (2010) evaluated the effect of allyl isothiocyanate (AITC) on the radical scavenging capacity and fruit decay of blueberries caused by *Colletotrichum acutatum* and *Alternaria* spp. The fruits were placed into polystyrene containers and the AITC was tested at a concentration of 5 μL L^{-1}. The authors found that AITC was effective in retarding blueberry decay during storage at 10°C. In addition to reducing decay, fruit treated with AITC showed lower amounts of total phenolics, and anthocyanins, as well as low antioxidant activities.

The biological activities of isothiocyanates are due to their ability to react with the amino group of amino acids forming thioureas and with the thio groups giving rise to N-allylthio-carbamoyl (or dithiocarbamates) derivatives. Experimental evidence suggests that isothiocyanates activity may be due to possible non-specific and covalent interaction of the isothiocyanate group (-N=C=S) with sulfhydryl groups (-SH), amino groups and disulfide bonds in proteins and amino acids such as lysine, cysteine and the phenolic group of tyrosine residues of proteins (Murthy and Rao, 1986; Kawakishi and Kaneko, 1987). After reaction, the isothiocyanates remain covalently bound to the protein which brings changes in the tertiary structure of the protein leading to partial or total loss of enzymatic activity (Rawel et al., 1998; Yang et al., 2000; Nakamura, 2009).

Although there is evidence indicating that the alteration of proteins is implicated in the antifungal effects of ITCs, there is scarce information about the effects of these compounds on specific microbial proteins or enzymes involved in the metabolic functions of pathogens. To the best of our knowledge, the only work carried out about the effects of ITCs on microbial proteins was carried out by Kojima and Ogawa (1971). The authors evaluated the oxygen uptake of three yeast strains treated with allyl, methyl, and phenylthyl isothiocyanate, finding that cytochrome c oxidase was inhibited mainly by allyl isothiocyanate, suggesting that the ITCs may act as uncouplers of oxidative phosphorylation, which may in turn account for the high susceptibility of the strictly aerobic fungi to these compounds.

Chitosan

Chitosan (poly β-(1-4)N-acetyl-d-glucosamine) is considered a biodegradable polymer, non-toxic, bioactive, that has fungicidal properties, and can induce defense mechanism in plants (Bautista-Baños et al., 2004; Devlieghere et al., 2004; Terry and Joyce, 2004). It has been proposed for applications either at pre- or postharvest to prevent postharvest fruit decay.

The antifungal activity of chitosan has been tested against several microorganisms (Bautista-Baños et al., 2006; Velázquez-del Valle et al., 2012) and it has been shown to inhibit the growth of *A. alternata* (Sánchez-Domínguez et al., 2011). In previous studies, it was shown that application of chitosan affects growth, morphology, and toxin production by *A. alternata* f. sp. *lycopersici*, the causal agent of black mold rot of tomato (Reddy et al., 1998). Similarly, Reddy et al. (2000) reported that stem scar application of chitosan inhibited the development of *A. alternata* on tomatoes stored at 20°C for 28 days by 34%, and reduces production of pathogenic factors by the fungus, such as cell wall-degrading enzymes (polygalacturonase, pectate lyase, and cellulose), organic acids, and host-specific toxins responsible for fungal penetration and host tissue damage. In addition, the authors found that chitosan induces phytoalexin production in the host as a natural defense. In another study, it was reported that chitosan with medium molecular weight at 2.5% (w/v), incorporated in potato dextrose agar inhibited mycelial growth of *A. alternata* up to 50.6%, under *in vitro* conditions (Sánchez-Domínguez et al., 2007). In a similar study, López et al. (2013) found that the use of chitosan of low molecular weight at a concentration of 1.0% reduced mycelial growth and sporulation of *A. alternata* up to 70% and completely inhibited conidial germination under *in vitro* conditions. In addition, the authors found that there was no fungicidal effect of chitosan when applied directly to the fruit.

The antifungal effects of chitosan are related to its deacetylation level, concentration and polymerization grade, among other factors. This polymer has different mechanisms of action to inhibit phytopathogenic fungi and to prevent fungal diseases (Meng et al., 2010). Several works described the morphological changes induced by chitosan on fungal hyphae and reproductive structures. Scanning electron examination of ultrasections of the hyphae and conidia of chitosan-treated *A. alternata* revealed marked alterations in the cell wall. The chitosan-treated mycelium showed predominantly loosened cell walls and, in some areas, lysis was observed. The conidia exposed to chitosan were intensely damaged, usually eroded and with broken cell walls with no cytoplasm in some cases (Sánchez-Domínguez et al., 2007). In other microscopical studies, Sánchez-Domínguez et al. (2011) observed great damage on fungal cells such as cell disintegration, plasma membrane retraction, a remarkable increase in vacuolization, release of the apical portion of the conidia and lysis of the cells, among other effects. Similar results were reported by López et al., (2013), who found an intense and broad vacuolization throughout mycelia and conidia, cytoplasmic content leakage and presence of fibrillar material in *A. alternata* strains with chitosan of low molecular weight applied at a concentration of 1.0%

The antimicrobial activity of chitosan seems to rely on electrostatic interactions between positive chitosan charges and the negatively charged plasma membrane phospholipids that integrate the fungal membrane. It was observed that chitosan first binds to the target membrane surface and covers it and, in a second step, after a threshold concentration has been reached, chitosan causes membrane permeabilization and release of cellular contents (Palma-Guerrero et al., 2010).

Elicitors of Natural Defense Mechanisms

Fruits display a wide range of physical and biochemical strategies to defend themselves from the attack of pathogenic microorganisms. These responses can be induced in fruit during the postharvest period by a variety of abiotic and biotic inducers or elicitors (Terry and Joyce, 2004; García-Brugger et al., 2006; Thakur and Sohal, 2013; Wan and Pentecost, 2013).

Although induced resistance for the control of fungal postharvest diseases has been shown to be effective in the laboratory and against different fungi, there is scarce information about the induction of defense mechanisms to control postharvest decay in fruits caused by *A. alternata*. For example, different elicitors (salicylic acid, oxalic acid, calcium chloride and the antagonist yeast *Cryptococcus laurentii*) were tested on pear fruit to induce the defense responses and control the disease caused by this fungus. The authors found that all the elicitors tested could significantly enhance defense-related enzyme activities, such as β-1,3-glucanase, phenylalanine ammonia lyase, peroxidase, and polyphenol oxidase activity, and reduce the disease incidence caused by *A. alternata* (Tian et al., 2005). In another study, the effect of salicylic acid dipping on *A. alternata* and disease resistance was tested in jujube fruit during postharvest storage. It was observed that 2 and 2.5 mg L^{-1} of salicylic acid dipping significantly inhibited the disease incidence and lesion area. In addition, the authors observed that salicylic acid enhanced activities of phenylalanine ammonia-lyase, peroxidase, chitinase and β-1,3-glucanase in fruit during storage (Cao et al., 2013).

Although interesting results have been obtained using UV-C irradiation, and heat treatments on the control of fungal decay in several stored commodities (Adikaram et al., 1988; Boulet et al., 1989; Rodov et al., 1992; Mercier et al., 1993), by the induction of disease resistance (Spott and Chen, 1987; Kim et al., 1991; Terry and Joyce, 2004), there is no information available about the effect of these treatments on the induction of disease resistance against *A. alternata*. Thus, it would be interesting to conduct studies to evaluate whether UV-C irradiation and heat treatments can stimulate resistance responses in fruits against the development of *Alternaria* rot.

CONCLUDING REMARKS

Based on the recent and most relevant data concerning the identification, biology of the infection, and postharvest treatments to control the disease caused by *A. alternata* in fruits and vegetables, it is clear that this fungus is a necrotrophic and destructive fungus that causes black spot in postharvest life of many fruits and vegetables. It is a latent fungus that develops during the cold storage of the fruits and is visible during the marketing period. This pathogen gains entry into the fruit via wounds or natural openings, and remains quiescent until the fruit ripens, and the conditions are more favorable for disease development. Also, as well as fruits, it has been found that this fungus can attack other plant tissue like seeds, leaves, stem and flowers. In the development of strategies for postharvest disease control, it is important to consider the production and postharvest handling systems. Many preharvest and postharvest factors directly and indirectly influence the development of

Alternaria's postharvest disease. Traditionally, fungicides have played a central role in postharvest disease control, however, trends towards reduced chemical usage in horticulture are forcing the development of new strategies, like biological control, natural compounds such as chitosan, essential oils, isothiocyanates, elicitors of defense mechanism, physical methods like heat treatments, UV light, ozone, among others. Therefore, it is possible it could become easier to control the postharvest disease caused by *A. alternata* using these novel alternatives that can be more economical, environmentally friendly and safe.

Despite its importance, a protocol to identify unambiguously *A. alternata* from other species belonging to same genus is something that needs to be developed and a tool that will help to identify the pathogen during infection of fruits and vegetables and to test the effectiveness of the future protocols designed to control the disease caused by this fungus.

REFERENCES

Abdolahi, A., Hassani, A., Ghosta, Y., Javadi, T., Meshkatalsadat, M.H., 2010. Essential oils as control agents of postharvest *Alternaria* and *Penicillium* rots on tomato fruits. J. Food Saf. 30, 341–352.

Abo-El-Seoud, M.A., Sarhan, M.M., Omar, A.E., Helal, M.M., 2005. Bioside formulation of essential oils having antimicrobial activity. Arch. Phytopathol. Plant Prot. 38, 175–184.

Adachi, Y., Watanabe, H., Tababe, K., Doke, N., Nishimura, S., Tsuge, T., 1993. Nuclear ribosomal DNA as a probe for genetic variability in the Japanese pear pathotype of *Alternaria alternata*. Appl. Environ. Microbiol. 59, 3197–3205.

Adikaram, N.K.B., Brown, A.E., Swinburne, T.R., 1988. Phytoalexin induction as a factor in the protection of *Capsicum annuum* L. fruits against infection by *Botrytis cinerea* Pers. J. Phytopathol. 122, 267–273.

Al-Sadi, A.M., Al-Said, F.A., Al-Kaabi, S.M., et al., 2011. Occurrence, characterization and management of fruit rot of immature cucumbers under greenhouse conditions in Oman. Phytopathol. Med. 50, 421–429.

Amin, M., Malik, A.U., Khan, A.S., Javed, N., 2011. Potential of fungicides and plant activator for postharvest disease management in mangoes. Int. J. Agric. Biol. 13, 671–676.

Anand, T., Bhaskaran, R., Raguchander, T., Samiyappan, R., Prakasam, V., Gopalakrishnan, C., 2009. Defence responses of chilli fruits to *Colletotrichum capsici* and *Alternaria alternata*. Biol. Plantarum 53, 553–559.

Andersen, B., Thrane, U., 1996. Differentiation of *Alternaria infectoria* and *Alternaria alternata* based on morphology, metabolite profiles, and cultural characteristics. Can. J. Microbiol. 42, 685–689.

Andersen, B., Kroger, E., Roberts, R.G., 2001. Chemical and morphological segregation of *Alternaria alternata*, *A. gaisen* and *A. longipes*. Mycol. Res. 15, 291–299.

Andersen, B., Hansen, M.E., Smedsgaard, J., 2005. Automated and unbiased image analyses as tools in phenotypic classification of small-spored *Alternaria* spp. Phytopathology 95, 1021–1029.

Andrew, M., Peever, T.L., Pryor, B.M., 2009. An expanded multilocus phylogeny does not resolve morphological species within the small-spored *Alternaria* species complex. Mycologia 101, 95–109.

Aradhya, M.K., Chan, H.M., Parfitt, D.E., 2001. Genetic variability in the pistachio late blight fungus, *Alternaria alternata*. Mycol. Res. 105, 300–306.

Barkai-Golan, R., 1973. Postharvest heat treatment to control *Alternaria tenuis* rot in tomato. Phytopathol. Med. 12, 108–111.

Bashar, M.A., Shamsi, S., Hossain, M., 2012. Fungi associated with rotten fruits in Dhaka Metropolis. Bangladesh J. Bot. 41, 115–117.

Bautista-Baños, S., Hernández-López, M., Bosquez-Molina, E., 2004. Growth inhibition of selected fungi by chitosan and plant extracts. Mex. J. Phytopathol. 22, 178–186.

Bautista-Baños, S., Hernández-Lauzardo, A.N., Velázquez-del Valle, M.G., et al., 2006. Chitosan as a potential natural compound to control pre and postharvest diseases of horticultural commodities. Crop Prot. 25, 108–118.

Begum, M.F., Rahman, M.A., Alam, F., 2010. Biological control of *Alternaria* fruit rot of chili by *Trichoderma* species under field conditions. Mycobiology 38, 113–117.

Belisario, A., Maccaroni, M., Corazza, L., Balmas, V., Valier, A., 2002. Occurrence and etiology of brown apical necrosis on Persian (English) walnut fruit. Plant Dis. 86, 599–602.

Ben-Arie, R., Zutkhi, Y., 1992. Extending the storage life of 'Fuyu' persimmon by modified-atmosphere packaging. HortScience 27, 811–813.

Ben-Arie, R., Klein, J., Zutkhi, Y., 1991. Long-term storage of persimmons. Postharvest Biol. Technol. 4, 169–179.

Boulet, M., Arul, J., Verret, P., Kane, O., 1989. Induced resistance of stored mango (*Mangifera indica* L.) fruits to mold infection by treatment with *Colletotrichum gloeosporioides* L. cell wall hydrolysate. Can. Inst. Food Sci. Technol. J. 22, 161–168.

Canihos, Y., Peever, T.L., Timmer, L.W., 1999. Temperature, leaf wetness, and isolate effects on infection of Minneola tangelo leaves by *Alternaria* sp. Plant Dis. 83, 429–433.

Cao, J., Yan, J., Zhao, Y., Jiang, W., 2013. Effects of postharvest salicylic acid dipping on *Alternaria* rot and disease resistance of jujube fruit during storage. J. Sci. Food Agric. 93, 3252–3258.

Cappellini, R.A., Ceponis, M.J., Lightner, G.W., 1988. Disorders in apricot and papaya shipments to the New York market, 1972–1985. Plant Dis. 72, 366–369.

Carvalho, D.D.C., Alves, E., Batista, T.R.S., Camargos, R.B., Lopes, E.A.G.L., 2008. Comparison of methodologies for conidia production by *Alternaria alternata* from citrus. Braz. J. Microbiol. 39, 792–798.

Ceponis, M.J., Cappellini, R.A., 1978. Relationship of chemical application time to postharvest disease control of blueberries. Plant Dis. Rep. 62, 1005–1007.

Ceponis, M.J., Cappellini, R.A., Lightner, G.W., 1987. Disorders in fresh pepper shipments to the New York market, 1972–1984. Plant Dis. 71, 1162–1165.

Chen, Y., Zhang, C., Cong, P., 2012. Dynamics of growth regulators during infection of apple leaves by *Alternaria alternata* apple pathotype. Australas. Plant Pathol. 41, 247–253.

Choi, M.O., Kim, S.G., Hyun, I.H., et al., 2010. First report of black spot caused by *Alternaria alternata* on grafted cactus. Plant Pathol. J. 26, 80–82.

Cole, J.S., Gates, L.F., Stephen, R.C., 1961. The effect of nitrogen, phosphorus, and potassium on susceptibility of tobacco to *Alternaria longipes*. Rhodesia Agric. J. 58, 354–361.

Conner, R.L., Carefoot, J.M., Bole, J.B., Kozu, G.C., 1992. The effect of nitrogen fertilizer and irrigation on black point incidence in soft white spring wheat. Plant Soil 140, 41–47.

Cota, I.E., Troncoso-Rojas, R., Sotelo-Mundo, R., Sánchez-Estrada, A., Tiznado-Hernández, M.E., 2007. Chitinase and β-1,3-Glucanase enzymatic activities in response to infection by *Alternaria alternata* evaluated in two stages of development in different tomato fruit varieties. Sci. Horticulturae 112, 42–50.

Couey, H.M., Hayes, C.F., 1986. A quarantine system for papayas using fruit selection and a two-stage hot-water treatment. J. Econ. Entomol. 79, 1307–1314.

de Hoog, G.S., Horré, R., 2002. Molecular taxonomy of the *Alternaria* and *Ulocladium* species from humans and their identification in the routine laboratory. Mycoses 45, 259–276.

Dehpour, A.A., Alavi, S.V., Majd, A., 2007. Light and scanning electron microscopy studies on the penetration and infection processes of *Alternaria alternata*, causing brown spot on Minneola Tangelo in the West Mazandaran-Iran. World Appl. Sci. J. 2, 68–72.

de Oliveira, E.N., de Melo, I.S., Franco, T.T., 2012. Changes in hyphal morphology due to chitosan treatment in some fungal species. Braz. Arch. Biol. Technol. 55, 637–646.

Devlieghere, F., Vermeulen, A., Debevere, J., 2004. Chitosan: antimicrobial activity, interactions with food components and applicability as a coating on fruits and vegetables. Food Microbiol. 21, 703–714.

Doster, M.A., Michaillides, T.J., 2007. Fungal decay of first-crop and main-crop figs. Plant Dis. 91, 1657–1662.

Droby, S., Prusky, D., Dinoor, A., Barkai-Golan, R., 1984. *Alternaria alternata*: A new pathogen on stored potatoes. Plant Dis. 68, 160–161.

Eckert, J.W., Ogawa, J.M., 1988. The chemical control of postharvest diseases: deciduous fruits, berries, vegetables and root/tuber crops. Ann. Rev. Phytopathol. 26, 433–469.

Elena, K., 2006. *Alternaria* brown spot of Minneola in Greece; evaluation of citrus species susceptibility. Eur. J. Plant Pathol. 115, 259–262.

El-Katatny, M.H., Emam, A.S., 2012. Control of postharvest tomato rot by spore suspension and antifungal metabolites of *Trichoderma harzianum*. J. Microbiol. Biotechnol. Food Sci. 1, 1505–1528.

Esguerra, E.B., Chavez, S.M., Traya, R.V., 2006. A modified and rapid heat treatment for the control of postharvest diseases of mango (*Mangifera indica* Linn. cv. Carabao) fruits. Int. J. Postharvest Technol. 1, 170–177.

Eshel, D., Lichter, A., Dinoor, A., Prusky, D., 2002. Characterization of *Alternaria alternata* glucanase genes expressed during infection of resistant and susceptible persimmon fruits. Mol. Plant Pathol. 3, 347–358.

Espinoza-Verduzco, M.D.A., Santos-Cervantes, M.E., Fernandez-Herrera, E., et al., 2012. First report of *Alternaria alternata* (Fr.) Keissler causing inflorescence blight in *Jatropha curcas* in Sinaloa, Mexico. Can. J. Plant Pathol. 34, 455–458.

Faid, M., Charai, M., Mosaddak, M., 1996. Chemical composition and antimicrobial activities of two aromatic plants: *Origanum majorana* L. and *O. compactum* Benth. J. Essent. Oil Res. 8, 957–964.

Fallik, E., Grinberg, S., Alkalai, S., Lurie, S., 1996. The effectiveness of postharvest hot water dipping on the control of grey and black moulds in sweet red pepper (*Capsicum annuum*). Plant Pathol. 45, 644–649.

Fallik, E., Aharoni, Y., Copel, A., et al., 2000. Reduction of postharvest losses of Galia melon by a short hot-water rinse. Plant Pathol. 49, 333–338.

Fatima, N., Batool, H., Sultana, V., Ara, J., Ehteshamul-Haque, S., 2009. Prevalance of post-harvest rot of vegetables and fruits in Karachi, Pakistan. Pak. J. Bot. 41, 3185–3190.

Fehr, M., Pahlke, G., Fritz, J., et al., 2009. Alternariol acts as an affecting topoisomerase poison, preferentially the IIα isoform. Mol. Nutr. Food Res. 53, 441–451.

Feng, W., Zheng, X., 2007. Essential oils to control *Alternaria alternata in vitro* and *in vivo*. Food Control 18, 1126–1130.

Filajdič, N., Sutton, T.B., 1991. Identification and distribution of *Alternaria mali* on apples in North Carolina and susceptibility of different varieties of apples to *Alternaria* blotch. Plant Dis. 75, 1045–1048.

Fourie, P.H., du Preez, M., Brink, J.C., Schutte, G.C., 2009. The effect of runoff on spray deposition and control of *Alternaria* brown spot of mandarins. Australas. Plant Pathol. 38, 173–182.

Gao, L.L., Zhang, Q., Sun, X.Y., et al., 2013. Etiology of moldy core, core browning, and core rot of fuji apple in China. Plant Dis. 97, 510–516.

García-Brugger, A., Lamotte, O., Vandelle, E., et al., 2006. Early signaling events induced by elicitors of plant defenses. Mol. Plant Microbe Int. 19, 711–724.

Gilchrist, D.G., Grogan, R.G., 1976. Production and nature of a host-specific toxin from *Alternaria alternata* f. sp. *lycopersici*. Phytopathology 66, 165–171.

Golmohammadi, M., Andrew, M., Peever, T.L., Peres, N.A., Timmer, L.W., 2006. Brown spot of tangerine hybrid cultivars Minneola, Page and Fortune caused by *Alternaria alternata* in Iran. Plant Pathol. 55, 578.

González, L.A., Valenzuela, Q.A.I., 2007. The postharvest use of synthetic fungicides, implications on human health. In: Troncoso Rojas, R., Tiznado Hernández, M.E., González León, A. (Eds.). Recent Advances in Alternative Postharvest Technologies to Control Fungal Diseases in Fruits & Vegetables, Kerala, India, pp. 1–20.

Grogan, R.G., Kimble, K.A., Misaghi, I., 1975. A stem canker disease of tomato caused by *Alternaria alternata* f. sp. *lycopersici*. Phytopathology 65, 880–886.

Guo, L.D., Xu, L., Zheng, W.H., Hyde, K.D., 2004. Genetic variation of *Alternaria alternata*, an endophytic fungus isolated from *Pinus tabulaeformis* as determined by random amplified microsatellites (RAMS). Fungal Divers. 16, 53–65.

Hadizadeh, I., Pivastegan, B., Hamzehzarghani, H., 2009. Antifungal activity of essential oils from some medicinal plants of Iran against *Alternaria alternata*. Am. J. Appl. Sci. 6, 857–861.

Halfon-Meiri, A., Rylski, I., 1983. Internal mold caused in sweet pepper by *Alternaria alternata*: Fungal ingress. Phytopathology 73, 67–70.

Harteveld, D.O.C., Akinsanmi, O.A., Drenth, A., 2013. Multiple *Alternaria* species groups are associated with leaf blotch and fruit spot diseases of apple in Australia. Plant Pathol. 62, 289–297.

Hidalgo, P.J., Ubera, J.A., Santos, F., et al., 2002. Essential oils in *Calamintha sylvatica* Bromf. ssp. *ascendens* (Jorden) P.W. Ball: Wild and cultivated productions and antifungal activity. J. Essent. Oil Res. 14, 68–71.

Hong, S.G., Liu, D., Pryor, B.M., 2005. Restriction mapping of the IGS region in *Alternaria* spp. reveals variable and conserved domains. Mycol. Res. 109, 87–95.

Hubballi, M., Nakkeeran, S., Raguchander, T., Rajendran, L., Renukadevi, P., Samiyappan, R., 2010. First report of leaf blight of noni caused by *Alternaria alternata* (Fr.) Keissler. J. Gen. Plant Pathol. 76, 284–286.

Jia, Ch., Zhang, L., Liu, L., Wang, J., Li, Ch, Wang, Q., 2013. Multiple phytohormone signaling pathways modulate susceptibility of tomato plants to *Alternaria alternata* f. sp. *lycopersici*. J. Exp. Bot. 64, 637–650.

Jiang, Y.M., Zhu, X.R., Li, Y.B., 2001. Postharvest control of litchi fruit rot by *Bacillus subtilis*. Lebensmittel Wissenschaft Technol. 34, 430–436.

Johnson, R.D., Johnson, L., Kohmoto, K., Otani, H., Lane, C.R., Kodama, M., 2000. A polymerase chain reaction-based method to specifically detect *Alternaria alternata* apple pathotype (*A. mali*), the causal agent of *Alternaria* blotch of apple. Phytopathology 90, 973–976.

Jones, E.B.G., 1994. Fungal adhesion. Mycol. Res. 98, 961–981.

Kai, W.H., Yu, Z.T., Meng, Z., 2001. Application of sequencing of 5.8S rDNA, ITS1 and ITS2 on identification and classification of *Alternaria* at species level. Mycosystema 20, 168–173.

Kawakishi, S., Kaneko, T., 1987. Interaction of proteins with allyl isothiocyanate. J. Agric. Food Chem. 35, 85–88.

Kim, J.J., Ben-Yehoshua, S., Shapiro, B., Henis, Y., Carmeli, S., 1991. Accumulation of scoparone in heat-treated lemon fruit inoculated with *Penicillium digitatum* Sacc. Plant Physiol. 97, 880–885.

Kobiler, I., Shalom, Y., Roth, I., et al., 2001. Effect of 2,4-dichlorophenoxy acetic acid on the incidence of side and stem end rots in mango fruits. Postharvest Biol. Technol. 23, 23–32.

Kohmoto, K., Itoh, Y., Shimomura, N., et al., 1993. Isolation and biological activities of two host-specific toxins from the tangerine pathotype of *Alternaria alternata*. Phytopathology 83. 495–450.

Kojima, M., Ogawa, K., 1971. Studies on the effects of isothiocyanates and their analogues on microorganisms. J. Fermentation Technol. 49, 740–746.

Kunkel, B.N., Brooks, D.M., 2002. Cross talk between signaling pathways in pathogen defense. Curr. Opin. Plant Biol. 5, 325–331.

Kwon, J.H., Cheon, M.G., Kim, J., Kwack, Y.B., 2011. Black rot of kiwifruit caused by *Alternaria alternata* in Korea. Plant Pathol. J. 27. 298–298.

Lagopodi, A.L., Thanassoulopoulos, C.C., 1998. Effect of a leaf spot disease caused by *Alternaria alternata* on yield of sunflower in Greece. Plant Dis. 82, 41–44.

Lawrence, D.P., Gannibal, P.B., Dugan, F.M., Pryor, B.M., 2013. Characterization of *Alternaria* isolates from the infectoria species-group and a new taxon from *Arrhenatherum, Pseudoalternaria arrhenatheria* sp. nov. Mycolo. Prog.http://dx.doi.org/10.1007/s11557-013-0910-x.

Li, Y.C., Bi, Y., An, L.Z., 2007. Occurrence and latent infection of *Alternaria* rot of pingguoli pear (*Pyrus bretschneideri* Rehd. cv. Pingguoli) fruits in Gansu, China. J. Phytopathol. 155, 56–60.

López, M.L.I., Gutiérrez, M.P., Bautista, B.S., Jiménez, G.L.F., Zavaleta, M.H.A., 2013. Evaluation of the fungicidal activity of chitosan on *Alternaria alternata* and on the quality of Mango 'Tommy Atkins' during storage. Rev. Chapingo Serie Horticulturahttp://dx.doi.org/10.5154/r.rchsh.2012.07.038.

Lurie, S., Klein, J.D., 1991. Acquisition of low-temperature tolerance in tomatoes by exposure to high-temperature stress. J. Am. Soc. Horticultural Sci. 116, 1007–1012.

Mahdavi, S., Zakerin, A., Sadeghi, H., Niazmand, A.R., 2013. Antifungal effects of essential oils of three medicinal plants on post-harvest rot ofValencia oranges at normal and storage temperatures. Afri. J. Microbiol. Res. 7, 3831–3835.

Malathrakis, N.E., 1983. *Alternaria* stem canker of tomato in Greece. Phytopathol. Med. 22, 33–38.

Marín, J.E., Fernández, H.S., Peres, N.A., Andrew, M., Peever, T.L., Timmer, L.W., 2006. *Alternaria* Brown spot of citrus caused by *Alternaria alternata* in Peru. Plant Dis. 90, 686.

Masunaka, A., Otani, K., Peever, T.L., et al., 2005. An isolate of *Alternaria alternata* that is pathogenic to both tangerine and rough lemon and produces two host selective toxins, ACT- and ACR-toxins. Phytopathology 95, 241–247.

McColloch, L.P., Worthington, J.T., 1952. Low temperature as a factor in the susceptibility of mature-green tomatoes to *Alternaria* rot. Phytopathology 42, 425–427.

Meath, M.B., 2000. Residues of mancozeb and iprodione in mango fruits treated for control of stem end rot. Pak. J. Agric. Res. 16, 35–39.

Meir, S., Rosenberger, I., Aharon, Z., Grinberg, S., Fallik, E., 1995. Improvement of the postharvest keeping quality and colour development of bell pepper (cv. 'Maor') by packaging with polyethylene bags at a reduced temperature. Postharvest Biol. Technol. 5, 303–309.

Mendgen, K., Hahn, M., Deising, H., 1996. Morphogenesis and mechanism of penetration by plant pathogenic fungi. Ann. Rev. Phytopathol. 34, 364–386.

Meng, X., Yang, L., Kennedy, J.F., Tian, S., 2010. Effects of chitosan and oligochitosan on growth of two fungal pathogens and physiological properties in pear fruit. Carbohydr. Polymers 81, 70–75.

Mercier, J., Arul, J., Ponnampalam, R., Boulet, M., 1993. Induction of 6-methoxymellein on resistance to storage pathogens in carrots slices by UV-C. J. Phytopathol. 137, 44–54.

Mersha, Z., Zhang, S., Bartz, J.A., 2012. A postharvest fruit disease – black mold rot caused by *Alternaria* sp on imported plum tomatoes in South Florida. The Vegetarian Newsletter. Horticultural Sciences at University of Florida. Issue No. 579 htpp://host.ufl.edu/newsletters/vegetarian/issue-no-579.

Mirkova, E., Konstantinova, P., 2003. First report of *Alternaria* leaf spot on gerbera (*Gerbera jamesonii* H. Bolux ex J.D. Hook) in Bulgaria. J. Phytopathol. 151, 323–328.

Misaghi, I.J., Grogan, R.G., Duniway, J.M., Kimble, K.A., 1977. Influence of environment and culture media on spore morphology of *Alternaria alternata*. Phytopathology 68, 29–34.

Mmbaga, M.T., Shi, A., Kim, M.S., 2011. Identification of *Alternaria alternata* as a causal agent for leaf blight in *Syringa* species. Plant Pathol. J. 27, 120–127.
Mohsan, M., Intizar-ul-Hassan, M., Ali, L., 2011. Chemotherapeutic management of *Alternaria* black spot (*Alternaria alternata*) in mango fruits. J. Agric. Res. 49, 499–506.
Morris, P.F., Connolly, M.S., St Clair, D.A., 2000. Genetic diversity of *Alternaria alternata* isolated from tomato in California assessed using RAPDs. Mycol. Res. 104, 286–292.
Murthy, K.K.N.V., Rao, M.S.N., 1986. Interaction of phytate with mustard 12S protein. J. Agric. Food Chem. 32, 493–498.
Nakamura, Y., 2009. Chemoprevention by isothiocyanates: molecular basis of apoptosis induction. In: Food Factors for Health Promotion, Forum Nutrition, 61, , pp. 170–181. Karger, Basel.
Nemsa, I., Hernandez, M.A., Lacasa, A., et al., 2012. Pathogenicity of *Alternaria alternata* on fruits and leaves of 'Fortune' mandarin (*Citrus clementina* × *Citrus tangerina*). Can. J. Plant Pathol. 34, 195–202.
Nishimura, S., Kohmoto, K., 1983. Host-specific toxins and chemical structures from *Alternaria* species. Ann. Rev. Phytopathol. 21, 87–116.
Ojeda-Contreras, A.J., Hernández-Martínez, J., Domínguez, Z., et al., 2008. Utilization of caffeic acid phenethyl ester to control *Alternaria alternata* rot in tomato (*Lycopersicon esculentum* Mill.) fruit. J. Phytopathol. 156, 164–173.
Ortega, X., Polanco, R., Castañeda, P., Perez, L.M., 2002. Signal transduction in lemon seedlings in the hypersensitive response against *Alternaria alternata*: participation of calmodulin, G-protein and protein kinases. Biol. Res. 35, 373–383.
Ozcan, M., 1999. Antifungal properties of propolis. Grasas Aceites 50, 395–398.
Palma-Guerrero, J., Lopez-Jimenez, J., Pérez-Berna, A., et al., 2010. Membrane fluidity determines sensitivity of filamentous fungi to chitosan. Mol. Microbiol. 75, 1021–1032.
Pearson, R.C., Hall, D.H., 1975. Factors affecting the occurrence and severity of black mold of ripe tomato fruit caused by *Alternaria alternata*. Phytopathology 57, 1352–1359.
Peever, T.L., Ibanez, A., Akimitsu, K., Timmer, L.W., 2002. Worldwide phylogeography of the citrus brown spot pathogen, *Alternaria alternata*. Phytopathlogy 92, 794–802.
Peever, T.L., Su, G., Carpenter-Boggs, L., Timmer, L.W., 2004. Molecular systematics of citrus-associated *Alternaria* species. Mycologia 96, 119–134.
Peres, N.A.R., Agostini, J.P., Timmer, L.W., 2003. Outbreaks of *Alternaria* brown spot of citrus in Brazil and Argentina. Plant Dis. 87, 750.
Phillips, C.A., Laird, K., Allen, S.C., 2012. The use of Citri-V™® – An antimicrobial citrus essential oil vapour for the control of *Penicillium chrysogenum*, *Aspergillus niger* and *Alternaria alternata in vitro* and on food. Food Res. Int. 47, 310–314.
Pose, G.N., Ludemann, V., Fernandez, D., Segura, J.A., Pinto, V.F., 2010. *Alternaria* species associated with 'moldy heart' on peaches in Argentina. Trop. Plant Pathol. 35, 174–177.
Pratella, G.C., Mari, M., 1993. Effectiveness of *Trichoderma*, *Gliocladium* and *Paecilomyces* in postharvest fruit protection. Postharvest Biol. Technol. 3, 49–56.
Prusky, D., 1996. Pathogen quiescence in postharvest diseases. Ann. Rev. Phytopathol. 34, 413–434.
Prusky, D., Ben-Arie, R., 1981. Control by imazalil of fruit storage rots caused by *Alternaria alternata*. Ann. Appl. Biol. 98, 87–92.
Prusky, D., Ben-Arie, R., Guelfat-Reich, S., 1981. Etiology and histology of *Alternaria* rot of persimmon fruit. Phytopathology 71, 1124–1128.
Prusky, D., Fuchs, Y., Kobilar, I., et al., 1998. Effect of hot water brushing, prochloraz treatment and waxing on the incidence of black spot decay caused by *Alternaria alternata* in mango fruits. Dep. Postharvest Sci. Fresh Produce Inst. Technol. Storage Agric. Prod.. ARO. The Volcanic Centre, P.O. Box 6, Bet Dagan 50250, Israel.

Prusky, D., Shalom, Y., Kobiler, I., Akerman, M., Fuchs, Y., 2002. The level of quiescent infection of *Alternaria alternata* in mango fruit at harvest determines the postharvest treatment applied for the control of rots during storage. Postharvest Biol. Technol. 25, 339–347.

Prusky, D., Alkan, N., Mengiste, T., Fluhr, R., 2013. Quiescent and necrotrophic lifestyle choice during postharvest disease development. Ann. Rev. Phytopathol. 51, 155–176.

Pryor, B.M., Gilbertson, R.L., 2000. Molecular phylogenetic relationship amongst *Alternaria* species and related fungi based upon analysis of nuclear ITS and mtSSU rDNA sequences. Mycol. Res. 104, 1312–1321.

Pryor, B.M., Michailides, T.J., 2002. Morphological, pathogenic, and molecular characterization of *Alternaria* isolates associated with *Alternaria* late blight of pistachio. Phytopathology 92, 406–416.

Pusz, W., 2009. Morpho-physiological and molecular analyses of *Alternaria alternata* isolated from seeds of *Amaranthus*. Phytopathology 54, 5–14.

Qiang, S., Wang, L., Wei, R., et al., 2010. Bioassay of the herbicidal activity of AAC-toxin produced by *Alternaria alternata* isolated from *Ageratina adenophora*. Weed Technol. 24, 197–201.

Qin, G.Z., Tian, S.P., 2004. Biocontrol of postharvest diseases of jujube fruit by Cryptococcus laurentii combined with a low dosage of fungicides under different storage conditions. Plant Dis. 88, 497–501.

Qin, G.Z., Tian, S.P., Xu, Y., 2004. Biocontrol of postharvest diseases on sweet cherries by four antagonistic yeasts in different storage conditions. Postharvest Biol. Technol. 31, 51–58.

Ramm, C., von, Lucas, G.B., 1962. Influence of temperature and light on brown spot of tobacco. J. Elisha Mitchell Sci. Soc. 78, 94.

Rathod, G., 2012. A review on biological control of post-harvest fungal diseases of fruits. Curr. Bot. 3, 05–07.

Rawel, H.M., Kroll, J., Riese-Schneider, B., Haebel, S., 1998. Physicochemical and enzymatic properties of benzyl isothiocyanate derivatized proteinases. J. Agricu. Food Chem. 46, 5043–5051.

Reddy, M.V., Arul, J., Ait-Barka, E., Angers, P., Richard, C., Castaigne, F., 1998. Effect of chitosan on growth and toxin production by *Alternaria alternata* f. sp. *lycopersici*. Biocontrol Sci. Technol. 8, 33–43.

Reddy, B.M.V., Angers, P., Castaigne, F., Arul, J., 2000. Chitosan effect on black mold rot and pathogenic factors produced by *Alternaria alternata* in postharvest tomatoes. J. Am. Soc. Horticultural Sci. 125, 742–747.

Reis, R.F., de Goes, A., Mondal, S.N., Shilts, T., Brentu, F.C., Timmer, L.W., 2006. Effect of lesion age, humidity, and fungicide application on sporulation of *Alternaria alternata*, the cause of brown spot of tangerine. Plant Dis. 90, 1051–1054.

Riley, E.A., 1949. Pathological and physiological studies on the brown spot disease of tobacco caused by *Alternaria longipes* (Ell. & Ev.) Mason. Mem. Dep. Agric. South Rhode Island 3, 1–32.

Roberts, R.G., Reymond, S.T., Andersen, B., 2000. RAPD fragment analysis and morphological segregation of small-spores *Alternaria* species and species group. Mycol. Res. 104, 151–160.

Roberts, R.G., Bischoff, J.F., Reymond, S.T., 2012. Differential gene expression in *Alternaria gaisen* exposed to dark and light. Mycol. Progress 11, 373–382.

Rodov, V., Ben-Yehoshua, S., Kim, J.J., Shapiro, B., Ittah, Y., 1992. Ultraviolet illumination induces scoparone production in kumquat and orange fruit and improves decay resistance. J. Am. Soc. Horticultural Sci. 117, 788–792.

Romanazzi, G., Lichter, A., Mlikota Gabler, F., Smilanick, J.L., 2012. Recent advances on the use of natural and safe alternatives to conventional methods to control postharvest gray mold of table grapes. Postharvest Biol. Technol. 63, 141–147.

Rotem, J., 1998. The genus of Alternaria: Biology, Epidemiology and Pathogenicity. The American Phytopathological Society, USA.

Ruelas, C., Tiznado-Hernández, M.E., Sánchez-Estrada, A., Robles-Burgueño, M.R., Troncoso-Rojas, R., 2006. Changes in phenolic acid content during *Alternaria alternata* infection in tomato fruit. J. Phytopathol. 154, 236–244.

Sánchez-Domínguez, D., Bautista-Baños, S., Castillo-Ocampo, P., 2007. Efecto del quitosano enel desarrollo y morfología de *Alternaria alternata*(Fr.:Fr.) Keissl. Anales de Biología 29, 23–32.

Sánchez-Domínguez, D., Ríos, M.Y., Castillo-Ocampo, P., Zavala-Padilla, G., Ramos-Garcíía, M., Bautista-Baños, S., 2011. Cytological and biochemical changes induce by chitosan in the pathosystem *Alternaria alternata*-tomato. Pesticide Biochem. Physiol. 99, 250–255.

Sawamura, K., 1972. Studies on apple Alternaria blotch caused by *Alternaria mali* Roberts. Bull. Faculty Agric. Hirosaki Univ. 18, 152–325.

Scott, P.M., 2001. Analysis of agricultural commodities and foods for *Alternaria* mycotoxins. J. AOAC Int. 84, 1809–1817.

Serdani, M.M., Kang, J.C., Andersen, B., Crous, P.W., 2002. Characterisation of *Alternaria* species-groups associated with core rot of apples in South Africa. Mycol. Res. 106, 561–569.

Sharma, R.R., Singh, D., Singh, R., 2009. Biological control of postharvest diseases of fruits and vegetables by microbial antagonists: A review. Biol. Control 50, 205–221.

Sheard, L.B., Tan, X., Mao, H., et al., 2010. Jasmonate perception by inositolphosphate-potentiated COI1-JAZ co-receptor. Nature 468, 400–405.

Shtienberg, D., 2012. Effects of host physiology on the development of core rot, caused by *Alternaria alternata*, in red delicious apples. Phytopathology 102, 769–778.

Sid, A., Ezziyyani, M., Egea-Gilaert, C., Candela, M.E., 2003. Selecting bacterial strains for use in the biocontrol of diseases caused by *Phytophthora capsici* and *Alternaria alternata* in sweet pepper plants. Biol. Plantarum 47, 569–574.

Simmons, E.G., 1999. *Alternaria* themes and variations (236-243). Mycotaxon 70, 325–369.

Simmons, E.G., Roberts, R.G., 1993. *Alternaria* themes and variations. Mycotaxon 48, 109–140.

Singh, J., Tripathi, N.N., 1999. Inhibition of storage fungi of black gram (*Vigna mungo* L.) by some essential oils. Flavour Fragrance J. 14, 42–44.

Singh, P.C., Singh, D., 2006. *In vitro* evaluation of fungicides against *Alternaria alternata*. Ann. Plant Prot. Sci. 14, 500–502.

Sitton, J.W., Pierson, C.F., 1983. Interaction and control of *Alternaria* stem decay and blue mold of d'Anjou pears. Plant Dis. 67, 904–907.

Sivakumar, D., Zeeman, K., Korsten, L., 2007. Effect of a biocontrol agent (*Bacillus subtilis*) and modified atmosphere packaging on postharvest decay control and quality retention of litchi during storage. Phytoparasitica 35, 507–518.

Snowdon, A.L., 1990. A Color Atlas of Post-harvest Disease & Disorders of Fruits & Vegetables, vol. 1, CRC Press, Inc. Boca Raton, Florida. pp. 30–57.

Solel, Z., Kimchi, M., 1998. Histopathology of infection of Mineola tangelo by *Alternaria alternata* pv. *citri* and the effect of host and environmental factors in lesion development. J. Phytopathol. 146, 557–561.

Sommer, N., 1985. Role of controlled environments in suppression of postharvest diseases. Can. J. Plant Pathol. 7, 331–339.

Spalding, D.H., 1980. Control of *Alternaria* rot of tomatoes by postharvest application of Imazalil. Plant Dis. 64, 169–171.

Spalding, D.H., King, L.R., 1980. Inhibition of *Alternaria* rot of tomatoes and bell peppers by postharvest treatment with CGA-6425 I or imazalil. Proc. Florida State Horticultural Soc. 93, 307–308.

Spotts, R.A., Chen, P.M., 1987. Prestorage heat treatments for control of decay of pear fruit. Phytopathology 77, 1578–1582.

Stavely, J.R., Main, C.E., 1970. Influence of temperature and other factors on initiation of tobacco brown spot. Phytopathology 60, 1591–1596.

Stevenson, R.E., Pennypacker, S.P., 1988. Effect of radiation, temperature, and moisture on conidial germination of *Alternaria solani*. Phytopathology 78, 926–930.

Suzuki, T., Shinogi, T., Narusaka, Y., 2003. Infection behavior of *Alternaria alternata* Japanese pear pathotype and localization of 1,3-D-glucan in compatible and incompatible interactions between the pathogen and host plants. J. Gen. Plant Pathol. 69, 91–100.

Swart, A.E., Lennox, C.L., Holz, G., 1995. Infection of table grape bunches by *Alternaria alternata*. South Afri. J. Enol. Viticulture 16, 1–4.

Szopinska, D., Tylkowska, K., Deng, C.J., Gao, Y., 2012. Comparison of modified blotter and agar incubation methods for detecting fungi in *Zinnia elegans* seeds. Seed Sci. Technol. 40, 32–42.

Taba, S., Takara, A., Nasu, K., Miyahira, N., Takushi, T., Moromizato, Z., 2009. *Alternaria* leaf spot of basil caused by *Alternaria alternata* in Japan. J. Gen. Plant Pathol. 75, 160–162.

Teitel, D.C., Barkai-Golan, R., Aharnoi, Y., Copel, A., Davidson, H., 1991. Toward a practical postharvest heat treatment for 'Galia' melons. Sci. Horticulturae 45, 339–344.

Terry, L.A., Joyce, D.C., 2004. Elicitors of induced disease resistance in postharvest horticultural crops: a brief review. Postharvest Biol. Technol. 32, 1–13.

Thakur, M., Sohal, S., 2013. Role of elicitors in inducing resistance in plants against pathogen infection: A review. . ISRN Biochemistry http://dx.doi.org/10.1155/2013/762412.

Thomas, H.R., 1948. Effect of nitrogen, phosphorus, and potassium on susceptibility of tomatoes to *Alternaria solani*. J. Agric. Res. 76, 289–306.

Tian, S.P., Qin, G.Z., Xu, Y., 2005. Synergistic effects of combining biocontrol agents with silicon against postharvest diseases of jujube fruit. J. Food Prot. 68, 544–550.

Timmer, L.W., Darhower, H.M., Zitko, S.E., Peever, T.L., Ibanez, A.M., Bushong, P.M., 2000. Environmental factors affecting the severity of *Alternaria* brown spot of citrus and their potential use in timing fungicide applications. Plant Dis. 84, 638–643.

Tiznado-Hernández, M.E., Troncoso-Rojas, R., 2006. Control of fungal diseases with isothiocyanates. Stewart Postharvest Rev.. Online ISSN www.stewartpostharvest.com.

Tohamy, M.R.A., Helal, G.A., Ibrahim, K.I., Abd-El-Aziz, S.A., 2004. Control of postharvest tomato fruit rots: II. Using heat treatments. Egypt. J. Phytopahtol. 32, 129–138.

Tran-Dinh, N., Hocking, A., 2006. Isolation and characterization of polymorphic microsatellite markers for *Alternaria alternata*. Mol. Ecol. Notes 6, 405–407.

Tripathi, P., Dubey, N.K., 2004. Exploitation of natural products as an alternative strategy to control postharvest fungal rotting of fruit and vegetables. Postharvest Biol. Technol. 32, 235–245.

Troncoso, R., Espinoza, C., Sanchez-Estrada, A., Tiznado, M.E., Garcia, H.S., 2005a. Analysis of the isothiocyantes present in cabbage leaves extract and their potential application to control *Alternaria* rot in bell peppers. Food Res. Int. 38, 701–708.

Troncoso-Rojas, R., Sanchez-Estrada, A., Ruelas, C., Garcia, H.S., Tiznado-Hernandez, M.E., 2005b. Effect of benzyl isothiocyanate on tomato fruit infection development by *Alternaria alternata*. J. Sci. Food Agric. 85, 1427–1434.

Troncoso-Rojas, R., Tiznado-Hernández, M.E., 2007. Natural compounds to control fungal disease in fruits and vegetables. In: Troncoso-Rojas, R., Tiznado-Hernández, M.E., González León, A. (Eds.), Recent Advances in Alternative Postharvest Technologies to Control Fungal Diseases in Fruits and Vegetables. Transworld Research Network, Kerala, India, pp. 127–156.

Troncoso-Rojas, R., Corral-Acosta, Y., Sanchez-Estrada, A., et al., 2009. Postharvest treatment of isothiocyanates to control *Alternaria* rot in netted melon. Phytoparasitica 37, 445–451.

Troncoso-Rojas, R., Báez-Flores, M.E., Pryor, B., García, H.S., Tiznado-Hernández, M.E., 2013. Inter simple sequence repeat polymorphism in *Alternaria* genomic DNA exposed to lethal concentrations of isothiocyanates. Afri. J. Microbiol. Res. 7, 838–852.

Valiuškaitė, A., Kviklienė, N., Kviklys, D., Lanauskas, J., 2006. Post-harvest fruit rot incidence depending on apple maturity. Agronomy Res. 4, 427–431.

Velázquez-del Valle, M.G., Hernández-Lauzardo, A.N., Guerra-Sánchez, M.G., Mariaca-Gaspar, G.I., 2012. Chitosan as an alternative to control phytopathogenic fungi on fruits and vegetables in Mexico. Afri. J. Microbiol. Res. 6, 6606–6617.

Vicent, A., Armengol, J., Sales, R., García-Jiménez, J., Alfaro-Lassala, F., 2000. First report of *Alternaria* brown spot of citrus in Spain. Plant Dis. 84, 1044.

Vincent, A., Armengol, J., García-Jiménez, J., 2007. Rain fastness and persistence of fungicides for control of *Alternaria* brown spot of citrus. Plant Dis. 91, 393–399.

Wall, M.M., Biles, C.L., 1993. *Alternaria* fruit rot of ripening chile peppers. Phytopathology 83, 324–328.

Wan, J., Pentecost, G., 2013. Potential application of chitin signaling in engineering broad-spectrum disease resistance to fungal and bacterial pathogens in plants. Adv. Crop Sci. Technol. 1. doi:10.417/acst.1000e103.

Wang, S.Y., Chen, C.T., 2010. Effect of allyl isothiocyanate on antioxidant enzyme activities, flavonoids and post-harvest fruit quality of blueberries (*Vaccinium corymbosum* L., cv. Duke). Food Chem. 122, 1153–1158.

Wang, Y., Yu, T., Li, Y., et al., 2009. Postharvest biocontrol of *Alternaria alternata* in Chinese winter jujube by *Rhodosporidium paludigenum*. J. Appl. Microbiol. 107, 1492–1498.

Wang, Y., Xu, Z., Zhu, P., et al., 2010. Postharvest biological control of melon pathogens using *Bacillus subtilis* EXWB1. J. Plant Pathol. 92, 645–652.

Weir, T.L., Huff, D.R., Christ, B.J., Romaine, C.P., 1998. RAPD-PCR analysis of genetic variation among isolates of *Alternaria solani* and *Alternaria alternata* from potato and tomato. Mycologia 90, 813–821.

Wells, J., Uota, M., 1970. Germination and growth of five fungi in low oxygen and high carbon dioxide atmospheres. Phytopathology 60, 50–53.

Whiteside, J.O., 1976. A newly recorded *Alternaria*-induced brown spot disease in Dancy tangerines in Florida. Plant Dis. Rep. 60, 326–329.

Xu, Y., Gao, C.W., Li, X.H., et al., 2013. *In vitro* antifungal activity of silver nanoparticles against ocular pathogenic filamentous fungi. J. Ocul. Pharmacol. Ther. 29, 270–274.

Yamagishi, D., Akamatsu, H., Otani, H., 2006. Pathological evaluation of host-specifi AAL-toxins and fumonisin mycotoxins produced by *Alternaria* and *Fusarium* species. J. Gen. Plant Pathol. 72, 323–327.

Yang, S., Jiang, S.S., Van, R.C., Hsiao, Y.Y., Pan, R.L., 2000. A lysine residue involved in the inhibition of vacuolar H+pyrophosphatase by fluorescein 5'-isothiocyanate. Biochim. et Biophys. Acta Bioenerg. 1460, 375–383.

Yang, D.M., Bi, Y., Chen, X.R., Ge, Y.H., Zhao, J., 2006. Biological control of postharvestdiseases with *Bacillus subtilis* (B1 strain) on muskmelons (*Cucumis melo* L. cv. Yindi). Acta Horticulturae 712, 735–739.

Yoon, J.T., Lee, J.T., Choi, D.U., Shon, S.G., 1989. Inhibitory effects of calcium compounds on the outbreak of *Alternaria alternata* f. sp. *mali*. Korean J. Plant Pathol. 5, 306–311.

Zaker, M., Mosallanejad, H., 2010. Antifungal activity of some plant extracts on *Alternaria alternata*, the causal agent of Alternaria leaf spot of potato. Pak. J. Biol. Sci. 13, 1023–1029.

Zhao, Y., Wang, R., Tu, K., Liu, K., 2011. Efficacy of preharvest spraying with *Pichia guilliermondii* on postharvest decay and quality of cherry tomato fruit during storage. Afri. J. Biotechnol. 10, 9613–9622.

CHAPTER 6

Penicillium expansum (Blue Mold)

Deena Errampalli
Agriculture and Agri-Food Canada*, SCPFRC, Vineland Station, ON, Canada

Contents

OVERALL IMPORTANCE OF THE FUNGUS

Penicillium, especially the species *Penicillium expansum*, has been thoroughly studied because of its impact on human culture. Some of the *Penicillia* are beneficial to humans and the environment. The antibiotic penicillin is the most significant and well-known product of the genus. In 1928, Alexander Fleming first discovered the antibiotic and named it Penicillin, the most efficacious life-saving drug in the world that altered forever the treatment

* Mention of a trademark or proprietary product does not constitute a guarantee or warranty of the product by AAFC or imply approval to the exclusion of other products that may also be suitable.

Postharvest Decay
http://dx.doi.org/10.1016/B978-0-12-411552-1.00006-5

of bacterial infections (Houbraken et al., 2011). Many species of *Penicillium* are involved in recycling of organic matter in the environment, while production of cheeses (e.g. Brie, Camembert and Roquefort) relies on members of this genus. Certain species of this fungus biodeteriorate commodities and some cause rot in fruits and vegetables and produce mycotoxins. This ubiquitous fungus, *P. expansum*, has a worldwide distribution and can be found on decomposing organic substances and is a common airborne fungus capable of inducing allergic reaction in sensitive individuals.

Blue mold rot or soft rot on many economically important fruit and vegetable crops is caused by *P. expansum* (Rosenberger, 1990; Snowden, 1990, 1992). The host range includes, apples, pears, peaches, nectarines, apricots, plums and cherries, grapes, strawberries, raspberries, persimmons, mango, passion fruit, avocado, cucurbits, carrots, onions and tomato (Rosenberger, 1990; Lugauskas et al., 2005; Kwon et al., 2006). In pome fruits, *P. expansum* is the principal cause of blue mold. The disease is called blue mold on account of the color of the conidia produced by *P. expansum*. The disease symptoms on fruit at first appear as soft, watery lesions that are light brown in color and, as the lesion ages, the conidia turn blue green. Because of the soft, watery consistency of the rotted tissue, this disease is often referred as soft rot. The lesions have a very sharp margin between diseased and healthy tissues. Decayed tissue can be readily separated from healthy tissue, leaving it like a 'bowl'. The surface of older lesions may be covered with blue green spores and conidial tufts or coremea of *P. expansum*. Decayed fruit has an earthy, musty odor. The presence of blue-green spore masses at the decayed area and associated musty odor are the positive diagnostic indication of blue mold. *Penicillium* spp. including *P. expansum* are also associated with two important core diseases of apple fruit, moldy core and core rot in Australia, Canada, China, New Zealand, South Africa, The Netherlands, the UK, and the USA (Spotts, 1990; van der Walt et al., 2010; Gao et al., 2013).

Food losses take place at production, postharvest and processing stages in the food supply chain (Parfitt et al., 2010). Economic losses during the postharvest phase are estimated to range from 10 to 30% per year in fresh fruit and vegetables and the losses caused postharvest are greater than generally realized because the value of fresh fruits increases many fold from the field to the consumer (Harvey, 1978; Eckert and Ogawa, 1988). Postharvest losses up to 50% were reported in developing countries in which sanitation and refrigeration are lacking (Kader and Rolle, 2004). On a world scale, *P. expansum* is one of the most important pathogens of pome fruit and infection by *P. expansum* may occur during the growing season, at harvest, postharvest in storage or transport (Dennis, 1983).

Mycotoxins pose a threat to public health, agriculture and economics (Drusch and Ragab, 2003). The filamentous fungi belonging to the genera *Penicillium*, *Aspergillus* and *Alternaria* cause diseases in fruits and produce mycotoxins, which are a chemically diverse group of toxic secondary metabolites. The mycotoxins are of health concern to animals and humans and the toxins cause acute and chronic effects including reduced growth efficiency, vomiting, reproduction loss, cancer and immunosuppression (Coulombe, 1993). In addition to the effect on animals and humans, the mycotoxins are responsible for significant financial losses to the food industry that harvests, stores, processes and uses fruits and/or their products. The species *P. expansum* is believed to be the major fungal species that produces the mycotoxin in apples and apple products. Strains of *P. expansum* and patulin were frequently found in apple, pear and other fresh fruits affected by blue mold rot, as well as in derived juices and jams (Logrieco et al., 2003). Contamination with *P. expansum* and patulin commonly occurs during postharvest and in storage. Patulin is most commonly found in apples and contributes to the vast majority of patulin in the human diet (FDA, 2001). In terms of carcinogenicity, patulin is classified as being of 'no evidence in humans' (group 3). The toxin exhibits mutagenic properties and is strongly antibiotic to many human disease-causing bacteria, including *Mycobacterium tuberculosis*. Patulin is considered to be immunosuppressive, and produces ulceration, congestion, and hemorrhagic lesions. The mode of action of patulin is that it affects enzyme alkylation and inhibits DNA translation/transcription. The co-occurrence of mycotoxins (for example, ochratoxin A, also produced by P. *expansum*) can enhance toxicity, and non-toxic fungal metabolites can act synergistically with toxic ones (Paterson et al., 2004). Patulin is heat resistant, particularly in an acidic environment and, as a result, the toxin persists in shelf-stable apple products such as juices, concentrates, jellies, baby foods etc. Since the juices and baby foods are consumed by infants and children, many countries including the USA and Canada, and the European Union have set regulatory limits of patulin as $50\,\mu g\ L^{-1}$ or less in apple juices, including the apple juice portion of any juice blends or drinks, and unfermented apple cider (FDA, 2001; HealthCanada, 2013).

TAXONOMY AND MORPHOLOGY

The genus *Penicillium* was first described in the literature by John H.F. Link (1767–1851), a mycologist and a professor of Botany in Rostock, Breslau and Berlin in 1809. In the paper, Link described three species of genus *Penicillium*, *P. candidum* (Link), *P. expansum* (Link), and *P. glaucum* (Link), all of which produced brush-like conidiophores (Link, 1809). The species

P. expansum, which causes blue mold in apple was selected as the type species, to which the name of the species is permanently linked.

Since its first description by Link in 1809, researchers have been refining the taxonomy of the genus *Penicillium*. Historically, the identification of species in the genus *Penicillium* relied on the morphorological characteristics, the branching of the conidiophores. Conidiophores are the dense brush-like conidia-bearing structures. The conidiophores are simple or branched and are terminated by clusters of flask-shaped phialides. The conidia are produced in dry chains from the tips of the phialides, with the youngest spore at the base of the chain, and are nearly always green. Based on the type of spore-bearing systems (penicilli) *P. expansum* was classified as terverticillate, branches that bear a second order of branches, bearing in turn a cluster of phialides. A polyphasic approach to the taxonomy based on morphology, growth pattern, ecology, extrolites, and partial β-tubulin sequences, placed *P. expansum*, *P. marinum* and *P. sclerotigenum* in the section *penicillium* and series *expansa* (Samson and Pitt, 2000; Frisvad and Samson, 2004). Partial *RPB1*, *RPB2* (RNA polymerase II genes), *Tsr1* (putative ribosome biogenesis protein) and *Cct8* (putative chaperonin complex component TCP-1) gene sequences were applied to distinguish three families in Trichocomacea (Houbraken and Samson, 2011). Identification of species of *Penicillium*, including *P. expansum* that produce mycotoxins is critical and, in spite of the availability of modern methods, its identification remains difficult (Paterson et al., 2004). Proficiency testing is rare and conventional identification methods do not inform reliably as to whether mycotoxins were detected/produced. Paterson et al. (2004) described a solution which consists of identifying a *Penicillium* strain as terverticillate and then undertaking mycotoxin analysis. As the new technologies in DNA and secondary metabolite research characterize new identifiers, the refining of the classification of *Penicillium* continues (Table 6.1).

Table 6.1 Taxonomy of *Penicillium expansum* (Link)

Kingdom	Fungi
Phylum	Ascomycota
Order	Eurotiales
Class	Eurotiomycetes
Family	Trichocomaeae
Genus	*Penicillium*
Subgenus	*Penicillium*
Series	*expansa*
Speices	*expansum*

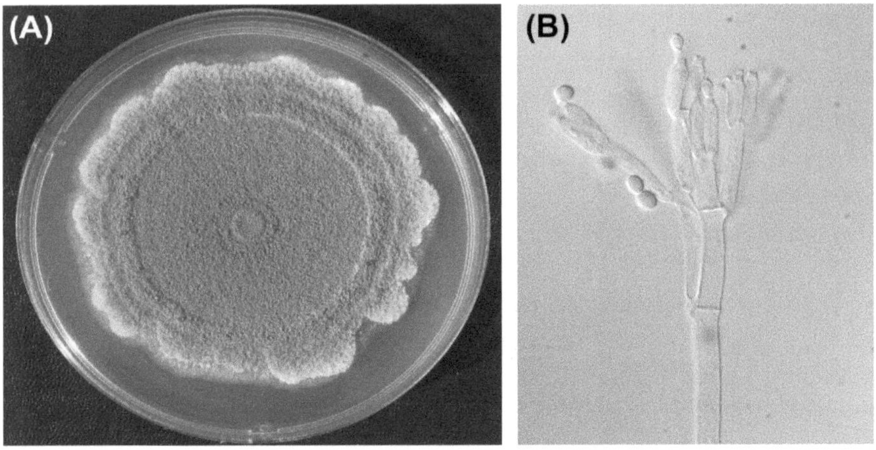

Figure 6.1 (A) Colony of *Penicillum expansum* on potato dextrose agar medium, and (B) photomicrograph of conidiophore *of P. expansum* showing terverticillate conidial heads and conidia. *(Photo (B) courtesy K. Seifert, ECORC, AAFC, Ottawa)* ×*100.*

Morphological characterization of *P. expansum* was based on the macroscopic study, growth rate, color texture and topography of the colony using two standard media namely potato dextrose agar (PDA) and Czapek Dox and microscopic study, observation of hyphae and conidia under a light microscope. Molecular characterization was based on molecular markers, (RAPD and ITS) and DNA sequence of β-tubulin gene (Peterson, 2000; Czerwinski et al., 2009). On potato dextrose agar at 25°C, the colonies of *P. expansum* grow rapidly forming a velvety carpet of surface filamentous mycelium up to 2 mm deep. The colonies are initially white becoming dull yellow green to blue green in 3–4 days (Fig. 6.1A). The plate reverse is usually pale to yellowish. As to micromorphology, the conidial heads are asymmetric and once or twice branched (terverticillate) and conidiophores are smooth, 400–700 μm long and with smooth elliptical conidia at 4–5 × 2.5–3.5 μm (Fig. 6.1B).

BIOLOGY OF *P. EXPANSUM* INFECTION PROCESS

Pathogen Presence

The microbiological flora on apples depends on the geographic area, climate, fungicide treatments, cultivars and the presence of competitive microorganisms on the fruit, fruit harvest methods, postharvest treatments and storage conditions. Species of *Penicillium* including *P. expansum* are ubiquitous soil inhabitants and are generally present on the decaying vegetative matter on the orchard floor or soil. Although the pathogen may accumulate as airborne spores on fruit surfaces during the growing season, it is likely that the majority of the propagules are carried in soil and debris on the fruit bins, equipment and spores are transmitted to wounds via water handling systems. Contamination can also occur from the air in the storage rooms. Some of the factors for a successful infection by a pathogen include time, wounds on the surface of the fruit and on the relative susceptibility of the fruit.

Spore germination is the initial step for fungal pathogens to infect host plants. The conidia are the asexual spores produced by actively growing mycelium of *P. expansum*. The conidia require O_2 or reduced partial pressures of O_2, relative humidity as in the form of a film of free moisture or moist environment on the fresh wound, and a source of metabolizable organic compounds in the wounds or natural secretions of the host (Sommer et al., 1992). The three distinct phases of the asexual life cycle of *Penicillium* include conidial germination, vegetative growth, and conidiation. Under favorable conditions, the conidium germinates and develops into an individual. In a study on the relationship between conidial enzymes and germination of *P. expansum*, it was shown that xylanase and pectinase activities were detected in the condia of the pathogen but not cellulose or amylase. The rate of conidial germination was dependent on the type of carbon source or the enzyme levels in conidia. Activities of xylanase and pectinase influenced the elongation of the germ tubes (Kimura et al., 2002; Kim et al., 2006). Li et al. (2010) have shown that ambient pH stress impaired protein synthesis and folding and inhibited spore germination in *P. expansum* (Li et al., 2010). pH values, 2.0 and 8.0 in growth media inhibited spore germination. Proteomic analysis demonstrated that 17 proteins were related to protein synthesis and folding in *P. expansum* and most of them were down-regulated at pH 2.0 and 8.0. Ambient pH affected intercellular pH and ATP level in conidia, as well.

Role of Wounds

Penicillium requires wounds to infect the host, and therefore is known as a wound pathogen. Unlike other fungi, *Penicillium* does not produce appressoria and therefore is unable to penetrate the intact epidermis of fruits and vegetables. The pathogen enters the fruit through the damaged skin caused by birds, insects or damaged areas near the lenticels (Amiri and Bompeix, 2005). The results from the sulfur dioxide test, which was used to study the number of lenticels, showed that the wounds on pome fruit occur during harvesting and wound parasites gain access into the fruit through open wounds. The rare reports of lenticel infection by *P. expansum* may probably occur through more susceptible larger, more open lenticels. *Penicillium expansum* was also reported to cause disease through stem ends (Rosenberger et al., 2006). The evaluation of 'Empire' apples artificially inoculated by placing conidia of *P. expansum* on the ends of stems and incubated in cold storage (1.1°C) or conrolled atmosphere (CA) storage (1.1°C, 1.6% O_2 and 2.2% CO_2), for 7–9 months showed that 27–47% of fruit developed decay during CA storage whereas less than 1% of similarly inoculated fruit decayed during cold-air storage.

Although the wounds provide opportunities for *P. expansum* infection, fruit's reaction to wounds can form a barrier to fungal infection. Aged wounds became more resistant to *P. expansum* infection than fresh wounds and an increase in tannins, lignins and other phonologic substances was observed in the cell walls adjacent to wounds in apple fruit (Vilanova et al., 2012). 'Red Delicious', 'Gala' and 'Empire' apple fruit inoculated 3–7 days after wounding had a lower infection rate than fresh, 0-day old wounds. Similarly, in 'Golden Delicious', 'Granny Smith', and 'Red Delicious' apple fruit, 1 or 3 day-old puncture wounds had less blue mold severity and incidence from *P. expansum* than in 0-day-old puncture wounds (Filonow, 2005).

Interaction between Pathogen Enzymes and Host Response

Multiple enzymes are involved in the degradation of different cell wall components and breakdown of host tissues. The plant cell wall is the first effective physical barrier against attack by pathogens and most phytopathogenic microorganisms produce enzymes that are capable of degrading cell wall polymers such as pectins and hemicelluloses. Polygalacturonase (PG), produced by phytopathogens, has been implicated in facilitating the invasion and colonization of host tissue, particularly in diseases characterized by tissue maceration or soft rot. Purified PG produced by *P. expansum* during postharvest decay of 'Anjou' pear was found to be different from the PG produced by the *P. solitum* pathogen in 'Golden Delicious' apples (Jurick Ii et al., 2010). The authors also found that the highest PG activity was observed on an apple pectin medium inoculated with *P. expansum*.

In fruit, polygalacturonase-inhibiting proteins (PGIP), which are induced in response to early stages of pathogen infection, prevent cell wall degradation, inhibit the action of PG and limit fungal growth and colonization (Yao et al., 1999). Compared to the uninoculated controls, very high PGIP transcript levels were detected in decayed areas and the tissue adjacent to the inoculation sites of *P. expansum* and *Botyritis cinerea* on apple fruit. In addition, the induction of PGIP in response to wounding and cold storage suggests that this enzyme may have multiple roles during fruit development and stress response. The PGIP proteins vary in the specificity of inhibition among the plant species.

Sánchez-Torres and González-Candelas (2003) reported that 18 genes showed a higher expression level during 'Golden Delicious' apple and *P. expansum* interaction than in the control apple tissues. Among the 12 genes expressed differentially, one fungal gene encoding an unknown protein was expressed during the infection process in apple tissue. Several upregulated *P. expansum* genes seem to mediate adaptive responses to the host environment. In the infected apple tissue, *P. expansum* secreted total d-gluconic acid (GLA). The analysis of two *P. expansum* glucose oxidase (GOX)-encoding genes, GOX1 and GOX2, showed that the GOX activity and GLA accumulation were strongly related to GOX2 expression, which increased with pH to a maximum at pH 7.0 (Barad et al., 2012). The authors have shown that pH modulation, mediated by d-gluconic acid (GLA) accumulation, is an important factor in generating the initial signal or signals for fungal development leading to host-tissue colonization by *P. expansum*.

Environmental Factors

Although the contamination of *P. expansum* can occur before harvest, it is commonly found postharvest and in storage and therefore, this section will focus on the environmental factors affecting the infection of fruit and vegetables in storage. Cold chain systems, in which produce is kept at a low temperature from harvest to retail, is an effective mechanism in managing postharvest diseases. Cooling of the produce immediately after harvest and low storage temperature are the two critical factors that reduce postharvest pathogen growth. All other control methods are sometimes described as supplemental to refrigeration (Sommer, 1985). Temperate fruits are stored at low temperatures, −1–4°C, for apples and −1.1–−0.5°C for pears. Unfortunately *P. expansum* can grow slowly at very low temperatures (Pitt and Hocking, 1997; Frisvad and Samson, 2004; Lahlali et al., 2005). Survival and growth of this pathogen at −1.1–4°C shows that additional control methods are required (Errampalli, 2006). Relative humidity over 95% in the storage environment is required to minimize the loss of moisture and prevent shrivelling of fresh produce during long-term storage (Kader, 2005). Studies show that 99–100% moisture is required for spore germination and successful infection by fungal pathogens. Air composition, oxygen (O_2) and carbon dioxide (CO_2) concentrations, in cold storage are some of the factors that affect postharvest blue mold. In general, increasing CO_2 levels over 5% and decreasing O_2 levels below 5% suppresses respiration, and maintains fruit quality and resistance (Sitton and Patterson, 1992). Oxygen and CO_2 ratios vary for cultivars of apple and pears.

AGRONOMIC AND ENVIRONMENTAL FACTORS FAVORING *P. EXPANSUM* INFECTION

Consumers prefer to purchase fruit according to its appearance, aroma, taste and shelf-life thus making fruit quality the most important factor. Market studies show that the flavor of the fruit is the quality parameter that brings the consumers back for more. Fruit quality parameters include color, juiciness, sweetness, flavor, sweet/acid balance, size and shelf-life. Postharvest quality of fruit is determined during its development on the tree or plant. Cultivar selection is a critical starting point and it is important to select the most suitable cultivar for a given location. Breeding programs and genetics are critical in developing new varieties which will behave differently in various locations under different irrigation and nutritional practices. Production locality is another agronomic factor to consider. Late spring frosts can damage flowers and fruit and it is important to avoid planting in areas where late frosts may occur after budburst (rainfall and temperature). Orchard or field management issues such as irrigation, plant nutrition, pruning and other factors will affect fruit quality and good agronomic practice is essential to maximize yield and quality. Application of the nutrient calcium (Ca) improves the strength of the fruit skin, which makes it less susceptible to postharvest injuries and, as a result, makes it less prone to invasion by *P. expansum*, a wound pathogen. Poor quality fruit with surface defects, such as wounds caused by insect or storm damage are the most common entry points for *P. expansum* in the orchard during the preharvest period. Improper fertilization, excessive rain or drought and poor soil conditions are some of the factors that contribute to mineral imbalances in fruit. Calcium is considered to be a critical element for fruit quality as this cation helps to maintain apple fruit firmness and decreases the incidence of physiological disorders such as water core, bitter pit and internal breakdown. Bramlage et al. (1985) reported that Ca was more important than magnesium, potassium, nitrogen, and phosphorus to maintain apple fruit quality. It has been demonstrated that foliar sprays of Ca during the growing season can reduce preharvest infection in apple fruit. It was shown that preharvest sprays of calcium chloride ($CaCl_2$) or other commercially available Ca mixtures did not influence firmness in 'McIntosh' apples (Bramlage et al., 1985). Similarly, preharvest $CaCl_2$ sprays do not affect the Ca content or fruit firmness of 'Wellspur Delicious' apples. On the contrary, 'Cox's Orange Pippin', 'Golden Delicious', and 'Delicious' apples receiving Ca foliar applications have been shown to be firmer than the respective nontreated apples (Peryea et al., 2007). It was suggested that combining preharvest application to fruit trees through foliar sprays and supplemental postharvest application of Ca is the best strategy to prevent losses due to low fruit Ca (Conway et al., 2002). A combination of a fungicide, pyrimethanil, with an additional application of $CaCl_2$ demonstrated a synergistic effect resulting in increased postharvest blue mold control (Pirgozliev et al., 2006a, 2006b).

POSTHARVEST FACTORS FAVORING *P. EXPANSUM* INFECTION

The physiological stage of maturity of fruit refers to the attainment of the final stage of biological function and development. The fruit is considered to be at a mature stage when, after harvesting, postharvest handling and storage, it will ripen to an acceptable quality. The fruits that are physiologically mature will continue ontogeny even if detached from the tree. In contrast, horticultural ripeness refers to fruit that is ready to eat. Physiological maturity at harvest is one of the important factors that determines storage life and final fruit quality for fruits in long-term storage, especially for apples and pears. Some of the quality standards required for fresh fruit include objective indices of maturity such as soluble solid concentration, titratable acidity, a ratio of soluble solid concentration and titratable acidity and flesh firmness. Fruit picked too early or too late is prone to fruit decay and infection by pathogens. Fruit ripening is associated with increased susceptibility to postharvest diseases. The blue mold pathogen, *P. expansum*, mostly affects climacteric fruit. Climacteric fruits continue to ripen after harvest. During the ripening process, the fruits emit ethylene along with an increased rate of respiration. Therefore, climacteric fruit that was harvested at an unripe stage will ripen to an acceptable quality. This characteristic enables the fruit to be stored for months, for example, apples can be stored for up to 12 months in controlled atmosphere storage. In contrast, nonclimacteric fruit do not exhibit a rapid increase in ethylene production or respiration during the ripening process and are harvested when they are ready to eat.

The maturity stage of fruit was found to be an important factor in resistance of fruit to *P. expansum*. Immature fruit are generally highly resistant to the blue mold pathogen. A slower rate of blue mold lesion development was observed in immature 'Golden Smoothee' apples in comparison to the optimal and late harvest fruits both at 20 and 0°C storage temperatures. In immature apples, lignin accumulation in the tissues surrounding the *P. expansum* inoculated wound suggests a role of liginin in resistance to pathogen invasion (Vilanova et al., 2012). Similarly, immature 'D Anjou' pears were highly resistant to postharvest pathogens including *P. expansum* than pears at and near harvestable maturity (Sugar, 2002). Apples harvested 7 days after commercial harvest were more susceptible to *P. expansum* than apples harvested 7 days before commercial harvest (Torres et al., 2003). Based on the evaluation of tolerance to blue mold disease in 127 apple germplasm accessions and cultivars, it was shown that harvest date was the most important factor, as the firm fruit had small lesions and the fruit that lost firmness during storage had larger disease lesion size (Ahmadi-Afzadi et al., 2013).

CONTROL

Conventional Alternatives

Chemical Fungicides

Chemical fungicides have been used to control postharvest diseases of fruits and vegetables, to prevent infection and suppress development of pathogens on the fruit or vegetable surface, to increase their quality and yield as well as to extend storage life. Application of fungicides may begin in the orchard or in the field, with some applications closer to harvest (preharvest) or may be applied after harvest (postharvest) only. Postharvest treatment may be applied immediately after harvest and prior to storage, or immediately before fruit are packed into boxes for final storage or sale.

Preharvest Fungicides for Control of Blue Mold

Research indicates that there is an increasing interest in the application of antifungal preharvest treatments to reduce field populations of *P. expansum* and/or induce resistance as a part of integrated disease management (IDM). The preharvest treatments may reduce the fungal inoculum on the surface of the fruit, residues of fungicides on the fruit surface may protect the wounds that are created at harvest and handling and may control latent infections (Xiao et al., 2011). A preharvest application of cyprodinil (belongs to the anilinopyramidine class of fungicides), 3 weeks prior to harvest controlled postharvest blue mold in stored apples. In another study, preharvest application of pyrimethanil, an anilinopyramidine, at 800 μg mL^{-1}, applied 20 and 14 days before harvest controlled postharvest blue and gray mold on wounded and inoculated 'Jonagold' and 'Gala' apples, respectively, for up to 6 months in storage (Sholberg et al., 2005a). Pirgozliev et al. (2006a) demonstrated that preharvest spray application of pyrimethanil at 800 μg mL^{-1} on 'McIntosh' and 'Empire' apples was more effective in reducing blue mold incidence caused by *P. expansum* for up to 6 months in cold storage, if it was applied 10 days before harvest rather than 28 days (Pirgozliev et al., 2006b). A significant control of blue mold was observed in pyrimethanil treated apples for up to 6 months in cold storage. Combining pyrimethanil with an additional application of $CaCl_2$ demonstrated a synergistic effect resulting in increased postharvest disease control. Preharvest application of pyrimethanil gave control of postharvest blue mold in wounded and inoculated 'Empire' apples for 4 months in cold storage (Errampalli and Wainman, 2010d, 2011). Pristine, a premix of pyraclostrobin and boscalid, is another reduced risk fungicide registered for preharvest application on apples. The two components of the fungicide, pyraclostrobin and boscalid, have different modes of action and this fungicide has a broad-spectrum activity. The preharvest application of pristine controlled blue mold in 'Fuji' and 'Red Delicious' apples (Xiao and Boal, 2009). An application of pyrimethanil preharvest spray gave significant control of blue mold in 'Empire' apples for up to 4 months compared to untreated controls, however, a combination of preharvest pristine treatment with a postharvest application of fludioxonil or pyrimethanil gave complete control of the disease for up to 6 months in storage (Errampalli and Wainman, 2009, 2010d, 2011).

Postharvest Fungicides

Postharvest dips/drenches with fungicides are used to protect fruits and vegetables against postharvest diseases. Postharvest diseases of fruits and vegetables have been controlled by synthetic postharvest fungicides (Eckert and Ogawa, 1988). Fungicides were more effective when pathogen populations were low, environmental conditions for infection development were least favorable and the fruit had some intrinsic resistance to infection. The benzimidazole derivatives were the first systemic fungicides that became available for the control of postharvest pathogens (Delp, 1988). The fungicide, 2(4-thiazol-4-yl) benzimidazole (thiabendazole, TBZ), belonging to the chemical class of benzimidazoles, remains stable in an aqueous suspension throughout the storage life of treated fruit. TBZ is one of the few fungicides approved for postharvest use on apples, pears, and citrus as a drench, dip, or line spray. Following harvest, prestorage chemical control is commonly applied to reduce the incidence of blue mold in apples that are stored for long periods of time in CA storages in Canada, Australia, and the USA (Eckert and Ogawa, 1988). In commercial operations, some of the cultivars of apples destined for long-term storage are also treated with an antioxidant and scald inhibitor, diphenylamine (DPA). In pome fruits, postharvest fungicide treatments are applied as a postharvest and prestorage drench, with or without the DPA. Following storage, fungicides are applied as a spray just before packing and sometimes the fungicide is combined with wax. The effectiveness of benzimidazole fungicides, including TBZ, promoted increased fruit production because long-term storage became less risky (Sugar, 2002). In a 2-year study, TBZ applied to fruit without artificial wounding or inoculation effectively reduced decay when applied within 3 weeks or 6 weeks (Sugar and Basile, 2008). The TBZ controlled blue mold in apples for up to 6 months in cold storage (Eckert and Ogawa, 1988; Errampalli, 2006). The TBZ fungicide was highly effective in controlling blue mold caused by TBZ-sensitive isolates of *P. expansum* on 'Empire' apples for up to 105 days in CA storage, however, it was not effective against TBZ-resistant *P. expansum* (Errampalli, 2006). Three fungicides, imazalil, TBZ, benomyl, controlled postharvest blue mold in apples (Louzeiro et al., 2002).

The extensive application of benzimidazole fungicides in the orchard and in packing houses over many years has led to the development of fungicide resistance in major postharvest pathogens of pome fruit. In the past three decades, resistance to benomyl and TBZ in *Penicillium* spp. has been reported in apple packing houses in the USA and in Canada (Rosenberger and Meyer, 1979; Chand-Goyal and Spotts, 1997; Errampalli, 2004, 2006; Lozynska et al., 2010; Karaoglanidis et al., 2011). Resistance to TBZ in *P. expansum* ranged between 25% in the early packing season to about 75% in the later packing season in the packing houses in Oregon and Washington (Chand-Goyal and Spotts, 1997). DPA provided partial control of benzimidazole resistant isolates, but some isolates have developed resistance to both TBZ and DPA (Errampalli et al., 2006). Resistance to benzimidazole fungicide in *P. expansum* is linked to the mutation at codon 198 of the β-tubulin gene in the pathogen (Pianzzola et al., 2004; Sholberg et al., 2005b; Czerwinski et al., 2009; Malandrakis et al., 2013; Yin and Xiao, 2013).

New Postharvest Fungicides

The development of resistance in *P. expansum* to both TBZ and DPA has led to searches for alternative chemical control strategies. The detection of the benzimidazole resistance was cited as a critical factor in initiating research into developing broad-spectrum activity of fungicides that are less prone to cause fungicide resistance in plant pathogens and alternative control measures, such as biological as well as integrated diseases control measures (Sugar et al., 1998; Sugar, 2002). The main aim of the fungicide resistance management strategy is to reduce or eliminate fungicide resistant fungal populations by alternating or combining fungicides with unrelated modes of action so that the resistant populations selected by one fungicide may be supressed by others. In 2004, two new reduced risk postharvest fungicides were registered for the control of postharvest diseases of fruits, including pome fruit in the USA and subsequently in other countries. Fludioxonil, belonging to the chemical class of phenylpyrroles and pyrimethanil, belonging to an anilinopyramidine class, are rated with a reduced risk toxicology by the United States Environmental Protection Agency (USEPA, 2006), and were shown to exhibit broad-spectrum activity against plant pathogenic fungi. Both fungicides were found to be effective against blue mold of apples (Errampalli, 2004, 2006; Errampali and Crnko, 2004; Errampalli et al., 2005b, 2006; Sholberg et al., 2005a; Li and Xiao, 2008; Karaoglanidis et al., 2011). Fludioxonil and pyrimethanil are particularly useful where the existence of *P. expansum* strains resistant to TBZ was observed, since these fungicides have modes of action different from that of TBZ (Errampalli and Crnko, 2004; Xiao and Kim, 2010). Application of TBZ at harvest followed by an application of fludioxonil 3 weeks later was superior to the application of TBZ at harvest alone (Sugar and Basile, 2008). Research is being conducted on two fungicides for the postharvest control of blue mold. A new fungicide, Philabuster, 400 SC, which contains 200 g L^{-1} imazalil and 200 g L^{-1} pyrimethanil was reported to control blue mold in apples and pears (Vorstermans et al., 2005). A triazole fungicide, difenoconazole was found to be effective against blue mold in 'McIntosh', 'Empire', and 'Silken' for up to 140 days in cold storage (Errampalli and Wainman, 2010a, b, c). Today, there are very few fungicides registered for the control of postharvest blue mold. There is a need for continued research in identifying and developing fungicides with different modes of action for the effective management of postharvest blue mold in fruits and vegetables.

Host Resistance

The use of resistant plants to control plant diseases can be regarded as a form of biological control. However, because of its importance, host resistance will be considered in this section separately from microbial antagonists. The use of resistant varieties is an important strategy in IDM. Breeding for disease and insect resistance has always been considered an important part of apple breeding programs, but the focus has been on field production and little attention has been devoted to postharvest disease resistance and associated decays (Janick et al., 1996). Very few reports on the variability of resistance to postharvest decays in apple cultivars have been published. Fruit cultivars differ in their susceptibility to blue mold disease caused by *P. expansum*. Baker and Heald (1936) reported that 'Jonathan' and 'Delicious' are the most susceptible to blue mold while 'Winesap' is the least susceptible of the three. 'McIntosh', 'Femeuse', 'York Imperial Stayman Winesap', 'Winesap', 'Rome Beauty', and 'Baldwin' were affected by blue mold. 'Winter Nelis', 'd Anjou' and 'Cornice' pears are highly susceptible while 'Columbia Doyenne', 'D Alencin' and 'Seckel' were reported to be immune to blue mold. 'Royal Gala' and 'Granny Smith' apples were resistant to *P. expansum* and 'Fuji' and 'Jonagold' apple were generally susceptible to blue mold disease and the mean blue mold infection severity values ranged from 34.5 for Royal Gala to 62.5 for 'Fuji' apple (Spotts et al., 1999). Because generally *P. expansum* requires a wound to initiate infection, resistance of the epidermis to breakage may be an important factor in resistance of apple cultivars to blue mold rot. Of the 83 apple accessions with center of origin in Kazakhstan, two accessions were classified as immune (no decay at 10^3 and 10^4 ml L^{-1}), four as resistant (no decay at 10^3 ml L^{-1}), 53 as moderately resistant (lesions less than 10 mm at 10^3 ml L^{-1}), and the remaining 24 as susceptible to *P. expansum* and blue mold disease (Janisiewicz et al., 2008). Some of the reports indicate differences in susceptibility but relatively little on actual resistance to postharvest decay was published (Spotts et al., 1999; Janisiewicz and Peterson, 2004). In a cold storage study, at 1°C for 2 months, decay incidence was 0% on 'Pink Lady', 8.3% on 'Ace Spur Delicious', and 41% on 'Empire' on fruit with stem pulls (Janisiewicz and Peterson, 2004). In 'Starking' apples, isolate NRRL 2034 produced blue mold disease and also patulin at 3, 6 and 17°C, but only trace amounts of patulin were detected at 0 or 25°C (Paster et al., 1995). 'Fuji' and 'Golden Delicious' apple fruits were susceptible and 'Granny Smith' and 'Red Delicious' were resistant to blue mold (Konstantinou et al., 2011). It was suggested that even a slight improvement in fruit resistance to postharvest diseases could increase the effectiveness of alternatives to postharvest fungicides and result in an additive or even synergistic effect in controlling postharvest diseases in apples (Janisiewicz et al., 2008).

Sanitizers

The aim of sanitary practices is to reduce the supply of pathogen spores to the lowest possible point for any given environment. Sanitary practices during harvest are critical in controlling postharvest diseases of fruits and vegetables (Spotts and Cervantes, 1994; Paster et al., 1995). The spores of *Penicillium* easily survive from season to season on contaminated storage bins in long-term storages, where the fungus can grow and produce numerous amounts of spores. Contamination of packing and storage bins with orchard soil and rotting fruit from the previous season that adhered to the bottom of the storage bins can be reduced by washing and sanitizing bins before use. Steam was found to be effective on wooden and plastic bins. Sanitizing compounds such as chlorine compounds, sodium-o-phenylphenate (SOPP) and quaternary ammonia compounds were also effective in reducing the *Penicilum* spore loads (Spotts and Cervantes, 1994; Sholberg and Conway, 2004). Fallen fruit in the orchard must be discarded as significantly higher levels of patulin were detected in cider made from apples fallen on the ground as compared to the fruit harvested from the trees (Jackson et al., 2003). It was suggested that fallen fruits might have cracks and wounds on the apple fruit surface that enabled the invasion of *P. expansum* spores from the soil.

Non-Conventional Alternatives

Physical Control Methods

Heat Treatment

Heat treatment is effective in sanitizing the fruit and enhancing the wound curing process. It has the added benefit of improving fruit color but does not lead to softening, since it inhibits the synthesis of cell wall hydrolytic enzymes in the apple fruit and reduces ethylene production (Lurie et al., 1998). Hot water rinsing and brushing at 55°C for 15 s significantly reduced decay development in *P. expansum*-inoculated apple fruit after 4 weeks at 20°C, or in naturally infected fruit after prolonged storage of 4 months at 1°C plus 10 days at 20°C (Fallik et al., 2001). Heat treatment controlled postharvest decay in pears (Spotts and Chen, 1987). Although the heat treatment provides control against postharvest decay, the effect is not the same as that of fungicides (Barkai-Golan and Phillips, 1991).

Irradiation

Low doses of gamma irradiation were used for insect disinfestation, extension of shelf-life and disease control in fruit and vegetables in several countries. Moy (1983) reviewed the potential of irradiation for disease control. Gamma irradiation may be used as a single treatment for quarantine purposes or in combination with other treatments such as high- or low-temperature storage to achieve the desired efficacy. Partial decay control was reported in irradiated mangoes (Johnson et al., 1990). Currently, the USFDA allows 1.0 kGy or less on fruits and vegetables (FDA, 1999). Blue mold disease incidence in 'Fuji' and 'Granny Smith' apples caused by *P. expansum* was reduced after gamma irradiation exposure between 0.30 and 0.90 kGy (Drake et al., 1999). Gamma radiation had no effect on the decay of 'Anjou' pear fruit naturally infected with *P. expansum* and *B. cinerea*, however, a reduction in decay was observed in naturally infected 'Bosc' pear fruits treated with 0.90 kGy. The loss of firmness due to irradiation was cultivar dependent and no loss of titratable acidity was observed 'Fuji' or 'Granny Smith' apples (Drake et al., 1999). Foaad et al. (2002) demonstrated that a dose of gamma radiation at 1.0–3.0 kGy inhibited the growth of *P. expansum* and patulin production. Morphological changes, the disappearance of phialids and the loss of ability to form conidia were observed in mycelium treated with 1.0–3.0 kGy and higher. *In vivo* study on pre- and postinoculation of both irradiated and non-irradiated fungal spores on irradiated and non-irradiated apples showed that radiation was effective on the apple mycoflora but not on the apple itself. Low irradiation doses (0.3 and 0.6 kGy) combined with cold storage minimized quality losses in *P. expansum* inoculated 'Red Delicious' apples during a 9-month storage period and at a dose range of 0.9–1.2 kGy, a significant decrease in phenolic content and antioxidant activity was observed (Mostafavi et al., 2012). Different morphological mutations were detected in the conidia of *P. expansum* isolate PO88 that were exposed to UV radiation for 30 min and some of the conidia repaired after the exposure (Ulevičius et al., 2008). The new irradiation technology, pulsed light (pulsed UV light, pulsed white light), which uses short-time pulses of intense broad spectrum, rich in UV-C light was introduced to decontaminate fruits and vegetables (Gómez-López et al., 2007).

Complementary Physical Methods

Cold storage, CA storage or modified atmosphere (MA) technologies considerably reduce the metabolic activity of the host, delay senescence, provide fungistatic effect by delaying the growth and development of postharvest pathogens and maintain fruit quality for a longer duration than storage at ambient temperature and can be considered as complementary methods for the management of postharvest diseases of fruits and vegetables. In CA storage, generally, O_2 is decreased from the normal 21% in air to 1–3% and CO_2 is increased from 0.04% to 1.5– 5% (Kupferman, 1992). Low oxygen (LO) and ultralow oxygen (U-LO) conditions refer to both the percentage of O_2 and CO_2 in balance with N_2, and O_2/CO_2 partial pressures are 3/3% for LO conditions and 3/1% for U-LO condition. Sitton and Patterson (1992) reported that *P. expansum* is resistant to high CO_2 concentrations and that even at concentrations of 12% the pathogen is not completely inhibited. It was reported that *P. expansum* has a very low requirement for O_2 and is virtually unaffected by levels of O_2 as low as 2.1% (Pitt and Hocking, 1997). The growth of *P. expansum* was not reduced under CA conditions (Paster et al., 1995; Errampalli, 2006). After 2 months of CA storage, the blue mold lesions on apples in U-LO storage were larger than those in LO (Morales et al., 2007). Similarly, Nunes et al. (2002b) also reported larger lesion diameters in apples stored during 3 months under U-LO conditions. Storage of *P. expansum* inoculated apples in modified atmosphere bags controlled the progress of fruit decay and markedly controlled patulin accumulation (Suzaki et al., 2008).

Biological Control

Biological control of blue mold can be achieved by using microorganisms that are antagonistic to pathogen growth and development on host tissue. Over the past 25 years substantial progress has been made in developing antagonistic microorganisms for the control of postharvest diseases (Droby et al., 2009). Different strains of yeasts, e.g. *Candida ciferrii* (Vero et al., 2002), *C. guilliermondii* (Xu and Du, 2012), *C. oleophila* (Sugar and Spotts, 1999), *C. membranifaciens* (Farahani et al., 2012), *C. sake* (Teixidó et al., 1998; Nunes et al., 2002a), *Cryptococcus laurentii* (Yao and Tian, 2005), *C. infirmominiatus* (Chand-Goyal and Spotts, 1997), *Metschnikowia andauensis* (Manso and Nunes, 2011), *M. fructicola* (Spadaro et al., 2013), *M. pulcherrima* (Janisiewicz, 2006), *Kluyveromyces lactis* (Taczman-Brückner et al., 2005), *Pichia membranaefaciens* (Xu and Tian, 2008), *Rhodotorula glutinis* (Zhang et al., 2009), *Rhodosporidium kratochvilovae* (Lima et al., 2011), *Rhodotorula mucilaginosa* (Li et al., 2011) *Saccharomyces cerevisiae* (Scherm et al., 2003), *Torulaspora delbrueckii* (Ebrahimi et al., 2012a), bacteria, e.g. *Bacillus subtilis* (Lee et al., 2012), *Pantoea agglomerans* (Ritte et al., 2002), *Pseudomonas syringae* (Janisiewicz and Jeffers, 1997) and filamentous fungi e.g. *Aureobasidium pullulans* (Ippolito et al., 2000), *Burkholderia gladioli* (Scuderi et al., 2009), *Trichoderma virens* (Bordbar et al., 2010) have been identified and characterized for the control of postharvest diseases, including blue mold caused by *P. expansum* in fruits and vegetables. In a study, three yeast and one bacterial biocontrol agents (BCA) were reported to reduce decay in pear wounds inoculated with *P. expansum* for up to 14 days after inoculation with the pathogen, but were found to be ineffective when applied at 28 days after inoculation (Sugar and Basile, 2008). In *in vivo* assays, BCA *Pseudomonas syringae* significantly reduced blue mold caused by *P. expansum* in cold storage (Errampalli and Brubacher, 2006). Several researchers have published review articles in the area of biological control of postharvest diseases of fruits and vegetables (El Ghaouth et al., 2002; Janisiewicz and Korsten, 2002; Droby et al., 2003; Zhou et al., 2007; Sharma et al., 2009; Nunes, 2012).

In spite of a large volume of research published about postharvest control of blue mold, the commercial use of these products is limited. By the early 2010s, very few products were registered for use against postharvest diseases of pome fruits: Aspire (Ecogen, Inc., Langhorne, PA) containing *Candida oleophila* I-182; Yield Plus (Anchor Yeast, Cape Town, South Africa) containing *Cryptococcus albidus*; Biosave-10LP 110 (Village Farms LLC, FL), containing the bacterium, *Pseudomonas syringae*.

Natural Compounds

Plant Extracts, Essential Oils and Volatiles

Metabolites produced by plants are a promising alternative to synthetic fungicides because they are considered relatively safe, they are easily decomposed, environmentally friendly and non-phytotoxic (Cabral et al., 2013). Plants and plant extracts can play a significant role in several domains of agriculture and are recognized to be useful in food preservation, preventing molds and pest attacks that affect the quality of fresh fruits and vegetables. Many of the plant extracts are known to contain alkaloids, flavonoids, isoflavonoids, tannins, cumarins, glycosides, terpens, phenylpropanes, and organic acids. Essential oils are aromatic oily liquids, obtained by hydrodistillation from whole tissues or seeds of plants and are usually mixtures of several components. The presence of a variety of antimicrobial compounds in plant extracts confers them advantages such as having different modes of actions depending on the compound involved and thus hindering the development of resistance by the pathogen.

A lot of research has been conducted in this area in recent years. *Penicillium* spp. including *P. expansum* are one of the three commonly used fungal genera to test plant extracts and essentials oils. The antifungal properties of plant extracts from various plants have proven totally or partially to inhibit fungal spore germination, germ tube elongation and mycelial growth *in vitro* and *in vivo*. Powders, gels, and aqueous and organic solvent extracts of plants from different origins are reported to inhibit mycelial growth of *P. expansum* and, in some cases, reduce blue mold disease. Table 6.2 summarizes the results presented in this section.

The leaf extracts of *Adiantum capillus-veneris* were reported to be the most effective as a pre-infection dip treatment, followed by *Thuja occidentalis* and *Eucalyptus globulus* against *P. expansum* in 'Golden Delicious' apples and *Azadirachta indica* exhibited curative action against blue mold disease (Singh and Geeta, 2007). These bioactive compounds from 'Lampo' chili pepper (*Capsicum annum*) were found to be effective against *P. expansum* (Nazzaro et al., 2009). Capsaicin, an active component in *Capsicum* spp., is responsible for the pungent flavor of hot pepper, and its analogs exhibited antimicrobial activity against plant pathogenic bacteria and fungi and the minimum inhibitory concentration of capsaicin against *P. expansum* was 122.16 μg mL^{-1} (Luo et al., 2011). In 'Fuji' apples, blue mold disease incidence was not reduced by the capsaicin treatment, however, the growth of *P. expansum* was delayed for the first 14 days of the *in vivo* assay (Fieira et al., 2013). In *in vitro* studies, semicrude or crude derivatives of cayenne pepper *C. frutescens* completely inhibited conidial germination of *P. expansum* and the crude extract reduced blue mold on 'Golden Delicious' apples (Bautista-Baños and De Lucca, 2004). Aqueous leaf extracts of seven common invasive plant species in South Africa, *Cestrum laevigatum*, *Nicotiana glauca*, *Solanum mauritianum*, *Lantana camara*, *Datura stramonium*, *Ricinus communis* and *Campuloclinium macrocephalum*, at a concentration of 0.84 mg mL^{-1} were found to be most effective against *P. expansum in vitro* (Mdee et al., 2009). Extracts of grape pomace, a by-product of grape from the food processing industry, are known to have a high total phenol content and antioxidant capacity and have exhibited antifungal activity against *P. expansum* (Corrales et al., 2010). Plant extracts from garlic, thyme, lavender, ginkgo and dandelion, extracted with dichloromethane, exhibited antifungal activity against *P. expansum*, and garlic extracts at a concentration of 50 μL L^{-1} was effective on apple fruit against blue mold after 6 days and the vapor treatment of inoculated apples with garlic extract at a concentration of 1 mL L^{-1} with 72 h of exposure provided blue mold disease control (Ikeura et al., 2011). The ethyl acetate extract of seeds of atung (*Parinarium glaberrimum*) exhibited strong

Table 6.2 Plant Extracts and Essential Oils Tested for Antifungal Capacity Against *Penicillium Expansum In Vitro* or *In Vivo*

Plant (Scientific Name)	Plant (Common Name)	Fungus/Crop	References
Adiantum capillus-veneris	Maidenhair fern	*P. expansum*/apples	Singh and Geetha, 2007
Allium sativum	Garlic	*P. expansum*/apples	Ikeura et al., 2011
Anemarrhena asphodeloides	Zhi mu	*P. expansum*/apples	Paik and Chung, 1997
Azadirachta indica	Neem	*P. expansum*/apples	Singh and Geetha, 2007
Calotropis procera	Apple of Sodom, milk weed	*P. expansum*	Varahalarao and Naidu Chandrasekhar, 2010
Campuloclinium macrocephalum	Pom pom weed	*P. expansum*	Mdee et al., 2009
Capsicum spp., including *Capsicum annum*	Chili pepper	*P. expansum*/apples	Luo et al., 2011; Nazzaro et al., 2009; Fieira et al., 2013
Capsicum frutescens	Chili pepper	*P. expansum*/apples	Bautista-Baños and De Luca, 2004
Carum carvi	Caraway	*P. expansum*	Mohammadifar et al., 2012
Caryophillum aromaticus	N/A[a]	*P. expansum*	Mohammadifar et al., 2012
Cestrum laevigatum	Inkberry	*P. expansum*	Mdee et al., 2009
Carum carvi	Caraway	*P. expansum*	Zabka et al., 2009
Cinnamomum zeylanicum	Cinnamon	*P. expansum*	Čvek et al., 2010
Citrus limum	Lemon	*P. expansum*	Venturini et al., 2002
Coptis japonica	Japanese gold-thread	*P. expansum*/apples	Paik and Chung, 1997
Coriandrum sativum	Coriander	*P. expansum*	Matos et al., 2011
Cuminum cyminum	Cumin	*P. expansum*	Mohammadifar et al., 2012
Cymbopogan nardus	Citronella grass	*P. expansum*	Zabka et al., 2009
Datura stramonium	Jimson weed or Datura	*P. expansum*	Mdee et al., 2009
Eucalyptus globulus	Eucalyptus	*P. expansum*/apples	Singh and Geetha, 2007
Ginko biloba	Ginko	*P. expansum*	Ikeura et al., 2011
Lantana camara	Shrub verbenas or lantanas	*P. expansum*	Mdee et al., 2009

Continued

Table 6.2 Plant Extracts and Essential Oils Tested for Antifungal Capacity Against *Penicillium expansum In Vitro* or *In Vivo*—cont'd

Plant (Scientific Name)	Plant (Common Name)	Fungus/Crop	References
Mentha spicata	Mint	*P. expansum*/ apples	Singh and Geetha, 2007
Mentha piperita	Peppermint	*P. expansum*	Mohammadifar et al., 2012
Mentha pulegium	Mint	*P. expansum*	Matos et al., 2011
Mucuna pruriens	Velvet bean	*P. expansum*	Mastan et al., 2009
Nicotiana glauca	Tree tobacco	*P. expansum*	Mdee et al., 2009
Ocimum sanctum	Basil	*P. expansum*/ apples	Singh and Geetha, 2007
Origanum vulgare	Oregano	*P. expansum*/ pears	Matos et al., 2011
Parinarium glaberrimum	Autung	*P. expansum*	Moniharapon and Hashinga, 2004
Pelargonium roseum	Geranium	*P. expansum*	Zabka et al., 2009
Pimenta dioica	All spice/ Jamaica pepper	*P. expansum*	Zabka et al., 2009
Ricinus communis	Castor oil plant	*P. expansum*	Mdee et al., 2009
Satureia hortensis	Summer savory	*P. expansum*	Mohammadifar et al., 2012
Solanum mauritianum	Woolly nightshade	*P. expansum*	Mdee et al., 2009
Taraxacum officinale	Dandelion	*P. expansum*	Ikeura et al., 2011
Thuja occidentalis	White cedar	*P. expansum*/ apples	Singh and Geetha, 2007
Thymus kotschyanus	Thyme	*P. expansum*/ pears	Marandi et al., 2011
Thymus vulgaris	Thyme	*P. expansum*	Camele et al., 2012; Ikeura et al., 2011; Sellamuthu et al., 2013;Venturini et al., 2002; Zabka et al., 2009
Trachyspermum ammi	Ajwon/bishop's weed	*P. expansum*	Mohammadifar et al., 2012
Vitis vinifera	Grape	*P. expansum*	Corrales et al., 2010
Zataria multiflora	Savory	*P. expansum*	Mohammadifar et al., 2012

[a]Not available.

antimicrobial activity against *P. expansum* (Moniharapon and Hashinaga, 2004). The methanol extracts from *Mucuna pruriens* leaves exhibited antimicrobial activity against different species of bacteria and fungi including *P. expansum in vitro* (Mastan et al., 2009). The methanol extracts of two medicinal plants, *Coptis japonica* and *Anemarrhena asphodeloides*, inhibited the mycelial growth by 51.6% and spore germination by 100% in *P. expansum* and *in vivo*, disease lesions were not formed by the fungi at 5°C up to 2 weeks after inoculation in apples (Paik and Chung, 1997). Organic solvent extracts of stem, leaves and flowers of *Calotropis procera*, a medicinal plant from India, were found to be effective against the mycelium of *P. expansum* (Varahalarao and Naidu Chandrashekar, 2010). Plant extracts and essentials from *Carum carvi*, *Cymbopogan nardus*, *Pelargonium roseum*, *Pimenta dioica* and *Thymus vulgaris* inhibited mycelial growth of *P. expansum* (Zabka et al., 2009).

The essential oils thymol, menthol, peppermint oil, citral, cinnamol etc., are some of the phytochemicals that have been studied for antifungal activity. The major compounds in thyme, peppermint and citronella oils were thymol, menthol and R(+)citronella, respectively. Thyme oil vapor (5 μL/plate) was found to be the most effective treatment as it completely inhibited the mycelial growth of all five pathogens including *P. expansum in vitro* (Sellamuthu et al., 2013). Thyme oil from *T. kotschyanus* completely inhibited fungal growth of *P. expansum* at concentrations above 200 μL L^{-1}, and in *in vivo* studies, the inoculated pear fruit sprayed with 0, 100, 300 and 500 μL L^{-1} concentrations of *T. kotschyanus* oil, decreased the disease severity and disease incidence of blue mold (Marandi et al., 2011). The essential oils of *Mentha spicata* and *Ocimum sanctum* at concentrations lower than 1.0% significantly reduced blue mold (Singh and Geeta, 2007). Cinnaman oil, carvacol, citral and thymol showed the greatest inhibitory activities against *P. expansum* (Venturini et al., 2002; Čvek et al., 2010; Camele et al., 2012). The extracts or oils of aromatic plants, *Origanum vulgare*, *Coriandrum sativum* and *Mentha pulegium*, have inhibited mycelial growth of *P. expansum* and the extracts of *O. vulgare* and the standard compounds corresponding to the main constituents of aromatic plants, caprilic acid, benzoic acid and pulegone, were found to be effective against blue mold in 'Rocha' pears (Matos et al., 2011). A concentration of 500 μL L^{-1} of essential oil from each of the *Trachyspermum ammi*, *Zataria multiflora*, *Satureia hortensis*, *Caryophillum aromaticus*, *Menthe piperita* medicinal plants from Iran, exhibited fungicidal effects and inhibited mycelial growth and reduced spore germination by 80–90% in *P. expansum* (Mohammadifar et al., 2012).

Fumigation of fruit with volatiles, naturally occurring compounds, can reduce or control microorganisms present on fruit and vegetables and potentially reduce postharvest diseases. Vapors of short-chain organic acids, acetic, formic and propianic acid controlled *P. expansum* on cherry and pome fruit (Sholberg, 1998). Similarly, acetic acid at 8 μL L^{-1} completely inhibited spore germination of *P. expansum*, and reduced blue mold incidence by 44%. However, it caused phytotoxicity on fruits, indicating a potential use of acetic acid for reduction of inoculum of postharvest pathogens in the disinfestation of chambers (Osorio et al., 2013). Hexanal vapor is one of the volatiles that shows promise against *P. expansum in vitro* and reduction of blue mold disease on fruit. Blue mold control, ranging from 50 to 98%, and a reduction of patulin content were achieved when 'Abate Fetel', 'Bartlett', 'Conference' and 'Kaiser' pears and 'Golden Delicious' and 'Royal Gala' apples were fumigated with vapors from 12.5 μL $L^{-1}$of trans-2-hexenal 24 h after pathogen inoculation (Neri et al., 2006a, b). Fan et al. (2006) showed that hexanal vapor at a concentration of 40 μmol L^{-1} for 24 h reduced spore viability of *P. expansum* by 94%. In cold storage, at 4°C, apples treated with 5–7 μmol L^{-1} hexanal vapor also showed a significant reduction in lesion incidence and severity.

Over the past decade, the once popular fumigant methyl bromide, found to be depleting ozone in the atmosphere, has been phased out in many countries. Since then a search for an effective fumigant has been initiated by many researchers. Biofumigation refers to the use of volatile producing organisms or biomass to kill plant pathogens. One example is *Muscodor albus*, an endophytic fungus, represented by a number of isolates from tropical tree and vine species in several of the world's rain forests, that produces volatile organic compounds (VOCs) with antibiotic activity. The use of *M. albus* to release VOCs to control postharvest blue mold and gray mold of apples was reported (Strobel et al., 2001; Mercier and Jiménez, 2004). The fungus was isolated from cinnamon tree bark and was found to be related to the Xylareacea in Ascomycetes. The fungus was reported to produce 28 VOCs, mainly alcohol, ester, ketone and acid derivatives which inhibit or kill many postharvest pathogens including *P. expansum*. The most abundant VOCs are 2-methyl-1-butanol (MB), butanoic acid, isobutyric acid (IBA), ethyl propionate and phenythyl alcohol. Fumigation of apples for 7 days with culture of *M. albus* grown on autoclaved grain gave complete control of blue mold and gray mold in wound-inoculated fruits. In this experiment, the *M. albus* was not in direct contact with apples (Mercier and Jiménez, 2004). Two major *M. albus* VOCs, IBA and MB, suppressed or killed mycelium of

P. expansum at a concentration of 45 μL L^{-1} and 100 μL L^{-1}, respectively (Mercier and Jiménez, 2004; Ramin et al., 2005). Isolates of *M. albus*, including E-6, showed chemical, biological and structural characteristics that make them potentially useful in medicine, agricultural and industrial applications (Strobel et al., 2007). In 2012, AgraQuest received EPA and CAL-EPA Registration of a natural fumigant based on an isolate of *M. albus*, a safe, effective alternative for methyl bromide.

Induced Resistance

Alternatives to fungicides in fruit disease control have arisen with the discovery of inducible resistance in fruit against infection by pathogens, and the induction of fruit resistance was evaluated as a promising approach for control of postharvest diseases (Terry and Joyce, 2004; Liu et al., 2005, 2007). In plants, induction can be triggered by elicitors that are biological, chemical or physical, generally with salicylic acid (SA), a phenolic compound that naturally exists in plants, as a signal molecular in signal transduction pathway for systemic-induced resistance. Various elicitor compounds, including Acibenzolar-S-methyl (ASM), 2R-3R butanediol, methyl jasmonate (MeJA), salicylic acid, harpin have shown to activate inducible defenses in plants.

Chitin and Chitosan

Chitin, a long-chain polymer of N-acetylglucosamine, is the second most abundant biopolymer renewable source in nature after cellulose and is a primary component of the cell walls of fungi, the exoskeletons of crustaceans, and insect cuticles. Chitosan is an N-acetylated derivative of the polysaccharide chitin. The unique physiological and biological properties of chitin and chiotosan make them candidates for a wide range of medical and industrial applications and a review article on the effects of chitin and its derivative chitosan on postharvest decay of fruits has been published (Zhang et al., 2011). Chitosan is reported to create a semipermeable film that delays fruit ripening and prolongs shelf-life and inhibits fungal decay by direct antifungal activity (Bautista-Baños et al., 2006; Romanazzi et al., 2007) and by induction of host resistance to pathogens. Chitosan has been approved as a biopesticide by the US Environmental Protection agency (US EPA) and as a food additive by the US Food and Drug Administration (US FDA). Chitosan directly inhibited spore germination and germ tube elongation of postharvest fruit pathogens including *P. expansum*, and *B. cinerea* (Liu et al., 2007). Screening of 15 different chitosan types for their effect on mycelial growth of *P. expansum* showed that their antifungal activity increased with decreasing degrees of acetylation, and those with low degrees of acetylation and high degrees of polymerization showed the highest inhibitory activity against *P. expansum* (Oliveira Junior et al., 2012). In another study, Liu et al. (2007) demonstrated that chitosan inhibited spore germination, germ tube elongation, and mycelial growth of *P. expansum* (isolated from tomato) *in vitro*, and significantly reduced blue mold caused by *P. expansum* in tomato fruit stored at 25 and 2°C. A significant increase in the activities of polyphenoloxidase (PPO) and peroxidase (POD), enhanced the content of phenolic compounds observed in chitosan treated tomato fruit, suggesting that chitosan may be associated with the direct fungitoxic property against *P. expansum* and *B. cinerea*, and the elicitation of biochemical defense responses in tomato fruit (Liu et al., 2007). A 2% concentration of oligochitosan inhibited mycelial growth of *P. expansum* and reduced blue mold caused by *P. expansum* in apricot fruit (Jia et al., 2012). Chitosan alone or with a BCA *C. laurentii* effectively inhibited blue mold in apple fruit after 7 days of incubation at 20°C (Yu et al., 2007). A significant reduction of blue mold disease incidence and lesion diameter was reported in pear fruit treated with chitin alone or in combination with *Cryptococcus laurentii* which was cultivated in nutrient yeast dextrose broth (NYDB) amended with 1.0% chitin (Yu et al., 2008). Preharvest and postharvest applications of chitosan treatment reduced blue mold in table grapes in storage (Romanazzi et al., 2007). After 16 days following inoculation, blue mold incidence in 'Fuji' apple fruits immersed in 0.5% chitosan was reduced by around 40% (Canaver and Di Piero, 2011). Chitin and chitosan, extracted from fungal mycelia of *Aspergillus niger* and *Mucor rouxii* significantly reduced lesions caused by *P. expansum* in harvested apples. (Wu et al., 2005).

Methyl Jasmonate

The elicitor, methyl jasmonate (MeJA) is a highly volatile compound and has identical activity in vapor and liquid phases. Plants produce jasmonic acid (JA) and MeJA in response to many biotic and abiotic stresses (particularly herbivory and wounding), which build up in the damaged parts of the plant (Creelman and Mullet, 1997). Exogenous application of MeJA showed no effect on spore germination or on mycelial growth of *P. expansum*. However, the elicitor has been shown to extend shelf-life and suppress or reduce postharvest plant diseases or decrease microbial contamination in vegetables, fruits and cut flowers. MeJA treatment significantly reduced the postharvest diseases including blue mold in peaches. Increased levels of expression of defense enzymes including chitinase, β-1,3-glucanase, phenylalanine ammonia-lyase (PAL), polyphenol oxidase (POD) and peroxidase were observed in fruit treated with MeJA, and the level of total phenolics in MeJA-treated fruit was also higher than that in control fruit (Jin et al., 2009). Yao and Tian (2005) showed that MeJA in combination with BCA *Cryptococcus laurentii* (Kuffer) reduced postharvest disease lesions in 'Bayuecui' peach fruit by 50% and induced defense-related enzymes, chitinase, β-1,3-glucanase, PAL and POD. A combination of MeJA at 200 μM and the BCA *R. glutinis* significantly reduced blue mold disease incidence by 75% and lesion diameter in pears than the application of either MeJA or *R. glutinis* (Zhang et al., 2009). In a study, the BCA yeast, *Torulaspora delbrueckii*, in combination with MeJA controlled 45% of blue mold, caused by *P. expansum,* in apples (Ebrahimi et al., 2012b). Fumigation with ethylene inhibitor, 1- methylcyclopropene (1-MCP) had no effect on blue mold in apples (Errampalli et al., 2012).

Harpin

Harpin is a protein produced by *Erwinia amylovora* that is involved in its pathogenesis of apple. Morphological and cellular changes in apple fruit treated with harpin suggest that it may trigger or intensify cellular responses in apples (De Capdeville et al., 2008). The flavonoid quercetin induced resistance against *P. expansum* by acting on the transcription level of genes involved in several distinct metabolic processes and reduced blue mold in 'Golden Delicious' apples (Sanzani et al., 2010). Of the 88 unique gene transcripts obtained, several sequences revealed high similarities with different classes of pathogenesis-related proteins (RNase-like PR10 and PR8), or with proteins expressed under stress conditions.

Acibenzolar-S-Methyl

Acibenzolar-S-methyl (ASM) activates treated plants to produce pathogenesis-related proteins (PR-proteins) in intercellular spaces and causes systemic acquired resistance since it mimics the role of salicylic acid (Kessmann et al., 1994; Lawton et al., 1996). The application of ASM, 48 h prior to pathogen inoculation was ineffective against blue mold (Spadaro et al., 2004). Similarly, the resistance inducers, β-aminobutyric acid and MeJA, did not reduce the disease in apples (Quaglia et al., 2011). The 'Yali' pear (*Pyrus bretschneideri*) fruit that had been harvested from the trees treated with field applications of ASM had lower blue mold incidence and disease severity in storage (Cao et al., 2005). Higher activities of defense enzymes, phenylalanine ammonia-lyase (PAL), chitinase and β-1,3-glucanase, were observed in fruit from trees treated with ASM than in controls at harvest and the ASM spray significantly increased the H_2O_2 level and glutathione reductase activity, but reduced activities of catalase and ascorbate peroxidase in young pears (Cao et al., 2005; Cao and Jiang, 2006). The authors have suggested that the enhancement of disease resistance in harvested 'Yali' pear fruit may be the result of multiple effects of several factors related to plant defenses induced by ASM sprays on trees during fruit growth.

Integrated Management of Postharvest Blue Mold

During the past few decades, substantial research has been conducted in identifying non-fungicidal methods to control blue mold. However, these methods when used alone do not provide adequate control for postharvest blue mold. Therefore, integration of one or more of the different strategies, physical, chemical and biological control methods, are required for integrated disease management (IDM) of blue mold. Research indicates that there is increasing interest in the application of a combination of antifungal pre- and postharvest treatments, fungicides, BCA, fungicides or biological control with physical methods, biological and other methods to reduce the inoculum of *P. expansum* and or induce resistance in fruit and vegetables as a part of IDM (Vorstermans et al., 2007; Giraud and Bompeix, 2012). The combination treatments may result in reduced length of treatments or reduced dose of the antifungal active ingredient and improved control of *P. expansum* and blue mold disease.

Integration of Heat Treatment with other Control Methods

The combination of hot water with other controls may increase effectiveness of heat treatments against *P. expansum* and blue mold control. Heat, Ca infiltration and BCA provided control of blue mold in apples (Conway et al., 1999). Prestorage fumigation with 1-MCP, heat treatment at 38°C air for 4 days and CA storage decreased decay severity caused by wound-inoculated *P. expansum* (Saftner et al., 1997, 2003). Heat treatment at 40°C for 24 h combined with CA storage showed the best protective effect against bruising and fungal decay on apples (Tahir et al., 2009). A combination of heat treatment at 38°C air for 4 days and chitosan coating of apple fruit showed complete control of blue mold in 'Gala' apples (Shao et al., 2010). Apples heat treated at 38°C air for 4 days and wound inoculated with *P. expansum*, followed by the treatment with the BCA *Pseudomnas syringae* had better blue mold control than the untreated control (Leverentz et al., 2000).

Heat treatment with 1-MCP in combination with heat at 38°C for 4 days and the yeast antagonist, *M. pulcherrima*, maximized the control of blue mold in 'Golden Delicious' apples (Janisiewicz et al., 2003). The heat treatment of apples with 20% ethanol or 5% sodium bicarbonate before the application of BCA, BIO126 strain of *M. pulcherrima* and the application of BIO126 in 0.1% sodium bicarbonate gave significant control of blue mold on apples (Spadaro et al., 2004). Heat treatment and sodium bicarbonate significantly improved the efficacy of the BCA against blue mold with storage at 23°C for 5 days. Heat treatment at 37°C for 2 days and *C. laurentii* gave effective control of blue mold (Zhang et al., 2007). In sweet cherry, hot water treatment, two BAC, *Kloeckera apiculata* and *Metschnikowia fructicola*, in combination with 0.24 M sodium bicarbonate gave 50% and 47% control of blue mold, respectively (Karabulut et al., 2005). A combination of 10% ethanol and hot water treatment at 60°C for 30 s controlled 86% of decay in sweet cherry (Karabulut et al., 2004).

Integration of Chemical Control with other Control Methods

Postharvest application of benzimidazole fungicides, benomyl or TBZ in combination with hot water treatment at 54.4°C/4 min drip gave complete control of blue mold in apples (Spalding et al., 1969). Similar results were achieved in peaches and nectarines treated with a combination treatment of benomyl or TBZ with hot water treatment of 46°C/2.5 min drip gave complete control of blue mold (Smith and Anderson, 1975). A 2 or 4-min dip treatment of fludioxonil at 300 μg mL^{-1} at 20°C or a combination of fludioxonil and heat treatment at 50°C gave 100% control of blue mold disease over 10 days of incubation on artificially inoculated 'Coscia' pears (Schirra et al., 2009). A combination of heat and fludioxonil treatment was found to be effective against postharvest blue mold on artificially inoculated 'Precoce di Fiorano', 'Coscia', and 'Spadona estiva' summer pears (Schirra et al., 2008). Integration of CA storage and two fungicides, fludioxonil or pyrimethanil, and a combination of fludioxonil and cyprodinil, were very effective in reducing blue mold disease incidence in 'McIntosh' and 'Empire' apples and 'Bosc' pears (Errampalli, 2004, 2006; Errampalli et al., 2006, 2007, 2012). A combination of $CaCl_2$ and benzimidazole fungicides benomyl, methyl-thiophanate, TBZ significantly reduced fungal decay in apples (Maouni et al., 2007). A higher incidence of decay was observed in wounded plus 1-MCP treated 'McIntosh' apples than in wounded only apples after 90 and 120 days in air and in standard CA storage (Errampalli et al., 2005a). But 1-MCP had no effect on the control of blue mold with fludioxonil or pyrimethanil in cold and CA storages (Errampalli et al., 2012). 1-MCP had a variable effect on decay in 'Empire' apples, where higher decay incidence was observed in wounded plus 1-MCP treated apples than in wounded only apples at 90 days, but the reverse trend was found at 120 days after treatment (Errampalli et al., 2005a).

Integration of Biological Control with Fungicides

Co-treatment of BCA and fungicides may be more efficient than treatment with BCA alone and the fungicide residues are significantly reduced. A concentration of 3×10^7 CFU mL^{-1} of *Pseudomonas syringae* MA-4 and 20 $\mu g\ L^{-1}$ cypridinil gave 90% control of blue mold in 'Northern Spy' apples (Zhou et al., 2002). A combination of *Pseudomnas syringae* at 1×10^8 CFU mL^{-1} and fungicide at 20 $\mu g\ L^{-1}$ of cyprodinil gave 100% control of blue mold in 'Empire' apples for up to 30 days in cold storage and a 90% control in the subsequent shelf-life study at 20°C for 7 days (Errampalli and Brubacher, 2006). The BCA *Rhodosporidium kratochvilovae* LS11 and *Cryptococcus laurentii* LS28 in combination with 25% of the label dose of commercial formulations of boscalid or cyprodinil reduced 80% of blue mold disease in apples for up to 7 days in storage (Lima et al., 2011). Wan and Tian (2005) showed that a combination of BCA yeast *Rhodotorula glutinus* and 1 mM of ammonium molybdate reduced blue mold by 70% in pear fruits. With the ever present possibility of emergence of fungicide resistant populations, it is prudent to integrate BCA and only the 'low risk' fungicides.

Integration of Biological Control with other Control Methods

Most of the microbial BCA, when used alone, do not provide the level of control needed during the postharvest phase, which is more than 95% (Droby et al., 2009). A combination of both physical and biological treatment techniques could enhance the control of postharvest diseases on apples and pears (Vorstermans et al., 2007). The BCA *Candida oleophila* in combination with $CaCl_2$ solution controlled blue mold disease of apples (Wisniewski et al., 2001). Calcium chloride infiltration of apples followed by the application of the BCA *Pseudomonas syringae* enhanced the reduction of blue mold in 'Golden Delicious' apples as compared to the BCA or $CaCl_2$ infiltration only (Janisiewicz et al., 1998). In 'Golden Delicious' apples, the BCA *Pseudomonas syringae* pv. *syringae* and ASM reduced blue mold by 50% and induced significant increases in the levels of gene transcripts of the pathogenesis-related proteins PR-1a, PR-2, PR-5 and PR-8 in apple peel compared to controls (Quaglia et al., 2011). Strains of antagonistic yeasts,

Torulaspora spp., in combination with sodium bicarbonate reduced blue mold by 91% in 'Golden Delicious' and 'Red Delicious' apples (Soto-Munoz and Martinez-Peniche, 2009).

CONCLUDING REMARKS

Food production must increase significantly to meet the future demands of an increasing world population. To feed the world's expected 10 billion people within the next 30 years requires concerted effort of increased food production and improved efficiency in handling of harvested food, so that the food reaches the consumer with minimal loss (FAO, 2010; Gustavsson et al., 2011). Minimizing postharvest losses of fruits and vegetables is a very effective way of reducing the area needed for production and/or increasing food availability. Over the past three decades, many international organizations have been actively searching ways to reduce postharvest diseases of perishable crops. Prevention or reduction of postharvest losses of produce must become a priority in a world of increasing population and limited resources.

Blue mold and patulin produced by *P. expansum* cause considerable financial losses to apple growers and processors. Postharvest losses have been reduced traditionally through postharvest fungicides and, in recent years, the development of resistance to key fungicides in *P. expansum* is raising concern. The increase in demand for food free of pesticide residues by many consumers may increase the growth of niche markets in which alternative disease control methods are used. More research is needed to identify, by screening germplasm and commercial apple cultivars, resistance to blue mold, especially the cultivars that are stored for extended lengths in cold storage. Extensive research has been conducted worldwide, for many years, to identify and develop alternative control methods for blue mold control, but very few of these products are registered for commercial use. Several factors, such as host plant species, cultivar, sensitivity of pathogen, and storage conditions and practices influence the effectiveness and feasibility of disease control and research is needed to understand how these factors separately or together prevent blue mold. It is important to understand the fundamental basis underlying host–pathogen interactions and how the alternative control methods influence these interactions. Proteomic and genomic approaches in the area of host–pathogen–control method interaction will result in identifying durable and effective alternatives to synthetic fungicides.

Biological control is an important alternative to the chemical fungicides and the search for new antagonists that are effective against multiple postharvest and field pathogens in different crops must continue, similar to the searches for fungicides in the chemical industry. The majority of the microbial BCA that are being studied are epiphytes and there is a need to search for endophytes, preferably within the host where *P. expansum* causes disease. Pulsed light technology is one of the emerging technologies that is being used for postharvest produce and to my knowledge this technology has not been tested against blue mold caused by *P. expansum* in fruits and vegetables. Further research is required to determine the effects of UV-irradiation on pathogens, fruit and vegetable physiology, sensory quality and consumer acceptance.

The research indicates that physical, chemical and biological control approaches have potential in controlling *P. expansum* but any one of the alternative methods alone is not sufficient to provide blue mold control in postharvest produce that is stored for several months. Successful blue mold disease management can be achieved by integration of practices and conditions that affect disease dynamics. The combination of chemical and BCA with modes of action different from each other form complex mixtures to provide increased efficacy against pathogens and the possibility of resistance development will become unlikely. Integrated control of blue mold with conventional and other alternatives such as plant extracts, volatiles, BCA, and essential oils, are promising and there is a need for further research to make this approach a viable alternative to the use of synthetic fungicides.

REFERENCES

Ahmadi-Afzadi, M., Tahir, I., Nybom, H., 2013. Impact of harvesting time and fruit firmness on the tolerance to fungal storage diseases in an apple germplasm collection. Postharvest Biol. Technol. 82, 51–58.

Amiri, A., Bompeix, G., 2005. Diversity and population dynamics of *Penicillium* spp. on apples in pre- and postharvest environments: Consequences for decay development. Plant Pathol. 54, 74–81.

Baker, K.F., Heald, F.D., 1936. The effect of certain cultural and handling practices on the resistance of apples to *Penicillium expansum*. Phytopathol. 26, 471–477.

Barad, S., Horowitz, S.B., Moskovitch, O., Lichter, A., Sherman, A., Prusky, D., 2012. A *Penicillium expansum* glucose oxidase-encoding gene, GOX2, is essential for gluconic acid production and acidification during colonization of deciduous fruit. Mol. Plant Microbe Interact. 25, 779–788.

Barkai-Golan, R., Phillips, D.J., 1991. Postharvest heat treatment of fresh fruits and vegetables for decay control. Plant Dis. Rep. 75, 1085–1089.

Bautista-Baños, S., De Luca, A.J., 2004. Evaluation of the antifungal activity of natural compounds to reduce postharvest blue mould (*Penicillium expansum* Link.) of apples (*Malus domestica* Borkh.) during storage. Rev. Mex. de Fitopatologia 22, 362–369.

Bautista-Baños, S., Hernández-Lauzardo, A.N., Velázquez-Del Valle, M.G., et al., 2006. Chitosan as a potential natural compound to control pre and postharvest diseases of horticultural commodities. Crop Prot. 25, 108–118.
Bordbar, F., Etebarian, H., Sahebani, N., Rohani, H., 2010. Control of postharvest decay of apple fruit with *Trichoderma virens* isolates and induction of defense responses. J. Plant Prot. Res. 50, 146–152.
Bramlage, W.J., Drake, M., Weis, S.A., 1985. Comparisons of calcium chloride, calcium phosphate, and a calcium chelate as foliar sprays for 'McIntosh' apple trees. J. Am. Soc. Horticultural Sci. 110, 786–789.
Cabral, L.D.C., Pinto, V.F., Patriarca, A., 2013. Application of plant derived compounds to control fungal spoilage and mycotoxin production in foods. Int. J. Food Microbiol. 166, 1–14.
Camele, I., Altieri, L., de Martino, L., de Feo, V., Mancini, E., Rana, G.L., 2012. In vitro control of post-harvest fruit rot fungi by some plant essential oil components. Int. J. Mol. Sci. 13, 2290–2300.
Canaver, B.S., Di Piero, R.M., 2011. Chitosan and adjuvants for the control of blue mold in apples. Tropical Plant Pathol. 36, 419–423.
Cao, J., Jiang, W., 2006. Induction of resistance in Yali pear (*Pyrus bretschneideri* Rehd.) fruit against postharvest diseases by acibenzolar-s-methyl sprays on trees during fruit growth. Sci. Horticulturae 110, 181–186.
Cao, J., Jiang, W., He, H., 2005. Induced resistance in yali pear (*Pyrus bretschneideri* Rehd.) fruit against infection by *Penicillium expansum* by postharvest infiltration of acibenzolar-S-methyl. J. Phytopathol. 153, 640–646.
Chand-Goyal, T., Spotts, R.A., 1997. Biological control of postharvest diseases of apple and pear under semi-commercial and commercial conditions using three saprophytic yeasts. Biol. Control 10, 199–206.
Conway, W.S., Sams, C.E., Hickey, K.D., 2002. Pre- and postharvest calcium treatment of apple fruit and its effect on quality. Acta Horticulturae 594, 413–419.
Conway, W.S., Janisiewicz, W.J., Klein, J.D., Sams, C.E., 1999. Strategy for combining heat treatment, calcium infiltration, and biological control to reduce postharvest decay of 'Gala' apples. Hortscience 34, 700–704.
Corrales, M., Butz, P., Tauscher, B., Cabrejas, M.M., 2010. Antifungal effect of organic and conventional grape pomace extracts. Acta Horticulturae 856, 31–36.
Coulombe, J.R.A., 1993. Biological action of mycotoxins. J. Dairy Sci. 76, 880–891.
Creelman, R.A., Mullet, J.E., 1997. Biosynthesis and action of jasmonates in plants. Annual Rev. Plant Biol. 48, 355–381.
Čvek, D., Markov, K., Frece, J., Dragičević, T.L., Majica, M., Delaš, F., 2010. Growth inhibition of *Aspergillus ochraceus* ZMPBF 318 and *Penicillium expansum* ZMPBF 565 by four essential oils. Arhiv za Higijenu Rada i Toksikologiju 61, 191–196.
Czerwinski, M., Popovic, V., Errampalli, D., 2009. Multi-site mutations in beta-tubulin gene sequences are involved in thiabendazole resistance in *Penicillium* species collected from Southern Ontario. Can. J. Plant Pathol. 31, 52.
De Capdeville, G., Beer, S.V., Wilson, C.L., Aist, J.R., 2008. Some cellular correlates of harpin-induced resistance to blue mold of apples. Tropical Plant Pathol. 33, 103–113.
Delp, C.J., 1988. Resistance management strategies for benzimidazoles. In: Delp, C.J. (Ed.), Fungicide Resistance in North America. APS Press, St Paul, MN, pp. 41–43.
Dennis, C., 1983. Postharvest Pathology of Fruits and Vegetables. Academic Press, London.
Drake, S.R., Sanderson, P.G., Neven, L.G., 1999. Response of apple and winter pear fruit quality to irradiation as a quarantine treatment. J. Food Process. Preservation 23, 203–216.
Droby, S., Wisniewski, M., El-Ghaouth, A., Wilson, C., 2003. Biological control of postharvest diseases of fruits and vegetables: Current achievements and future challenges. Acta Horticulturae 628, 703–713.

Droby, S., Wisniewski, M., Macarisin, D., Wilson, C., 2009. Twenty years of postharvest biocontrol research: Is it time for a new paradigm? Postharvest Biol. Technol. 52, 137–145.

Drusch, S., Ragab, W., 2003. Mycotoxins in fruits, fruit juices, and dried fruits. J. Food Prot. 66, 1514–1527.

Ebrahimi, L., Aminian, H., Etebarian, H.R., Sahebani, N., 2012a. Control of apple blue mould disease with *Torulaspora delbrueckii* in combination with silicon. Arch. Phytopathol. Plant Prot. 45, 2057–2065.

Ebrahimi, L., Etebarian, H.R., Aminian, H., Sahebani, N., 2012b. Enhancement of biocontrol activity of *Torulaspora delbrueckii* with methyl jasmonate against apple blue mould disease. Arch. Phytopathol. Plant Prot. 45, 2355–2363.

Eckert, J.W., Ogawa, J.M., 1988. The chemical control of postharvest diseases: deciduous fruits, berries, vegetables and root/tuber crops. Annu. Rev. Phytopathol. 26, 433–469.

El Ghaouth, A., Wilson, C., Wisniewski, M., Droby, S., Smilanick, J.L., Korsten, L., 2002. Biological control of postharvest diseases of fruits and vegetables. Appl. Mycol. Biotechnol. 2, 219–238.

Errampalli, D., 2004. Effect of fludioxonil on germination and growth of *Penicillium expansum* and decay in apple cvs. Empire and Gala. Crop Protection 23, 811–817.

Errampalli, D., 2006. Postharvest control of blue mold of apples with reduced-risk fungicides together with anti-scald agent diphenylamine under cold and controlled atmosphere storage conditions. J. Food Agri. Envi. 4, 43–47.

Errampalli, D., Brubacher, N.R., 2006. Biological and integrated control of postharvest blue mold (*Penicillium expansum*) of apples by *Pseudomonas syringae* and cyprodinil. Biol. Control 36, 49–56.

Errampalli, D., Crnko, N., 2004. Control of blue mold caused by *Penicillium expansum* on apples 'Empire' with fludioxonil and cyprodinil. Can. J. Plant Pathol. 26, 70–75.

Errampalli, D., Wainman, L.I., 2009. Effect of preharvest application boscalid/pyraclostrobin (Pristine) and pyramethanil (Scala) for the control of postharvest blue and gray mold in 'Empire' apples, 2007-08. Pest Manag. Res. Rep. 47, 148–151.

Errampalli, D., Wainman, L.I., 2010a. Effect of difenoconazole protective drench on control of postharvest blue mold and gray mold in 'McIntosh' apples, 2008. Pest Manag. Res. Rep. 48, 44–46.

Errampalli, D., Wainman, L.I., 2010b. Effect of difenoconazole protective drench on control of postharvest blue mold and gray mold in 'Red Delicious' apples, 2009. Pest Manag. Res. Rep. 48, 75–76.

Errampalli, D., Wainman, L.I., 2010c. Effect of difenoconazole protective drench on control of postharvest blue mold and gray mold on 'Empire' apples, 2008. Pest Manag. Res. Rep. 48, 47–48.

Errampalli, D., Wainman, L.I., 2010d. Effect of preharvest application boscalid/pyraclostrobin (Pristine) and pyrimethanil (Scala) for the control of postharvest blue and gray mold in 'Empire' apples, 2008-09. Pest Manag. Res. Rep. 48, 40–43.

Errampalli, D., Wainman, L.I., 2011. Effect of preharvest boscalid/pyraclostrobin and pyrimethanil application for the control of postharvest gray mold in 'Empire' apples. 2009-10. Pest Manag. Res. Rep. 49.

Errampalli, D., DeEll, J.D., Murr, D., 2005a. Effect of 1-methylcyclopropene treatment on postharvest decay in apple cvs. Empire and McIntosh. Can. J. Plant Pathol. 27, 467.

Errampalli, D., Northover, J., Skog, L., Brubacher, N.R., Collucci, C.A., 2005b. Control of blue mold (*Penicillium expansum*) by fludioxonil in apples (cv Empire) under controlled atmosphere and cold storage conditions. Pest Manag. Sci. 61, 591–596.

Errampalli, D., Brubacher, N.R., DeEll, J.R., 2006. Sensitivity of *Penicillium expansum* to diphenylamine and thiabendazole and postharvest control of blue mold with fludioxonil in 'McIntosh' apples. Postharvest Biol. Technol. 39, 101–107.

Errampalli, D., Wainman, L., Chu, C.L., 2007. Evaluation of fludioxonil for the control of post-harvest gray mould and blue mould in pears in cold and controlled atmosphere storages. Int. J. Pest Manag. 53, 101–109.

Errampalli, D., Wainman, L.I., DeEll, J.R., 2012. Reduced risk control options for apple postharvest diseases in long-term storages. Acta Horticulturae 934, 313–318.

Fallik, E., Tuvia-Alkalai, S., Feng, X., Lurie, S., 2001. Ripening characterisation and decay development of stored apples after a short pre-storage hot water rinsing and brushing. Innovative Food Sci. Emerg. Technol. 2, 127–132.

Fan, L., Song, J., Beaudry, R.M., Hildebrand, P.D., 2006. Effect of hexanal vapor on spore viability of *Penicillium expansum*, lesion development on whole apples and fruit volatile biosynthesis. J. Food Sci. 71, M105–M109.

FAO., 2010. The state of food insecurity in the world. Addressing food insecurity in protracted crises. [Online]. Available at: http://www.fao.org/docrep/013/i1683e/i1683e.pdf.

Farahani, L., Etebarian, H.R., Sahebani, N., Aminian, H., 2012. Biocontrol of blue mold of apple by *Candida membranifaciens* in combination with silicon. Arch. Phytopathol. Plant Prot. 45, 310–317.

FDA., 1999. U. S. Regulatory Requirements for Irradiating Foods. [Online]. Available at: http://www.fda.gov/Food/GuidanceRegulation/GuidanceDocumentsRegulatoryInformation/IngredientsAdditivesGRASPackaging/ucm110730.htm.

FDA., 2001. Compliance policy guide. Compliance policy guidance for FDA staff. Sec. 510.150. Patulin in Apple Juice, Apple Juice Concentrates and Apple Juice Products – adulteration with patulin. [Online]. Available at: http://www.fda.gov/ora/compliancepref/cpg/cpgfod/cpg510–150.htm.

Fieira, C., Oliveira, F., Calegari, R.P., Machado, A., Coelho, A.R., 2013. In vitro and in vivo antifungal activity of natural inhibitors against *Penicillium expansum*. Food Sci. Technol. 33, 40–46.

Filonow, A.B., 2005. Wound type in apple fruits affects wound resistance to decay-causing fungi. J. Plant Pathol. 87, 233–238.

Foaad, M.A., Ayesh, A.M., Swailem, H.M.H., 2002. Efficiency of gamma irradiation on patulin elimination produced by apple mycoflora. Egypt. J. Food Sci. 30, 289–312.

Frisvad, J.C., Samson, R.A., 2004. Polyphasic taxonomy of *Penicillium* subgenus *Penicillium*. A guide to identification of food and air-borne terverticillate *Penicillia* and their mycotoxins. Stud. Mycol. Utrecht 49, 1–173.

Gao, L.L., Zhang, Q., Sun, X.Y., et al., 2013. Etiology of moldy core, core browning, and core rot of fuji apple in China. Plant Dis. 97, 510–516.

Giraud, M., Bompeix, G., 2012. Postharvest diseases of pome fruits in Europe: perspectives for integrated control. IOBC/WPRS Bull. 84, 257–263.

Gómez-López, V.M., Ragaert, P., Debevere, J., Devlieghere, F., 2007. Pulsed light for food decontamination: a review. Trends Food Sci. Technol. 18, 464–473.

Gustavsson, J., Cederberg, C., Sonesson, U., van Otterdijk, R., Meybeck, A., 2011. Global food losses and food waste: extent, causes and prevention: study conducted for the International Congress Save Food at Interpack 2011 Düsseldorf. Germany Food and Agriculture Organization of the United Nations, Rome.

Harvey, J.M., 1978. Reduction of losses in fresh fruits and vegetables. Annu. Rev. Phytopathol. 16, 321–341.

HealthCanada, 2013. Consultation on Health Canada's Proposed Standard (Maximum Level) for the Presence of the Mycotoxin Patulin in Apple Juice and Unfermented Apple Cider [Online]. Available at: http://www.hc-sc.gc.ca/fn-an/consult/max-mycotox-pat/index-eng.php.

Houbraken, J., Frisvad, J.C., Samson, R.A., 2011. Taxonomy of *Penicillium* section *Citrina*. Stud. Mycol. 70, 53–138.

Houbraken, J., Samson, R.A., 2011. Phylogeny of *Penicillium* and the segregation of Trichocomaceae into three families. Stud. Mycol. 70, 1–51.

Ikeura, H., Somsak, N., Kobayashi, F., Kanlayanarat, S., Hayata, Y., 2011. Application of selected plant extracts to inhibit growth of *Penicillium expansum* on apple fruits. Plant Pathol. J. 10, 79–84.

Ippolito, A., El Ghaouth, A., Wilson, C.L., Wisniewski, M., 2000. Control of postharvest decay of apple fruit by *Aureobasidium pullulans* and induction of defense responses. Postharvest Biol. Technol. 19, 265–272.

Jackson, L.S., Beacham-Bowden, T., Keller, S.E., et al., 2003. Apple quality, storage, and washing treatments affect patulin levels in apple cider. J. Food Prot. 66, 618–624.

Janick, J., Cummins, J.N., Brown, S.K., Hemmat, M., 1996. Apples. In: Janick, J., Moore, J.N. (Eds.), Fruit Breeding. Tree and Tropical Fruits, vol. 1. John Wiley & Sons, New York.

Janisiewicz, W.J., 2006. Biological control of postharvest decay of fruit using strains of *Metschnikowia* species. United States Dep. Agri. patents.

Janisiewicz, W.J., Jeffers, S.N., 1997. Efficacy of commercial formulation of two biofungicides for control of blue mold and gray mold of apples in cold storage. Crop Prot. 16, 629–633.

Janisiewicz, W.J., Korsten, L., 2002. Biological control of postharvest diseases of fruits. Annu. Rev. Phytopathol. 40, 411–441.

Janisiewicz, W.J., Peterson, D.L., 2004. Susceptibility of the stem pull area of mechanically harvested apples to blue mold decay and its control with a biocontrol agent. Plant Dis. 88, 662–664.

Janisiewicz, W.J., Conway, W.S., Glenn, D.M., Sams, C.E., 1998. Integrating biological control and calcium treatment for controlling postharvest decay of apples. Hortscience 33, 105–109.

Janisiewicz, W.J., Leverentz, B., Conway, W.S., Saftner, R.A., Reed, A.N., Camp, M.J., 2003. Control of bitter rot and blue mold of apples by integrating heat and antagonist treatments on 1-MCP treated fruit stored under controlled atmosphere conditions. Postharvest Biol. Technol. 29, 129–143.

Janisiewicz, W.J., Saftner, R.A., Conway, W.S., Forsline, P.L., 2008. Preliminary evaluation of apple germplasm from Kazakhstan for resistance to postharvest blue mold in fruit caused by *Penicillium expansum*. Hortscience 43, 420–426.

Jia, P., Liu, X., Lin, C., Zhu, X., 2012. Study on influencing factors of the antifungal activity on the main postharvest disease-producing fungus of apricot fruit by oligochitosan. Xinjiang Agri. Sci. 49, 290–295.

Jin, P., Zheng, Y., Tang, S., Rui, H., Wang, C.Y., 2009. Enhancing disease resistance in peach fruit with methyl jasmonate. J. Sci. Food Agri. 89, 802–808.

Johnson, G.I., Boag, T.S., Cooke, A.W., Izard, M., Panitz, Sangchote, S., 1990. Interaction of postharvest disease control treatments and gamma irradiation on mangoes. Annals Appl. Biol. 116, 245–257.

Jurick Ii, W.M., Vico, I., Gaskins, V.L., et al., 2010. Purification and biochemical characterization of polygalacturonase produced by *Penicillium expansum* during postharvest decay of 'Anjou' pear. Phytopathology 100, 42–48.

Kader, A.A., 2005. Increasing food availability by reducing postharvest losses of fresh produce. Acta Horticulturae 682, 2169–2175.

Kader, A.A., Rolle, R.S., 2004. The role of post-harvest management in assuring the quality and safety of horticultural produce. Food and Agriculture Organization of the United Nations, Rome.

Karabulut, O.A., Arslan, U., Ilhan, K., Kuruoglu, G., 2005. Integrated control of postharvest diseases of sweet cherry with yeast antagonists and sodium bicarbonate applications within a hydrocooler. Postharvest Biol. Technol. 37, 135–141.

Karabulut, O.A., Arslan, U., Kuruoglu, G., Ozgenc, T., 2004. Control of postharvest diseases of sweet cherry with ethanol and hot water. J. Phytopathol. 152, 298–303.

Karaoglanidis, G.S., Markoglou, A.N., Bardas, G.A., Doukas, E.G., Konstantinou, S., Kalampokis, J.F., 2011. Sensitivity of *Penicillium expansum* field isolates to tebuconazole, iprodione, fludioxonil and cyprodinil and characterization of fitness parameters and patulin production. Int. J. Food Microbiol. 145, 195–204.

Kessmann, H., Staub, T., Ligon, J., Oostendorp, M., Ryals, J., 1994. Activation of systemic acquired disease resistance in plants. Eur. J. Plant Pathol. 100, 359–369.

Kim, J.H., Mahoney, N., Chan, K.L., Molyneux, R.J., Campbell, B.C., 2006. Controlling food-contaminating fungi by targeting their antioxidative stress-response system with natural phenolic compounds. Appl. Microbiol. Biotechnol. 70, 735–739.

Kimura, S., Amachi, S., Ohno, N., Takahashi, H., Shinoyama, H., Fujii, T., 2002. Relationship between conidial enzymes and germination of the apple blue mold, *Penicillium expansum*. Biosci. Biotechnol. Biochem. 66, 1126–1129.

Konstantinou, S., Karaoglanidis, G.S., Bardas, G.A., Minas, I.S., Doukas, E., Markoglou, A.N., 2011. Postharvest fruit rots of apple in Greece: pathogen incidence and relationships between fruit quality parameters, cultivar susceptibility, and patulin production. Plant Dis. 95, 666–672.

Kupferman, E.M., 1992. Maturity and storage of apple varieties new to Washington State. Tree Fruit Postharvest J. 3, 9–15.

Kwon, J.H., Jeong, S.G., Hong, S.B., Chae, Y.S., Park, C.S., 2006. Occurrence of blue mold on sweet persimmon (Diospyros kaki) caused by *Penicillium expansum*. Res. Plant Dis. 12, 290–293.

Lahlali, R., Seehini, M.N., Jikakil, M.H., 2005. Studying and modelling the combined effect of temperature and water activity on the growth rate of *Penicillium expansum*. Int. J. Food Microbiol. 103, 315–322.

Lawton, K.A., Friedrich, L., Hunt, M., et al., 1996. Benzothiadiazole induces disease resistance in Arabidopsis by activation of the systemic acquired resistance signal transduction pathway. Plant J. 10, 71–82.

Lee, G.W., Ko, J.A., Oh, B.T., et al., 2012. Biological control of postharvest diseases of apples, peaches and nectarines by *Bacillus subtilis* S16 isolated from halophytes rhizosphere. Biocontrol Sci. Technol. 22, 351–361.

Leverentz, B., Janisiewicz, W.J., Conway, W.S., et al., 2000. Combining yeasts or a bacterial biocontrol agent and heat treatment to reduce postharvest decay of 'Gala' apples. Postharvest Biol. Technol. 21, 87–94.

Li, B., Lai, T., Qin, G., Tian, S., 2010. Ambient pH stress inhibits spore germination of *Penicillium expansum* by impairing protein synthesis and folding: A proteomic-based study. J. Proteome Res. 9, 298–307.

Li, H.X., Xiao, C.L., 2008. Baseline sensitivities to fludioxonil and pyrimethanil in *Penicillium expansum* populations from apple in Washington State. Postharvest Biol. Technol. 47, 239–245.

Li, R., Zhang, H., Liu, W., Zheng, X., 2011. Biocontrol of postharvest gray and blue mold decay of apples with *Rhodotorula mucilaginosa* and possible mechanisms of action. Int. J. Food Microbiol. 146, 151–156.

Lima, G., Castoria, R., De Curtis, F., Raiola, A., Ritieni, A., De Cicco, V., 2011. Integrated control of blue mould using new fungicides and biocontrol yeasts lowers levels of fungicide residues and patulin contamination in apples. Postharvest Biol. Technol. 60, 164–172.

Link, J.H.F., 1809. Observationes in ordines plantarum naturales: Dissertatio Ima complectens anandrarum ordines epiphytas, mucedines gastromycos et fungos [in Latin]. Der Gesellschaft Naturforschender Freunde zu Berlin Magazin für die neuesten Entdeckungen in der gesammten Naturkunde 3, 3–42.

Liu, H., Jiang, W., Bi, Y., Luo, Y., 2005. Postharvest BTH treatment induces resistance of peach (*Prunus persica* L. cv. Jiubao) fruit to infection by *Penicillium expansum* and enhances activity of fruit defense mechanisms. Postharvest Biol. Technol. 35, 263–269.

Liu, J., Tian, S., Meng, X., Xu, Y., 2007. Effects of chitosan on control of postharvest diseases and physiological responses of tomato fruit. Postharvest Biol. Technol. 44, 300–306.

Logrieco, A., Bottalico, A., Mulé, G., Moretti, A., Perrone, G., 2003. Epidemiology of toxigenic fungi and their associated mycotoxins for some Mediterranean crops. Eur. J. Plant Pathol. 109, 645–667.

Louzeiro, I.M., Ferrari, J.T., Nogueira, E.M.d.C., Santos, A.J.T.d, 2002. Control of postharvest diseases on apple fruits. Arquivos do Instituto. Biologico. 69, 225–226.
Lozynska, I., Van Driel, L., Errampalli, D., 2010. Classification of *Penicillium* spp. in apples collected from packinghouses in southern Ontario using PCR-RFLP. Can. J. Plant Pathol. 32, 426–427.
Lugauskas, A., Repečkiene, J., Novošinskas, H., 2005. Micromycetes, producers of toxins, detected on stored vegetables. Ann. Agri. Environ. Med. 12, 253–260.
Luo, X.J., Peng, J., Li, Y.J., 2011. Recent advances in the study on capsaicinoids and capsinoids. Eur. J. Pharmacol. 650, 1–7.
Lurie, S., Fallik, E., Klein, J.D., Kozar, F., Kovacs, K., 1998. Postharvest heat treatment of apples to control San Jose scale (*Quadraspidiotus perniciosus* Comstock) and blue mold (*Penicillium expansum* Link) and maintain fruit firmness. J. Am. Soc. Horticultural Sci. 123, 110–114.
Malandrakis, A.A., Markoglou, A.N., Konstantinou, S., Doukas, E.G., Kalampokis, J.F., Karaoglanidis, G.S., 2013. Molecular characterization, fitness and mycotoxin production of benzimidazole-resistant isolates of *Penicillium expansum*. Int. J. Food Microbiol. 162, 237–244.
Manso, T., Nunes, C., 2011. *Metschnikowia andauensis* as a new biocontrol agent of fruit postharvest diseases. Postharvest Biol. Technol. 61, 64–71.
Maouni, A., Lamarti, A., Aidoun, A., Khaddor, M., Badoc, A., 2007. Effect of benzimidazole fungicides and calcium chloride on *Alternaria alternata* and *Penicillium expansum* rot during storage of pears. Afr. J. Biotechnol. 6, 1289–1292.
Marandi, R.J., Hassani, A., Ghosta, Y., Abdollahi, A., Pirzad, A., Sefidkon, F., 2011. Control of *Penicillium expansum* and *Botrytis cinerea* on pear with *Thymus kotschyanus, Ocimum basilicum* and *Rosmarinus officinalis* essential oils. J. Med. Plants Res. 5, 626–634.
Mastan, S.A., Janaki Ramayya, P., Mutyala Naidu, L., Mallikarjuna, K., 2009. Antimicrobial activity of various extracts of *Mucuna pruriens* leaves. Biomed. Pharmacol. J. 2, 55–60.
Matos, O.C., Santos, M., Ramos, P., Barreiro, M.G., 2011. Aromatic plants and their bioactive products to control postharvest 'Rocha' pear diseases. Acta Horticulturae 925, 335–340.
Mdee, L.K., Masoko, P., Eloff, J.N., 2009. The activity of extracts of seven common invasive plant species on fungal phytopathogens. South Afr. J. Bot. 75, 375–379.
Mercier, J., Jiménez, J.I., 2004. Control of fungal decay of apples and peaches by the biofumigant fungus *Muscodor albus*. Postharvest Biol. Technol. 31, 1–8.
Mohammadifar, M., Norabadi, M.T., Hasanzadeh, M., Dashtipoor, S., Etebarian, H.R., Sahebani, N., 2012. Study of antifungal activities of seven essential oils from some Iranian medicinal plants against various postharvest phytopathogenic fungi. Arch. Phytopathol. Plant Prot. 45, 2046–2056.
Moniharapon, E., Hashinaga, F., 2004. Antimicrobial activity of atung (*Parinarium glaberrimum* Hassk) fruit extract. Pakistan J. Biol. Sci. 7, 1057–1061.
Morales, H., Sanchis, V., Rovira, A., Ramos, A.J., Marcn, S., 2007. Patulin accumulation in apples during postharvest: Effect of controlled atmosphere storage and fungicide treatments. Food Control 18, 1443–1448.
Mostafavi, H.A., Mirmajlessi, S.M., Mirjalili, S.M., Fathollahi, H., Askari, H., 2012. Gamma radiation effects on physico-chemical parameters of apple fruit during commercial postharvest preservation. Radiat. Physics. Chem. 81, 666–671.
Moy, J.H., 1983. Radurisation and radicidation: fruit and vegetables. In: Josephson, E.D., Peterson, M.S. (Eds.), Preservation of Food Ionizing Radiation, vol. 3. CRC Press, Boca Raton, FL, pp. 83–108.
Nazzaro, F., Caliendo, G., Arnesi, G., Veronesi, A., Sarzi, P., Fratianni, F., 2009. Comparative content of some bioactive compounds in two varieties of *Capsicum annuum* l. sweet pepper and evaluation of their antimicrobial and mutagenic activities. J. Food Biochem. 33, 852–868.

Neri, F., Mari, M., Menniti, A.M., Brigati, S., 2006a. Activity of trans-2-hexenal against *Penicillium expansum* in 'Conference' pears. J. Appl. Microbiol. 100, 1186–1193.

Neri, F., Mari, M., Menniti, A.M., Brigati, S., Bertolini, P., 2006b. Control of *Penicillium expansum* in pears and apples by trans-2-hexenal vapours. Postharvest Biol. Technol. 41, 101–108.

Nunes, C., Usall, J., Teixidó, N., Abadias, M., Viñas, I., 2002a. Improved control of postharvest decay of pears by the combination of *Candida sake* (CPA-1) and ammonium molybdate. Phytopathology 92, 281–287.

Nunes, C., Usall, J., Teixidó, N., Fons, E., Viñas, I., 2002b. Post-harvest biological control by *Pantoea agglomerans* (CPA-2) on Golden Delicious apples. J. Appl. Microbiol. 92, 247–255.

Nunes, C.A., 2012. Biological control of postharvest diseases of fruit. Eur. J. Plant Pathol. 133, 181–196.

Oliveira Junior, E.N., El Gueddari, N.E., Moerschbacher, B.M., Franco, T.T., 2012. Growth rate inhibition of phytopathogenic fungi by characterized chitosans. Braz. J. Microbiol. 43, 800–809.

Osorio, G.T., Oliveira, B.S., Di Piero, R.M., 2013. Effect of fumigants on blue and gray molds of apple fruit. Tropical Plant Pathol. 38, 63–67.

Paik, S., Chung, I., 1997. Effect of medicinal plant extracts on apple storage diseases. Korean J. Plant Pathol. 13, 57–62.

Parfitt, J., Barthel, M., Macnaughton, S., 2010. Food waste within food supply chains: quantification and potential for change to 2050. Philos. Trans. Royal Soc. B Biol. Sci. 365, 3065–3081.

Paster, N., Huppert, D., Barkai-Golan, R., 1995. Production of patulin by different strains of *Penicillium expansum* in pear and apple cultivars stored at different temperatures and modified atmospheres. Food Addit. Contam. 12, 51–58.

Paterson, R.R.M., Venâncio, A., Lima, N., 2004. Solutions to Penicillium taxonomy crucial to mycotoxin research and health. Res. Microbiol. 155, 507–513.

Peryea, F.J., Neilsen, G.H., Faubion, D., 2007. Start-timing for calcium chloride spray programs influences fruit calcium and bitter pit in 'Braeburn' and 'Honeycrisp' apples. J. Plant Nutr. 30, 1213–1227.

Peterson, S.W., 2000. Phylogenetic analysis of *Penicillium* species based on ITS and LSU-rDNA nucleotide sequences. In: Samson, R.A., Pitt, J.I. (Eds.), Integration of Modern Taxonomic Methods for *Penicillium* and *Aspergillus* Classification. Harwood Academic Publishers, Amsterdam, The Netherlands, pp. 163–178.

Pianzzola, M.J., Moscatelli, M., Vero, S., 2004. Characterization of *Penicillium* isolates associated with blue mold on apple in Uruguay. Plant Dis. 88, 23–28.

Pirgozliev, S.P., Errampalli, D., Sholberg, P.L., Stokes, S.C., DeEll, J.R., Murr, D.P., 2006a. Effect of pre-harvest pyrimethanil (scala sc, bayer cropscience Ltd) application for control of post-harvest gray and blue mold in 'Empire' apples, 2004-05. Pest Manag. Res. Rep. 44, 165–170.

Pirgozliev, S.P., Errampalli, D., Sholberg, P.L., Stokes, S.C., DeEll, J.R., Murr, D.P., 2006b. Effect of pre-harvest pyrimethanil (scala sc, bayer cropscience ltd) application for control of post-harvest gray and blue mold in 'McIntosh' apples, 2004-05. Pest Manag. Res. Rep. 44, 171–176.

Pitt, J.I., Hocking, A.D., 1997. Fungi and Food Spoilage. Blackie Academic and Professional., London.

Quaglia, M., Ederli, L., Pasqualini, S., Zazzerini, A., 2011. Biological control agents and chemical inducers of resistance for postharvest control of *Penicillium expansum* Link. on apple fruit. Postharvest Biol. Technol. 59, 307–315.

Ramin, A.A., Braun, P.G., Prange, R.K., DeLong, J.M., 2005. In vitro effects of *Muscodor albus* and three volatile components on growth of selected postharvest microorganisms. Hortscience 40, 2109–2114.

Ritte, E., Lurie, S., Droby, S., Ismailov, Z., Chet, I., Chernin, L., 2002. Biocontrol of postharvest fungal pathogens of peaches and apples by *Pantoae agglomerans* strain IC1270. Bulletin OILB/SROP 25, 199–202.

Romanazzi, G., Gabler, F.M., Santini, M., Landi, L., Karabulut, O.A., Smilanick, J.L., 2007. Advances in the use of chitosan to control postharvest decay of table grapes. In: Berolini, P. (Ed.). Proceedings of the COST 924, Novel approaches for the control of postharvest diseases and disorders. University of Bologna, Bologna, pp. 327–334.

Rosenberger, D.A., 1990. Blue mold. In: Jones, A.L., Aldwinckle, H.S. (Eds.), Compendium of Apple and Pear Diseases. American Phytopathological Society Press, St Paul, MN, pp. 54–55.

Rosenberger, D.A., Engle, C.A., Meyer, F.W., Watkins, C.B., 2006. *Penicillium expansum* invades apples trough stems during controlled atmosphere storage [Online]. Available at: http://www.plantmanagementnetwork.org/sub/php/research/2006/storage/.

Rosenberger, D.A., Meyer, F.W., 1979. Benomyl-tolerant *Penicillium expansum* in apple packinghouses in eastern New York. Plant Dis. Rep. 63, 37–48.

Saftner, R.A., Abbott, J.A., Conway, W.S., Barden, C.L., 2003. Effects of 1-methylcyclopropene and heat treatments on ripening and postharvest decay in 'Golden Delicious' apples. J. Am. Soc. Horticultural Sci. 128, 120–127.

Saftner, R.A., Conway, W.S., Sams, C.E., 1997. Effects of some polyamine biosynthesis inhibitors and calcium chloride on in vitro growth and decay development in apples caused by *Botrytis cinerea* and *Penicillium expansum*. J. Am. Soc. Horticultural Sci. 122, 380–385.

Samson, R.A., Pitt, J.I. (Eds.), 2000. Integration of Modern Taxonomic Methods for *Penicillium* and *Aspergillus* Classification. Harwood Academic, The Netherlands.

Sánchez-Torres, P., González-Candelas, L., 2003. Isolation and characterization of genes differentially expressed during the interaction between apple fruit and *Penicillium expansum*. Mol. Plant Pathol. 4, 447–457.

Sanzani, S.M., Nigro, F., Ippolito, A., De Girolamo, A., Schena, L., González-Candelas, L., 2010. Characterization of differentially expressed transcripts in quercetin-treated apples by suppression subtractive hybridization. Acta Horticulturae 877, 1691–1696.

Scherm, B., Ortu, G., Muzzu, A., Budroni, M., Arras, G., Migheli, Q., 2003. Biocontrol activity of antagonistic yeasts against *Penicillium expansum* on apple. J. Plant Pathol. 85, 205–213.

Schirra, M., D'Aquino, S., Migheli, Q., Pirisi, F.M., Angioni, A., 2009. Influence of postharvest treatments with fludioxonil and soy lecithin co-application in controlling blue and grey mould and fludioxonil residues in Coscia pears. Food Additives and Contaminants – Part A Chemistry, Analysis, Control. Expo. Risk Assess. 26, 68–72.

Schirra, M., D'Aquino, S., Mulas, M., et al., 2008. Efficacy of heat treatments with water and fludioxonil for postharvest control of blue and gray molds on inoculated pears and fludioxonil residues in fruit. J. Food Prot. 71, 967–972.

Scuderi, G., Bonaccorsi, A., Panebianco, S., Vitale, A., Polizzi, G., Cirvilleri, G., 2009. Some strains of *Burkholderia gladioli* are potential candidates for postharvest biocontrol of fungal rots in citrus and apple fruits. J. Plant Pathol. 91, 207–213.

Sellamuthu, P., Sivakumar, D., Soundy, P., 2013. Antifungal activity and chemical composition of thyme, peppermint and citronella oils in vapor phase against avocado and peach postharvest pathogens. J. Food Safety 33, 86–93.

Shao, X., Tu, K., Tu, S., Su, J., Zhao, Y., 2010. Effects of heat treatment on wound healing in Gala and Red Fuji apple fruits. J. Agri. Food Chem. 14 (58), 4303–4309.

Sharma, R.R., Singh, D., Singh, R., 2009. Biological control of postharvest diseases of fruits and vegetables by microbial antagonists: A review. Biol. Control 50, 205–221.

Sholberg, P.L., 1998. Fumigation of fruit with short-chain organic acids to reduce the potential of postharvest decay. Plant Dis. 82, 689–693.

Sholberg, P.L., Bedford, K., Stokes, S., 2005a. Sensitivity of *Penicillium* spp. and *Botrytis cinerea* to pyrimethanil and its control of blue and gray mold of stored apples. Crop Prot. 24, 127–134.

Sholberg, P.L., Conway, W.S., 2004. Postharvest pathology. In: Gross, K.C., Wang, C.Y., Saltveit, M. (Eds.), The Commercial Storage of Fruits, Vegetables, and Florist and Nursery Stocks, vol. 66. USDA, ARS Agricultural Handbook.
Sholberg, P.L., Harlton, C., Haag, P., Lávesque, C.A., O'Gorman, D., Seifert, K., 2005b. Benzimidazole and diphenylamine sensitivity and identity of *Penicillium* spp. that cause postharvest blue mold of apples using β-tubulin gene sequences. Postharvest Biol. Technol. 36, 41–49.
Singh, Y.P., Geeta, S., 2007. Efficacy of leaf extracts and essential oils of some plant species against *Penicillium expansum* rot of apples. Ann. Plant Prot. Sci. 15, 135–139.
Sitton, J.W., Patterson, M.E., 1992. Effect of high carbon dioxide and low oxygen controlled atmospheres on postharvest decay of apples. Plant Dis. 76, 992–995.
Smith, W.L., Anderson, R.E., 1975. Deacy control of peaches and nectarines during and after controlled atmospheres and air storage. J. Am. Soc. Horticultural Sci. 100, 84–86.
Snowden, A.L., 1990. A Color Atlas of Post-harvest Diseases and Disorders of Fruits and Vegetables. General Introduction and Fruits, vol. 1, CRC Press, Boca Raton, FL.
Snowden, A.L., 1992. A Color Atlas of Post-harvest Diseases and Disorders of Fruits and Vegetables, vol. 2, Vegetables. CRC Press, Boca Raton, FL.
Sommer, N.F., 1985. Role of controlled environments in suppression of postharvest diseases. Can. J. Plant Pathol. 7, 331–339.
Sommer, N.F., Fortlage, R.J., Edwards, D.C., 1992. Postharvest diseases of selected commqdities. In: Kader, A.A. (Ed.), Postharvest Technology of Horticultural Crops. University of California, USA. Division of agriculture and natural resources. 136 pp.
Soto-Munoz, L., Martinez-Peniche, R.A., 2009. Effect of antagonic yeasts and sodium bicarbonate on *Penicillium expansum* link in two apple varieties. Rev. Chapingo Serie Horticultura 15, 211–216.
Spadaro, D., Garibaldi, A., Gullino, M.L., 2004. Control of *Penicillium expansum* and *Botrytis cinerea* on apple combining a biocontrol agent with hot water dipping and acibenzolar-S-methyl, baking soda, or ethanol application. Postharvest Biol. Technol. 33, 141–151.
Spadaro, D., Lorè, A., Garibaldi, A., Gullino, M.L., 2013. A new strain of *Metschnikowia fructicola* for postharvest control of *Penicillium expansum* and patulin accumulation on four cultivars of apple. Postharvest Biol. Technol. 75, 1–8.
Spalding, D.H., Vaught, H.C., Day, R.H., Brown, G.A., 1969. Control of blue mold rot development in apples treated with heated and unheated fungicides. Plant Dis. Rep. 53, 738–742.
Spotts, R.A., 1990. Moldy core and core rot. In: Jones, A.L., Aldwinckle, H.S. (Eds.), Compendium of Apple and Pear Diseases. American Phytopathological Society, St Paul, MN, pp. 29–30.
Spotts, R.A., Cervantes, L.A., 1994. Contmination of harvest bins with pear decay fungi and evaluation of disinfectants on plastic and wood material. Acta Horticulturae 367, 419–425.
Spotts, R.A., Cervantes, L.A., Mielke, E.A., 1999. Variability in postharvest decay among apple cultivars. Plant Dis. 83, 1051–1054.
Spotts, R.A., Chen, P.M., 1987. Prestorage heat treatments for control of decay of pear fruit. Phytopathology 77, 1578–1582.
Strobel, G.A., Dirkse, E., Sears, J., Markworth, C., 2001. Volatile antimicrobials from *Muscodor albus*, a novel endophytic fungus. Microbiology 147, 2943–2950.
Strobel, G.A., Kluck, K., Hess, W.M., Sears, J., Ezra, D., Vargas, P.N., 2007. *Muscodor albus* E-6, an endophyte of *Guazuma ulmifolia* making volatile antibiotics: isolation, characterization and experimental establishment in the host plant. Microbiology 153, 2613–2620.
Sugar, D., 2002. Management of postharvest diseases. In: Michael, K. (Ed.), Fruit Quality and its Biological Basis. Wiley-Blackwell, pp. 225–252.
Sugar, D., Basile, S.R., 2008. Timing and sequence of postharvest fungicide and biocontrol agent applications for control of pear decay. Postharvest Biol. Technol. 49, 107–112.

Sugar, D., Richardson, D.G., Chen, P.M., Spotts, R.A., Roberts, R.G., Chand-Goyal, T., 1998. Advances in improving the postharvest quality of pears. Acta Horticulturae 475, 513–526.
Sugar, D., Spotts, R.A., 1999. Control of postharvest decay in pear by four laboratory-grown yeasts and two registered biocontrol products. Plant Dis. 83, 155–158.
Suzaki, K., Ito, T., Kanematsu, S., 2008. Occurrence and prevention of blue mold disease in apples. Mycotoxins 58, 137–141.
Taczman-Brückner, A., Mohácsi-Farkas, C., Balla, C., Klskó, G., 2005. Comparison of biocontrol activity of *Kluyveromyces lactis* with other yeast strains against *Penicillium expansum*. Acta Alimentaria 34, 71–80.
Tahir, I.I., Johansson, E., Olsson, M.E., 2009. Improvement of apple quality and storability by a combination of heat treatment and controlled atmosphere storage. Hortscience 44, 1648–1654.
Teixidó, N., Viñas, I., Usall, J., Magan, N., 1998. Control of blue mold of apples by preharvest application of *Candida sake* grown in media with different water activity. Phytopathology 88, 960–964.
Terry, L.A., Joyce, D.C., 2004. Elicitors of induced disease resistance in postharvest horticultural crops: A brief review. Postharvest Biol. Technol. 32, 1–13.
Torres, R., Valentines, M.C., Usall, J., Viñas, I., Larrigaudiere, C., 2003. Possible involvement of hydrogen peroxide in the development of resistance mechanisms in 'Golden Delicious' apple fruit. Postharvest Biol. Technol. 27, 235–242.
Ulevičius, V., Pečiulyte, D., Plauškaite, K., Špirkauskaite, N., 2008. Resistance of airborne fungal propagules to ultraviolet irradiation: Laboratory study. Lithuanian J. Physics 48, 265–273.
USEPA., 2006. U.S. Envoromental Protection Agency. Accomplishments under the Food Quality Protection Act (FQPA)-10th anniversary of the Food Quality Protectio Act. [Online]. Available at: http://www.epa.gov/pesticides/regulating/laws/fqpa/fqpa_accomplishments.htm.
van der Walt, L., Spotts, R.A., Visagie, C.M., Jacobs, K., Smit, F.J., McLeod, A., 2010. *Penicillium* species associated with preharvest wet core rot in South Africa and their pathogenicity on apple. Plant Dis. Rep. 94, 666–675.
Varahalarao, V., Naidu Chandrashekar, K., 2010. In vitro bioactivity of Indian medicinal plant *Calotropis procera* (AIT.). J. Global Pharma Technol. 2, 43–45.
Venturini, M.E., Blanco, D., Oria, R., 2002. In vitro antifungal activity of several antimicrobial compounds against *Penicillium expansum*. J. Food Prot. 65, 834–839.
Vero, S., Mondino, P., Burgueño, J., Soubes, M., Wisniewski, M., 2002. Characterization of biocontrol activity of two yeast strains from Uruguay against blue mold of apple. Postharvest Biol. Technol. 26, 91–98.
Vilanova, L., Viñas, I., Torres, R., Usall, J., Jauset, A.M., Teixidó, N., 2012. Infection capacities in the orange-pathogen relationship: Compatible (*Penicillium digitatum*) and incompatible (*Penicillium expansum*) interactions. Food Microbiol. 29, 56–66.
Vorstermans, B., Creemers, P., Bylemans, D., Garnier, A., 2005. A new post-harvest fungicide to control fruit rot on apple and pear. Commun. Agri. Appl. Biol. Sci. 70, 79–89.
Vorstermans, B., Laer, S.v., Creemers, P., Pujos, P., Jijakli, H., 2007. Improving control of storage diseases on apple by combining biological and physical post-harvest methods. Bulletin OILB/SROP 30, 423–426.
Wan, Y., Tian, S., 2005. Integrated control of postharvest diseases of pear fruits using antagonistic yeasts in combination with ammonium molybdate. J. Sci. Food Agri. 85, 2605–2610.
Wisniewski, M., Wilson, C., El-Ghaouth, A., Droby, S., 2001. Increasing the ability of the biocontrol product, Aspire, to control postharvest diseases of apple and peach with the use of additives. Bulletin OILB/SROP 24, 157–160.

Wu, T., Zivanovic, S., Draughon, F.A., Conway, W.S., Sams, C.E., 2005. Physicochemical properties and bioactivity of fungal chitin and chitosan. J. Agri. Food Chem. 53, 3888–3894.

Xiao, C.L., Boal, R.J., 2009. Preharvest application of a boscalid and pyraclostrobin mixture to control postharvest gray mold and blue mold in apples. Plant Dis. 93, 185–189.

Xiao, C.L., Kim, Y.K., 2010. Control of postharvest diseases in apples with reduced-risk fungicides. Stewart Postharvest Rev. 6, 1–6.

Xiao, C.L., Kim, Y.K., Boal, R.J., 2011. First report of occurrence of pyrimethanil resistance in *Penicillium expansum* from stored apples in Washington State. Plant Dis. Int. J. Appl. Plant Pathol. 95, 72.

Xu, L., Du, Y., 2012. Effects of yeast antagonist in combination with UV-C treatment on postharvest diseases of pear fruit. BioControl 57, 451–461.

Xu, X.B., Tian, S.P., 2008. Reducing oxidative stress in sweet cherry fruit by *Pichia membranaefaciens*: A possible mode of action against *Penicillium expansum*. J. Appl. Microbiol. 105, 1170–1177.

Yao, C., Conway, W.S., Ren, R., Smith, D., Ross, G.S., Sams, C.E., 1999. Gene encoding polygalacturonase inhibitor in apple fruit is developmentally regulated and activated by wounding and fungal infection. Plant Mol. Biol. 39, 1231–1241.

Yao, H.J., Tian, S.P., 2005. Effects of a biocontrol agent and methyl jasmonate on postharvest diseases of peach fruit and the possible mechanisms involved. J. Appl. Microbiol. 98, 941–950.

Yin, Y.N., Xiao, C.L., 2013. Molecular characterization and a multiplex allele-specific PCR method for detection of thiabendazole resistance in *Penicillium expansum* from apple. Eur. J. Plant Pathol. 136, 703–713.

Yu, T., Li, H.Y., Zheng, X.D., 2007. Synergistic effect of chitosan and *Cryptococcus laurentii* on inhibition of *Penicillium expansum* infections. Int. J. Food Microbiol. 114, 261–266.

Yu, T., Wang, L., Yin, Y., Wang, Y., Zheng, X., 2008. Effect of chitin on the antagonistic activity of *Cryptococcus laurentii* against *Penicillium expansum* in pear fruit. Int. J. Food Microbiol. 122, 44–48.

Zabka, M., Pavela, P., Slezakova, L., 2009. Antifungal effect of *Pimenta dioica* essential oil against dangerous pathogenic and toxinogenic fungi. Industrial Crops and Products. 30.

Zhang, H., Li, R., Liu, W., 2011. Effects of chitin and its derivative chitosan on postharvest decay of fruits: A review. Int. J. Mol. Sci. 12, 917–934.

Zhang, H., Ma, L., Turner, M., Xu, H., Dong, Y., Jiang, S., 2009. Methyl jasmonate enhances biocontrol efficacy of *Rhodotorula glutinis* to postharvest blue mold decay of pears. Food Chem. 117, 621–626.

Zhang, H., Wang, L., Zheng, X., Dong, Y., 2007. Effect of yeast antagonist in combination with heat treatment on postharvest blue mold decay and Rhizopus decay of peaches. Int. J. Food Microbiol. 115, 53–58.

Zhou, T., Northover, J., Schneider, K.E., Lu, X., 2002. Interactions between *Pseudomonas syringae* MA-4 and cyprodinil in the control of blue mold and gray mold of apples. Can. J. Plant Pathol. 24, 154–161.

Zhou, T., Yu, H., Errampalli, D., 2007. Strategies for biological control of fungal diseases of temperate fruits. In: Chincholkar, S.B., Mukerji, K.G. (Eds.), Biological Control of Plant Diseases. The Haworth Press, Binghampton, NY, pp. 191–221.

CHAPTER 7

Monilinia fructicola, *Monilinia laxa* (Monilinia Rot, Brown Rot)

Camilla Martini and Marta Mari
CRIOF, DipSA, University of Bologna, Cadriano, Bologna

Contents

OVERALL IMPORTANCE OF FUNGI

Worldwide, brown rot caused by *Monilinia laxa* (Aderhold & Ruhland) Honey, *Monilinia fructigena* (Aderhold & Ruhland) Honey and *Monilinia fructicola* (Winter) Honey is responsible for severe losses in important commercial crops such as peach and nectarine (*Prunus persica* (L.) Batsch), apricot (*P. armeniaca* L.), plum (*P. domestica* L.), sweet (*P. avium* L.) and sour cherry (*P. cerasus* L.), almond (*P. amygdalus* Batsch), apple (*Malus pumila* Mill.) and pear (*Pyrus communis* L.) (Batra, 1991), determining an important economic impact. *Monilinia fructigena* is prevalent on pome fruits and only a

Postharvest Decay
http://dx.doi.org/10.1016/B978-0-12-411552-1.00007-7

low percentage of isolates (less than 10%) derived from stone fruits. Direct yield losses result from blossom and twig blight, and from fruit rot before harvest and during the postharvest phase (Mari et al., 2012a). Losses depend on weather conditions and are especially severe if high humidity, warm temperatures and abundant rainfall prevail prior to harvest (Prusky, 1996; Bonaterra et al., 2003). Indirect losses are due to the cost of fungicide applications during the bloom and the pre- and postharvest periods (Thomidis and Michailides, 2010), however, the appearance of disease after fruit purchase, at consumer level, can negatively affect consumer satisfaction and propensity to consumption. Stone fruits are important in the agricultural economy of many countries; for example, the peach and nectarine world production reached 17 MT in 2011 (FAOSTAT, 2012) and, by calculating an underestimated average percentage of losses (10%) due to brown rot, *Monilinia* losses can account for 1.7 M€/year for peach and nectarine alone. In organic stone fruit orchards, *Monilinia* rot is a major yield-limiting factor that is difficult to control (Larena et al., 2005; Xu et al., 2007) and, in a wet season, for commercial organic orchards, it is not unusual to suffer 75% or more crop losses due to brown rot, despite the fact that copper-based fungicides are used. In some regions, organic cherry production is almost impossible because all attempts to control brown rot have failed.

Brown rot has been extensively studied in Europe for over 150 years, in North America for about 100 years, and more recently in other parts of the world, such as Australia, New Zealand, China and Japan. The first authentic description of the fungus causing brown rot on fruit was published in 1796 by Persoon. During the 19th century there was a greater awareness of fungal plant diseases in Europe and North America and the number of publications increased, including those concerned with brown rot on fruit. By the end of the 19th century it had been established that there are two brown rot species in Europe: *M. fructigena* and *M. laxa* (Byrde and Willetts, 1977). In North America, the fungus was initially thought to be one of the European species and only later was it found to be different, given the specific epithet of *fructicola*. Subsequently, *M. fructicola* was found in India, Japan, the Republic of Korea, Oceania, and North and South America, while it was considered a quarantine pathogen in Europe by the EPPO until 2001, when it was detected, for the first time, in peach orchards in France (Lichou at al., 2002), then in the Czech Republic (Duchaslavovà et al., 2007), in mummified peach fruit and trees in an orchard located in Lleida (Spain) (De Cal and Gell, 2009) and in stored nectarines in Italy (Pellegrino et al., 2009). Following these findings, *M. fructicola* was moved from A1 to A2 on the EPPO list. *Monilinia laxa* is a quarantine pathogen in China (though there is no finding) while in some states of North America it was detected for the first time in 2008 on 'Surefire' cherries in Niagara, in 2009 on cherries from Wayne, and in 2011 on sweet cherries and plums in New York, Rhode Island and Massachusetts by Cox et al. (2011). However, in a survey of California stone fruit in the first half of the 20th century, *M. laxa* was identified in over 79% of cases on 250 isolates of *Monilinia* spp. and the remaining 21% was represented by *M. fructicola* (Hewitt and Leach, 1939). In the early 1980s, both *M. fructicola* and *M. laxa* were widespread in plum and apricot growing areas of California (Michailides et al., 1987), suggesting that *M. fructicola* had partly displaced *M. laxa* in those orchards.

TAXONOMY AND MORPHOLOGY

The fungi determining brown rot are classified in the *Discomycetes*, a subdivision of the *Ascomycetes*; *Monilinia* is a genus of fungi in the family Sclerotiniaceae. The main diagnostic feature of the Discomycetes is that the apothecium or fruiting body originates from pseudosclerotia formed in mummified fruit partly or wholly buried in the soil, or in debris where the sexual spores (ascospores) are produced. The peculiar shape of the apothecium is cup- or disk-shaped and the inside or upper surface is lined with thousands of asci, containing eight single-celled ascospores, and interspersed with sterile hyphae. *M. laxa* apothecia are recorded less frequently than those of *M. fructicola*. The vegetative mycelium in *Monilinia* is well developed and is that of a typical Ascomycete. The hyphae are coenocytic and septate, with the septa perforated by a central opening. Hyphal walls and cytoplasm in infected tissues may be hyaline to variously pigmented, the pigments are usually light gray to tan or olive green, to nearly black in stromatic tissues.

Under warm, humid conditions, depending on the host, the mycelium produces chains of elliptical *Monilia*-type conidia on hyphal branches arranged in tufts (sporodochia). This type of arrangement, similar to beads on a string, is known as moniloid, from a Latin word, *monile*, necklace (Byrde and Willetts, 1977). The Discomycetes can also produce microconidia (spermatia) in a chain. The term microconidium refers to a small unicellular, usually globose or pyriform conidium-like, but non-germinative cell. It is not known whether the spermatia are functional; they seem to be involved in the fertilization of the fungus. The disease is known as 'brown rot fungi of fruits' since the three species present similar life cycles, symptoms, and host range (Byrde and Willetts, 1977; Batra, 1991).

The *Monilinia* species are difficult to distinguish from each other, especially between *M. laxa* and *M. fructicola*. The diagnostic protocol recommends isolation of *Monilinia* spp. from the host, followed by combining culture characteristics, such as growth rate, growth pattern and color, with morphological data like the conidial dimensions and the length of the germ tube (De Cal and Melgarejo, 1999). Following this protocol, sometimes atypical isolates of *M. fructicola* may be misidentified as *M. laxa* and *vice versa* (Martini, 2012).

Colonies of *M. laxa* have lower growth rates, about half that of *M. fructicola* on potato dextrose agar (PDA) at 22°C under 12–16 h near-UV light (320–380 nm) per 24 h (De Cal and Melgarejo, 1999). *Monilinia laxa* shows a markedly lobed colony margin and the colonies are rosette, and sporulation is sparse (Fig. 7.1). Characteristic black rings and arcs are associated with the petals of the rosettes in the colony. The bottom of the Petri dish shows black arcs or rings. The stromata of *M. laxa* are grayish or hazel in color. Under the conditions mentioned above, *M. fructicola* shows a growth rate ranging from 9 to 20 mm per 24 h with an average of approximately 13 mm (van Leeuwen and van Kesteren, 1998). The colony margin should be mostly entire, even and no rosettes with black arcs are present on the surface. There is considerable variation among isolates of *M. fructicola*, particularly in the amount of stromatal tissues on media. The stromata are grayish or hazel in color and have irregular crusts and discoid sclerotia that may develop on the agar surface or within the medium as colonies age. Abundant microconidia may be apparent macroscopically as black raised areas,

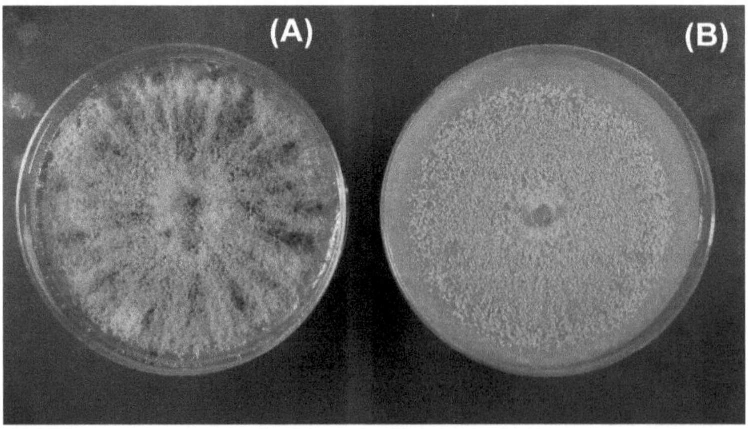

Figure 7.1 (A) *Monilinia fructicola* and (B) *Monilinia laxa* (right) on V8 agar.

particularly at the edge of the Petri dish (EPPO, 2009). The colonies of *M. fructigena* show lower growth rates, about half that of *M. fructicola* under the conditions mentioned above. Some isolates produce distinct concentric rings while others either do not sporulate or form only indistinct sporangeus bands (Byrde and Willetts, 1977). The colony color of *M. fructigena* is cream/yellow and the sporulation is sparse (EPPO, 2009).

Conidia can be useful for a preliminary *Monilinia* identification: conidia of *M. fructigena* are uniformly larger than those of *M. fructicola* and *M. laxa* (mostly 17–21 × 10–13 μm), and often form two germs tubes per conidia (van Leeuwen and van Kesteren, 1998). *Monilinia laxa* produces conidia similar in size to those of *M. fructicola*, germ tubes are single but short (150–350 μm) and twisted (EPPO, 2009) (Fig. 7.2A, B). However, conidial size is extremely variable and influenced by environmental factors and host species (EPPO, 2003) (Table 7.1). Conidia are blastic, formed in chains with the youngest spore at the distal end, occasionally they could be arthric, ellipsoid, ovoid or limoniform. Qualitative and quantitative characteristics are not sufficient to delineate the three brown rot species accurately, and other specific laboratory-based techniques are indispensable for correct identification. Willetts et al. (1977) found consistent differences in cell-wall-degrading enzyme patterns, in particular α-L-arabinofuranosidase, that could help in distinguishing between species of the brown rot fungi. van Leeuwen and van Kesteren (1998) proposed anastomiosis between germ tubes as an additional feature to distinguish *M. fructicola* from *M. laxa*; in fact, the first shows abundant anastomiosis, that rarely occurs in the second. Moreover, classic identification methods require at least 5–7 days of incubation so they are not adequate for the phytosanitary diagnosis, which

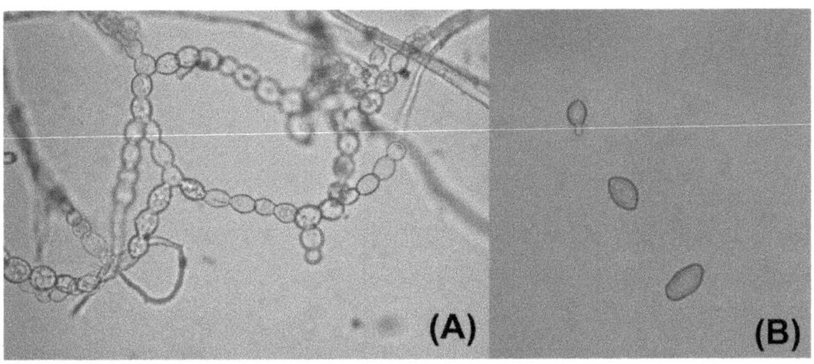

Figure 7.2 (A) Conidia chains of *Monilinia laxa* (×256), (B) conidia of *Monilinia fructicola* (×320).

Table 7.1 Main Morphological Differences Among *Monilinia* Species

Species	Conidial Dimension in Culture (μm)	Germ Tube	Sporulation[1]	Hyphal Diameter & Colony Shape[2]
M. fructicola	14.5–16 × 9.5–11	Long and branching far from spore	Abundant with microconidia and stromata	Relatively narrow and margin entire
M. laxa	11.5–17 × 8–11	Short and branching near spore	Abundant	Relatively narrow and lobed margin
M. fructigena	18–21 × 11.5–13	Long and branching far from spore	As *M. fructicola*, but much less abundant	Relatively wide and margin entire

[1]After 3 days in V8 Juice agar in the light
[2]After 12 h in nutrient medium
Source: EPPO, 2011; Martini, 2012 modified

should be rapid and reliable. Subsequently, several molecular diagnostic protocols have been developed for *Monilinia* species. The first methods were based on the use of SSU rDNA group I intron (Fulton and Brown, 1997, Snyder and Jones, 1999). Subsequently, other studies showed that these methods were not reliable and new polymerase chain reaction (PCR) primers were developed by Hughes et al. (2000), Ioos and Frey (2000), Coté et al. (2004) and Gell et al. (2008). To develop a more reliable method, Ioos and Frey (2000) designed specific primer pairs for the ribosomal internal transcribed spacer 1 (ITS1) region, the 5.8S rRNA gene, and the ITS2 region between the 18S and 28S rRNA genes. These primer pairs are specific for *M. laxa*, *M. fructicola*, and *M. fructigena* and can be used on conventional PCR assays. Subsequently, real-time PCR methods were developed by Luo et al. (2007) and van Brouwershaven et al. (2010). The last method is a TaqMan assay validated against *M. laxa*, *M. fructicola* and *M. fructigena* with the addition of *M. polystroma*, most similar to *M. fructigena*, at first found only in Japan (Coté et al., 2004) but later also detected in Hungary (Petroczy and Palkovics, 2009) and the Czech Republic (EPPO, 2011). The detection methods, using PCR assays, can discover and identify *Monilinia* spp. in at least 2 h. As few as 20 spores of *M. fructicola* were detected in air samples with real-time PCR but it is still difficult to detect small numbers of spores resting on the non-infected plant tissue. The plant tissues probably contain more PCR inhibitors than the air samples

(Ma et al., 2003). For this reason, the MALDI-TOF MS technique was recently proposed to differentiate between morphologically similar *Monilinia* spp. as a rapid, efficient and accurate method providing results in several minutes and requiring minimum sample handling. Since the MALDI analysis could be influenced by host origin of the fungi, the combination of this technique with capillary isoeletric focusing or capillary zone electrophoresis can solve the problem (Horka et al., 2012). These techniques obviously require specialized personnel and sophisticated equipment available in laboratories devoted to fungi diagnostic for pest alert; other methods that are simple but at the same time reliable have yet to be investigated.

BIOLOGY OF *MONILINIA FRUCTICOLA* AND *M. LAXA* INFECTION PROCESS

The pathogen infects aerial parts of host plants with a variety of symptoms, including blighting of blossoms, cankers on woody tissues and rotting of fruits. Blossom blight is the first symptom in the spring; it appears on the blossoms and develops when spores or conidia land on and penetrate flower stems of susceptible plants. The pathogen overwinters as mycelium in mummified fruit on the tree and in twig cankers or as pseudosclerotia in mummies on the ground. The new conidia are produced by mycelium on mummified fruit on the tree and in twig cankers, whereas the pseodiosclerotia in mummified fruit buried in the ground produce apothecia, which form asci and ascospores. *Monilinia laxa* conidia in mummified fruits or twig cankers and *M. fructicola* ascopores in mummies on the ground can cause the first infection. The conidia are windblown or carried on floral parts by rainwater splash or insects. The ascospores are discharged by the ascus, forming a whitish cloud over the apothecium. Any part of the flowers, stigma, stamens, petals or sepals can be an infection site. The infected tissue turns dark brown and the mycelium produces numerous conidial tufts on the rotten, shrivelled floral parts from which new masses of conidia are released. The mycelium advances rapidly into the blossom petioles and into the fruit spurs and the twigs, where a depressed, reddish-brown, shield-shaped canker forms. Some infected blossoms fall to the ground but others can remain attached to the tree for long periods and, during moist weather, the pathogen within the withered flowers produces tufts of spores over the infected tissues (van Leeuwen and van Kesteren, 1998; Hong and Michailides, 1999; Michailides et al., 2007; Ceredi and Mari, 2011) (Fig. 7.3A). The chains of conidia provide the

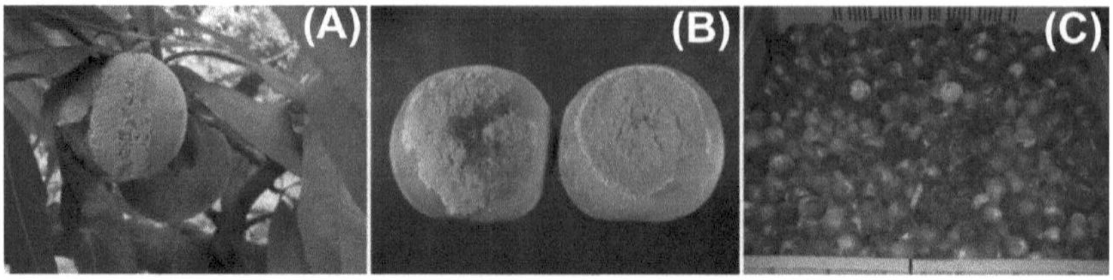

Figure 7.3 (A) *Monilina laxa* on peach on the field, (B) peaches infected by *Monilinia fructicola* (left) and *M. laxa* (right) and (C) *Monilia* rot on peaches after shelf-life.

inoculum for other parts of the same or neighboring trees. Infections of green fruit are not frequent, they show soft water-soaked, dark areas; rot first develops in clustered fruit, in fruit contact spots, and insect- or wind-damaged fruit, since clustered fruit are more favorable for disease development. For instance, fruit-to-fruit contact surfaces have microcracks, thinner cuticles, and microenvironmental conditions predisposing to fruit infections. Under moist conditions and on soft, ripe fruit, almost all the surface is covered with conidial tufts or vegetative mycelium, but when the relative humidity is low and/or the fruits are not ripe, no mycelium and little or no conidial tufts develop (Michailides et al., 2000).

Stem cankers usually develop from blighted twigs or fruit spurs by growth of mycelium in larger limbs of the tree. In the early stages of canker formation, the bark in the infected area dies, the tissue beneath the bark becomes sunken and discolored and, finally, an open wound develops. Gum is often exuded in the diseased area. Callus tissue usually develops around the canker and restricts further growth, so that the cankers heal up. If environmental conditions are suitable for sporulation, active cankers sometimes produce conidia and thereby provide a source of inoculum for fruit infection later in the season, when the fruit begins to ripen. Occasionally, the wounds continue to develop for several years and extensive damage of even death of the tree may be caused, but this is often the result of invasion by secondary organisms (Byrde and Willetts, 1977). The conidia can penetrate the fruits from wounds and lesions and, in some cases, they also gain access through stomata (EFSA, 2011). The fungus grows intercellularly at first, and secretes enzymes that cause maceration and browning of the infected tissues; it invades the fruit quite rapidly, also producing conidial tufts on the already rotted area. The new conidia can infect more fruits.

Infections occurring in the field can remain quiescent until the fruit reaches ripening, allowing *Monilinia* to overcome host defenses (Emery et al., 2000) and produce the disease (see Fig. 7.3B, C). These infections are often classified as 'quiescent' or 'latent' according to how the infection is initiated. There are different mechanisms involved in pathogen quiescence and several studies have reported that brown rot incidence during fruit development varies considerably: green fruits appear more resistant to the pathogen than later ripe fruits (Mari et al., 2003). *Monilinia* spp. use the ungerminated and germinated spores in different ways in correlation with cuticle and cell-wall thickness (Adaskaveg et al., 1991). The quiescent period lengthened as the cuticle and cell-wall thickness of peaches increased (Adaskaveg et al., 1989, 1991; Michailidis and Johnson, 1992) while, in nectarines, the development of *M. fructicola* quiescent infections was increased with decreasing fruit cuticle thickness (Michailidis and Johnson, 1992). Different hypotheses have been proposed to explain the association between the decrease in *Monilinia* susceptibility and the green fruit stage: (1) the higher mechanical resistance; (2) a different biochemical response between unripe and ripe fruits; and (3) the inhibitory substances abundant in the tissue of fruits at this stage. More recently, susceptibility of peach fruit to *M. laxa* has been characterized in more detail by monitoring it throughout fruit growth. It was found that, during the period corresponding to the pit hardening stage, fruit susceptibility drastically decreases (Northover and Biggs, 1990; Fourie and Holz, 2003; Mari et al., 2003; Keske et al., 2011). However, the molecular basis of this phenomenon is still not well understood and not yet elucidated; analyzing the transcript levels of the phenylpropanoid pathway key-gene expression in susceptible and resistant fruits found the presence of inhibitory compounds in the peach fruit peel during pit hardening was associated with their variation (Zubini, 2008).

Monolinia latent infections were documented in peach for the first time in 1971 (Kable, 1971); latent infections by *M. fructicola* on peach fruits were later described by Schlagbauer and Holz (1989) and in California by Michailides et al. (2000). Other authors also reported latent infection in apricot, plum, prune and cherry (Northover and Cerkauskas, 1994; Witting et al., 1997; Forsters and Adaskaveg, 2000).

ENVIRONMENTAL AND AGRONOMIC FACTORS FAVORING *MONILINIA FRUCTICOLA* AND *M. LAXA* INFECTION

High humidity and abundant rainfall are the environmental factors that favor epidemics of brown rot in stone fruit orchards (Prusky, 1996; Gell et al., 2008). Temperature plays an important role in *Monilinia* survival; in fact, under laboratory conditions, high temperatures (>25°C) reduced conidial viability, while high relative humidity (RH) (≥97%) reduced the rate of loss in

M. fructigena conidial viability (Xu et al., 2001). Climatic conditions can also affect the development of primary and secondary inocula; Villarino et al. (2010) found two significant correlations between the sources of primary inoculum and the climate conditions in the interval July–February in Spain: first, the mean temperature was significantly correlated with necrotic infected twigs and, second, the mean RH was negatively correlated with the number of overwintered non-abscised fruit. For *M. fructicola* apothecia (primary inoculum), soil moisture and adequate temperatures are necessary; mummies that were incubated at <90% of RH never produced apothecia when stored at 2°C and then transferred to warmer temperatures with light (Holtz et al., 1998). Non-abscised aborted fruit in the trees and thinned fruit on the orchard floor can be important sources of secondary inoculum for brown rot in humid and semi-arid climates (Landgraf and Zehr, 1982), while dry weather conditions suppressed the sporulation of *M. fructicola* on non-abscised aborted fruit in the trees during the growing season (Hong et al., 1997). However, the dry weather conditions require regular irrigation of the crop during the season and this practice could create a microclimate that encourages potential infection and sporulation of *Monilinia* spp. on overwintered aborted fruit during the growing season (Villarino et al., 2010).

The pathogen infects the aerial parts of host plants under moist conditions and temperatures between 20 and 25°C; at these optimum conditions, decay on ripe infected peaches may be visible within 48–72 h of infection. The optimal temperature range for *M. laxa, M. fructicola* and *M. fructigena* conidia germination is 15–30°C. The species are even able to germinate at suboptimal temperatures (0 and 35°C) but, at 38°C and less than 0°C, no germination was observed (Casals et al., 2010).

POSTHARVEST FACTORS FAVORING *MONILINIA FRUCTICOLA* AND *M. LAXA* INFECTION

Stone fruits are characterized by a relatively short postharvest life, passing quickly from ideal ripeness to over-mature fruits, depending in part on temperature and handling (Kader and Mitchell, 1989). In general, fruit maturity at harvest will determine postharvest life: a too high harvest maturity increases fruit susceptibility to decay and a too low harvest maturity decreases its potential flavor and texture qualities. In addition, peaches and nectarines are very responsive to high temperature exposure, such as delays between harvest and cooling. *Monilinia* rots routinely occur in the postharvest phase, during storage and transport, and can be typically more severe than in the preharvest phase, reaching high values (59%) (Larena et al., 2005). The losses can be very expensive because of the accumulated costs of growing, harvesting and storing these high-value fruits and they can be reduced by maintaining natural fruit resistance and by delaying senescence (Sanzani et al., 2009). In the last few decades, the use of appropriate postharvest technologies of storage, such as low temperature or a controlled atmosphere, has greatly reduced disease development, preventing fungal growth by postponing fruit ripening (Lurie, 2002). Maximum storage duration should not be exceeded, either in air or in a controlled atmosphere, to prevent the occurrence of senescence and the development of chilling injuries that increase fruit susceptibility to brown rot. In fact, when stone fruits are moved to a higher temperature than that used in storage to develop their typical flavor and texture and reach optimal eating quality, the pathogens resume growth and develop rapidly into decay lesions (Prusky, 1996; Neri, 2003; Gregori et al., 2008). It is important to stress that the management of diseases in each fruit species, but in particular for stone fruits, should be accomplished by integrated control programs that take into consideration the numerous factors influencing decay development, and above all fruit maturity at harvest and conditions applied during handling, storage and shelf-life of fruits. The management of *Monilinia* rots at retail and consumer sites, where the cold chain is often interrupted, is more critical. In addition, excessive handling in self-service displays or suboptimal disposal of fruits at retail stores often causes mechanical injuries (punctures, cuts or bruises) that create infection opportunities for pathogens that develop after purchase (Mari et al., 2009). Despite the use of the cited beneficial practices, modern storage facilities and techniques are not enough to prevent postharvest fruit losses due to *Monilinia* rots. In addition, as previously described, stone fruit infections by *Monilinia* spp. occur in the field and remain latent until harvest, therefore postharvest treatments to control brown rot may provide a curative effect for both established infections and potential conidia present on the surface of the fruit (Casals et al., 2012).

CONTROL

Conventional Approach

Cultural Practices

Some cultural practices are essential for brown rot management and they consist of a sanitization of the orchard by minimizing inoculum potential, reducing the risk of blossom and fruit infection. The removal of all remaining fruit from the tree, after the final picking, limits the infection of fruit peduncles and twigs, reducing the appearance of brown rot cankers. In addition, this practice prevents the formation of mummies within the tree immediately adjacent to susceptible blossoms in the spring. Another important aspect is the management of thinned fruits: in general, fruit thinned before pit hardening decomposes rapidly, whereas fruit thinned after pit hardening becomes infected on the orchard floor and serves as spore sources for the disease. It would thus be ideal to thin all cultivars before the pit hardening stage, although this practice is quite difficult since all cultivars reach this stage at the same time; alternatively, the thinned fruit should be removed from the orchard with additional costs. In addition, cultivation of the orchard floor and burying mummies to disrupt formation of apothecia is recommended; in fact, in orchard experiments, *M. fructicola* apothecia were observed only in plots with non-disturbed orchard floor vegetation, while no apothecia were found in either herbicide-treated or rototilled plots, showing the importance of these practices in inoculum reduction (Holtz et al., 1998). Excessive fertilization can influence the overcrowding of branches, and limiting the air circulation prevents a rapid drying, favoring the proliferation of pathogens, while preharvest calcium applications significantly reduced the incidence of brown rot infected fruit per tree at harvest and postharvest rots (Elmer et al., 2007). The integration of calcium foliar sprays into current brown rot management practices has been widely adopted by stone fruit growers in New Zealand as a practical tool to reduce brown rot losses. Disproportionate irrigation, particularly during the last days before harvest, can produce microlesions on fruit and allow *Monilinia* spores to penetrate into the fruit. In some peach production areas, removal of alternative hosts such as wild *Prunus* spp. could be an effective strategy (Byrde and Willetts, 1977). However, all these practices are partially successful but are not reliable for consistent, effective control. Until now there are no commercial peach cultivars resistant to brown rot. The 'Bolinha' peach is reported as resistant, but shows poor fruit quality, including small size, high susceptibility to bruising and extensive fruit dropping before harvest (Gradziel et al., 1998).

Fungicides

Control of *Monilinia* spp., especially *M. fructicola*, is critical in stone fruit production worldwide. Fungicide sprays are usually used to control brown rot in commercial orchards, however, the measures vary and depend on (1) the nature of inoculum sources, (2) the ontogeny of both the fungus and the host species concerned, (3) the modes of infection and the organs attacked, (4) the regional and local climate, and (5) economic and other social factors.

Effective brown rot control became possible after the introduction of captan and dichloran in the 1950s; the subsequent introduction of benzimidazole (MBC) fungicides in the 1970s dramatically improved the control of *Monilinia* spp. in the field. Two sprays of benomyl or thiophanate methyl during bloom were sufficient against brown rot in most environments, above all when rainfall was light. The wide and almost exclusive use of benomyl to control the disease produced the appearance of benomyl-resistant populations of *M. fructicola* in many countries only 5–6 years after its registration, in Michigan (Jones and Ehret, 1976), Australia (Whan, 1976), New York (Szkonik and Gilpatrick, 1977), then in South Carolina and California (Ogawa et al., 1981). The reason for the rapid development of resistance was also correlated with the mode of action of MBCs. They are, in fact, single site inhibitors of fungal microtubule assembly during mitosis and the resistance has been associated with a point mutation in the β-tubulin gene which results in altered amino acid sequences at the MBC binding site (Koenraadt et al., 1992). Two different MBC-resistance levels in *M. fructicola* strains were observed in California: low resistant (LR) isolates showed the EC_{50} values ranging from 0.61 to 0.96 μg mL^{-1}, while for high resistant (HR) isolates the EC_{50}s were greater than 50 μg of a.i./mL (Ma et al., 2003). Interestingly, the same authors reported the influence of temperatures on the HR isolates that were heat sensitive and did not show high resistance to benomyl when exposed to 31°C, while the LR isolates of *M. fructicola* were cold sensitive and did not show resistance to benomyl at temperatures below 15°C. Generally, in stone fruit production areas like Central Valley in California or in European countries like Spain, France and Italy, the temperatures during early bloom season could be approximately under 15°C and MBC fungicides may still be effective in controlling the blossom blight, since the LR isolates could seem not resistant to benomyl at these temperatures. Nevertheless, MBC fungicides cannot control the blossom blight caused by HR isolates of *M. fructicola*. In the case of *M. laxa*, HR isolates were detected in Greece on 37% of assayed samples (Malandrakis et al., 2012), increased with respect to a previous screening with a frequency ranging from 11 to 15% (Thomidis et al., 2009), while a low level of resistance to benomyl and thyophanate-methyl was reported in *M. laxa* isolates from California (Ma et al., 2005). HR isolates of *M. laxa* were found in both conventional and organic orchards where no fungicide sprays were applied for at least 8 years, showing that the high persistence in the field of these isolates, also after the interruption of fungicide treatments, was probably due to the elevated fitness of MBC-resistant isolates (Penrose, 1990).

After the appearance of *Monilinia* spp. MBC-resistant isolates, other new fungicides were registered and used in the control of brown rot, such as demethylation inhibitors (DMI, e.g. myclobutanil and fenexamide), succinate dehydrogenase inhibitors (SdhI, e.g. boscalid), dicarbossimides (e.g. iprodione) and quinine outer binding site inhibitors (QoI, e.g. pyraclostrobin and azoxystrobin). Although *Monilinia* species are classified by the FRAC (Fungicide Resistance Action Committee) as pathogens with a moderate risk of developing resistance to fungicides, the appearance of resistant isolates to the above mentioned fungicides in some regions after several years of exposure was reported (Elmer and Gaunt, 1994; Zehr et al., 1999; Luo et al., 2008). DMI fungicides are still an important component of anti-resistance strategies in view of the few effective alternatives and according to field testing in experimental orchards of some fungicides, such as captan, QoI, fenhexamid and DMI, in Georgia (Schnabel et al., 2004). The results obtained showed that the DMI fungicides were still among the most efficacious products for brown rot control. New products such as QoI fungicides may also be viable disease control alternatives. Azoxystrobin, pyraclostrobin (QoI) and boscalid (SdhI) are respiration inhibitors (RIs), both chemical classes are site specific and therefore prone to resistance development. A mixture of pyraclostrobin + boscalid was used for preharvest control of brown rot of stone fruit by Amiri et al. (2010). No cross-resistance was observed between the QoI and the SdhI fungicides, which implies that alternation or tank mixtures of these two chemical classes can be used as a resistance management strategy.

Across the world, recommendations for brown rot control with fungicide treatments vary depending on climatic conditions, *Monilinia* species presence and the legislative measures allowed in each country. In the USA, according to one of the suggested spraying programs, fungicides are applied at the following growth stages: blossom, immature fruit, ripening fruit, and mature fruit (in this program, only fungicides that have a minimum preharvest interval of three or fewer days are recommended). However, fungicides need not be applied to immature fruit unless wetness conditions are especially favorable for infection or injury caused by insects or cold has increased the likelihood of disease. Insect control may thus be important during the immature peach stage (Adaskaveg et al., 2008). In Brazil, disease control is based on protective treatments at the blooming and preharvest stages (Keske et al., 2011). In China, the most prevalent method of controlling brown rot is to combine postharvest treatments with dichloran and preharvest spraying with fungicides such as benomyl, captan or iprodione (Li and Yu, 2000). In

Europe, brown rot is controlled by fungicide spray programs in the field, and according to the guidelines for integrated production of stone fruit only two or three chemical treatments with DMI or QoI are recommended, one during the full bloom and one 7 or 10 days before harvest, while no chemical treatments are allowed on stone fruit after harvest (Casals et al., 2010). For *M. laxa*, eradicative sprays are also useful in late winter since such treatments destroy conidia on cankers (Byrde and Willetts, 1977).

Postharvest Practices

To prevent infections at harvest and during storage and transit, fruit should be picked and handled with the greatest care to avoid punctures and skin abrasions on the fruit, which enable the brown rot fungus to gain entrance more easily. All fruit with brown rot spots should be discarded; for instance, fruit-to-fruit contact surfaces have microcracks, thinner cuticles, and micro-environmental conditions predisposing fruit infections (Michailides et al., 2000). Protection from heating after harvest and rapid movement to cooling are thus important to avoid excessive flesh softening and also to reduce postharvest decay (Crisosto et al., 1995).

Alternative Control Means

Current control of fruit diseases requires a considerable reduction in the use of conventional fungicides. This is demanded by public concerns about environmental contamination and risks to human health associated with chemical residues in fruits, as well as loss of efficacy in pathogen control by some synthetic compounds as a result of the appearance of resistant isolates of pathogens. In addition, the requirements in sustainable agriculture, integrated crop management and organic production, which have increased in the last few decades, have resulted in the need to develop other methods to control brown rot (Nunes, 2012). Many studies conducted in recent years have defined three main research fields within the alternative methods for brown rot control: they are (1) biological control *sensu strictum* with microbial antagonists, (2) use of natural bioactive compounds and (3) use of physicochemical methods, all characterized by a great number of studies and widely reviewed (Droby et al., 2009; Sharma et al., 2009; Nunes, 2012).

Microbial Antagonists

The postharvest phase is suited to the application of biological control agents (BCAs). In restricted environments such as storage rooms, parameters like temperature, RH and gas composition can be altered to favor antagonist growth. Injuries caused during harvest and handling can be protected with a single application of BCA directly on the infection site, producing an intimate contact between antagonist and pathogen that improves its activity (Wilson and Wisniewski, 1989). The first study on biological control of *M. fructicola* in the postharvest phase with BCA appeared in 1984 (Pusey and Wilson, 1984); testing two bacteria (*Pseudomonas* and *Bacillus*) isolates, the authors demonstrated the possibility of controlling *M. fructicola* infections on peaches and laid the fundamentals of a new strategy to combat postharvest diseases. During these last 30 years, interesting results have been obtained using a wide range of antagonists: bacteria, yeasts and fungi. The isolates generally belonged to microflora present on the fruit surface and their selection may have ecological advantages since they seem better adapted to the environment than antagonists originating from microflora casually deposited on the fruit or residing mainly in the soil. In addition, they may be more acceptable as postharvest fruit treatment to the public and regulatory agencies, since they have been eaten with fruit for millennia without known adverse effects (Janisiewicz et al., 2011). Within bacteria, *Bacillus* spp. represents a promising candidate for biological control of *Monilinia* spp. due to its ability to produce a wide variety of antimicrobial compounds, including iturin (Gueldner et al., 1988) and lipopeptide antibiotics (Yañez-Mendizibal et al., 2011). However, the production of these substances was never detected on fruit, raising doubts on the effective role of antibiosis in postharvest disease control. Another bacterium, *Pantoea agglomerans*, assayed against natural infections of brown rot, showed a reduction of decayed fruits of over 50% (Bonaterra et al., 2003). The propensity for *Bacillus* in biological control derives from its high levels of resistance to extreme environmental conditions that make the bacterium adequate for a stable and efficient formulation. In fact, the *B. subtilis* CPA-8, active against *M. frcuticola*, can survive spray-drying formulation thanks to the heat-resistant endospore production, while the survival in the same formulation of a non-spore-forming BCA such as *P. agglomerans* was very low, less than 2% (Yañez-Mendizibal et al., 2011). Treatments based on yeasts were effective in reducing *Monilinia* spp. infections (Zhang et al., 2010; Mari et al., 2012a). Yeast and yeast-like microorganisms deserve particular attention for their use in the future as BCAs, since their activity does not generally depend on the production of toxic metabolites and they are easily cultivated with simple nutritional requirements for large scale production (Droby and Chalutz, 1994). Within yeasts, *Aureobasidium pullulans* isolates were investigated with success against many postharvest pathogens on different hosts such as citrus (Wilson and Chalutz, 1989), grape (Schena et al., 2003), pome fruits (Mari et al., 2012b), stone fruits (Zhang et al., 2010), and strawberries (Adikaram et al., 2002). *Aureobasidium pullulans*, strains L1 and L8, completely inhibited *Botrytis cinerea* infections on apple at a low concentration (10^7 cfu mL^{-1}); they were active on *Monilinia* spp. at low temperature (0°C) on peach (Mari et al., 2012a) and also on *Penicillium expansum* on apple under a controlled atmosphere (Mari, personal communication). The efficacy of this BCA against a wide range of pathogens, on different harvested commodities, also at low concentrations, and its compatibility with storage conditions make it a good candidate as an ideal antagonist, although other criteria such as genetic stability, easy formulation, capacity to survive under adverse environmental conditions, etc., also have to be satisfied. In the past, the control of brown rot with treatments based on fungal antagonists has been proposed. First, a field application of *Epicoccum nigrum*, a component of resident mycoflora of peach twigs and flowers, to control postharvest *Monilinia* rots was studied in three European countries; the results showed that two applications of the antagonist *E. nigrum* in the orchard (at bloom and preharvest) significantly reduced brown rot of peaches after harvest in five trials out of the seven tested (Larena et al., 2005). Similarly, a postharvest treatment with fresh cells of *E. nigrum* obtained a significant reduction of brown rot on nectarines both naturally and artificially infected (Mari et al., 2007). Other fungal biocontrol agents, such as *P. frequentans*, achieved control of *Monilinia* rot by more than 50% but a conidial viability of the antagonist, up to 40%, was necessary for this level of pathogen control (Guijarro et al., 2007), revealing the importance of an optimal formulation to achieve an acceptable disease reduction that Chalutz and Droby (1997) defined around 95–98% in the postharvest phase. Recently, endophytic fungi were isolated from plum (*Prunus domestica*) leaves and their antagonistic activity was tested against *M. fructicola* by volatile inverted-plate tests. The most frequently isolated species was *Phaeosphaeria nodorum* that produced inhibitory volatiles to *M. fructicola*. The volatiles were identified as ethyl acetate, 3-methyl-1-butanol, acetic acid, 2-propyn-1-ol, and 2-propenenitrile; they inhibited growth and reduced width of the hyphae, causing a disintegration of the hyphal content (Pimenta et al., 2012). A volatile compound (phenethyl-alcohol) was also produced from a strain of *P. expansum* (R82) and in *in vitro* trials it completely inhibited the germination and the mycelium growth of *M. laxa* at the concentrations naturally produced (0.726 μg mL^{-1}) (Rouissi et al., 2013).

Despite this intense production of data and good results on biological control of brown rot with BCAs, their development for postharvest disease control is a long and expensive process if compared with the relatively small potential market for the use of postharvest biofungicides (Nunes, 2012). In fact, only a few of the BCAs reported to control postharvest diseases of fruits under laboratory conditions are commercialized, and within these those active against *Monilinia* rots are even fewer; 'Biosave' 10LP and 110 based on *P. syringae* (strain 10LP and 110); 'Serenade' based on *B. subtilis*, are two biofungicides commercialized for brown rot control on stone fruits but not yet registered for postharvest treatments in Europe. Biofungicides are therefore not routinely applied in the packing houses, the main reasons also being the relative ineffectiveness compared to fungicide activity, the difficulty of obtaining adequate formulations and in controlling previously established infections, so-called 'latent infections' that, in the case of *Monilinia* spp., play an important role in the appearance of brown rot in the postharvest phase. The possibility of selecting new BCAs, analyzing the direct interactions between pathogen structures involved in latent infections (appressoria) and antagonists was therefore recently studied (Janisiewicz et al., 2011). These experiments could result in a new BCA generation that will improve postharvest *Monilinia* spp. control.

Natural Bioactive Compounds

Natural compounds derived from plants and animals have received attention as one of several non-chemical control options of postharvest fruit diseases and their potential has long been recognized (Ark and Thompson, 1959). Plants produce a myriad of secondary metabolites important for their interaction with the environment; many of them are associated with the defense system and can function as fungal inhibitors (Treutter, 2005). Some antimicrobials, generally stored in secretory glands, ducts or glandular trichomes, are constitutive and occur in healthy plants in their active forms. Others occur as inactive precursors and are activated in their biologically active form in response to pathogen or tissue damage.

Plant Derivates

Plant Extracts

Many results have been reported on the antimicrobial properties of plant extracts containing different classes of compounds like phenols that occur in plants as preformed compounds. Plant extracts from peach peel tissue at fruit pit hardening showed considerable inhibition of brown rot in TLC assays (Martini, 2012). These results can be associated with the variation of gene expression of phenylpropanoid pathway key genes as explained previously by Lee and Bostock (2007). Similarly chlorogenic acid and its isomer, neochlorogenic acid, were found to be the major phenolic compounds in the flesh and peel of three peach cultivars. The results provide evidence that the amount of phenolic compounds in the fruits could be causally linked to the differences in disease susceptibility of immature and mature fruits (Villarino et al., 2011).

Extracts obtained from nine wild edible herbaceous species were active against some important postharvest pathogens such as *M. laxa*. The inhibition efficacy of the extracts was ascribed to the presence of some caffeic acid derivatives and/or flavonoids (Gatto et al., 2011). Likewise, extracts from perennial Mediterranean weeds showed varying degrees of antifungal activities *in vitro* trials on colony growth and conidia germination of *M. laxa* and *M. fructigena* (Mamoci et al., 2011).

Volatiles

The defatted meal of *Brassica carinata* containing GLCs was activated by hydrolysis in order to obtain an isothyocianate (ITC), allyl-isothiocyanate (AITC). AITC vapors were active against *M. laxa* in *in vitro* tests and on peaches and nectarines, reducing the brown rot incidence by more than 89% (Mari et al., 2008). This evidence showed that some volatile substances such as ITCs could potentially be successfully employed in the postharvest phase as gaseous treatments in a process defined 'biofumigation'. In addition, the use of bio-based chemicals obtained from renewable natural resources fit well with the goals of the European technology Platform 'Plant for the Future'. Other volatile compounds, such as flavor compounds, can be perceived by the olfactory receptors and confer an odor. Some of them, defined 'character impact compounds', possess the typical aroma of the products that contain them, while other compounds only contribute to the overall odor of a product. Some flavor compounds were assayed against many postharvest pathogens including *Monilinia* spp. and were found to be active inhibitors of pathogen growth. The high electrophilic properties of the carbonyl group adjacent to the double bond make *trans*-2-hexanal particularly reactive with nucleophiles, such as protein sulfydryl and the amino groups of the pathogens. Postharvest exposure to *trans*-2-hexanal was found to reduce significantly the infections of *M. laxa* in stone fruits (Neri et al., 2007) only when fumigation was applied immediately (2 h) after pathogen inoculation. Carvacrol, a monoterpenoid phenol, making up 60–80% of oregano essential oil and its isomer, thymol, that occurs mainly in thyme were used for postharvest fumigation of stone fruits. After fumigation of plum with thymol, the compound was found deposited in large crystals on the surface of conidia and hyphae of *M. fructicola*. These results were partially confirmed by Svircev et al. (2007) who detected thymol crystallized on the outer surface of the fungal cell wall, while the cytoplasm of the thick-walled spores did not appear to differ from the untreated spores. The antimicrobial activity of carvacrol and thymol has been ascribed to the hydrophobicity of the compound, the presence of a phenolic hydroxyl group in the molecule and an adequate system of delocalized electrons that allows the OH group to release its proton (Ben Arfa et al., 2006). Both compounds were effective in controlling *M. fructicola* on cherry (Tsao and Zhou, 2000) and on apricots and plums (Liu et al., 2002). However, these treatments caused phytotoxic symptoms and off-flavors in cherries and firmer texture and phytotoxicity in apricots. Recently, results obtained in *in vitro* trials showed that the thyme oil in the vapor-phase at 5 mL/plate completely inhibited the radial mycelium growth of *M. fructicola* (Sellamuthu et al., 2013). Likewise, thymol fumigation (2 μg mL^{-1}) immediately affected *M. fructicola* conidia with a 50% reduction in viability (Svircev et al., 2007).

Jasmonates

Jasmonic acid (JA) and its methyl ester (MJ), collectively referred to as jasmonates, are naturally occurring plant regulators, known to regulate plant development and response to environmental stress. The direct inhibitory effect of jasmonates on spore germination and germ tube elongation of some fungal pathogens was found on *Colletotrichum acutatum* (Cao et al., 2008) and *P. expansum* (Yao and Tian, 2005) while no direct influence on *M. fructicola* was observed (Tsao and Zhou, 2000). These different results could be related to different sensitivity levels of fungal species to MJ, therefore jasmonates could be considered the signal molecules in fruits, inducing defense responses and increasing resistance (Ding et al., 2002). A large number of studies have demonstrated the importance of many metabolites as plant defense compounds against pest and pathogens including *Monilinia* spp. However, further research is needed to investigate the activity of the compounds in large-scale experiments, their mode of action and their degradation in organisms which are not still fully understood. The different responses found in many studies indicate cultivar specificity in the product–pathogen–compounds interaction and a barrier to the use of some plant bioactive compounds may not be efficacy, but rather the off-odors caused in fruits and/or phytotoxicity. Moreover, if natural bioactive compounds are present in low concentrations in human food as flavors, their application in higher doses for pathogen control could increase human exposure to these compounds and cause toxic effects. Possible phytotoxicity, effects on quality traits, human toxicity, and treatment residue in products should be examined closely in order to determine the practical and economic feasibility of this alternative mean for brown rot control.

Animal Derivatives

Chitosan

Among the natural compounds derived from animals, chitosan, a high molecular polymer, non-toxic, bioactive agent, has become a useful appreciated compound due to its fungicidal effects and elicitation of defense mechanisms in plant tissues (Terry and Joyce, 2004). Its effects against plant pathogens causing postharvest diseases have been reviewed by Bautista-Baños et al. (2006). The *in vitro* ability of chitosan (1%) to reduce mycelial growth of some postharvest fungal pathogens including *M. laxa* was comparable to that obtained with the synthetic fungicide fenhexamid (Feliziani et al., 2013). Chitosan could act directly on the microorganisms but it also has the potential to induce resistance in fruit. Li and Yu (2000) suggested that the superoxide dismutase activity in chitosan-treated peaches contributed to delaying the degeneration of *M. fructicola*-infected cells and, therefore, slowed down the spread of the disease. Moreover, the same authors reported a lower fungicidal effect of chitosan on peaches artificially inoculated with *M. fructicola* compared to the fungicide prochloraz.

Physicochemical Methods

The use of physicochemical methods to prevent brown rot in stone fruits has been widely investigated. Heat, radio frequency (RF) energy, and ultraviolet C (UV-C) irradiation are only some of the main physical methods assayed in recent years to replace synthetic fungicides.

Heat

Heat has been applied in different ways as hot water (HW) dips (Jemirc et al., 2011; Spadoni et al., 2013), hot air (curing) (Casals et al., 2010) and hot water brushing (Karabulut et al., 2002), and it effectively controlled *Monilinia* spp. in stone fruits. Conidia germination of *M. laxa*, *M. fructicola* and *M. fructigena* were completely inhibited with a dip in water at 55°C for 60 s, while lower temperatures required longer times of exposure to heat to obtain significant inhibition (Casals et al., 2010; Spadoni et al., 2013). *Monilinia fructicola* appeared more resistant than the other two species due probably to its greater ability to tolerate the intracellular ROS accumulation (Helmerhorst et al., 2001). In *in vivo* trials, the best results were obtained by dipping fruit for 20 s in water at 60°C, with a reduction of brown rot caused by *M. laxa* ranging from 51% (Casals et al., 2010) to 94.7% (Spadoni et al. 2013). Very short treatments (20–60 s) are optimal to accelerate fruit handling in packing houses and better for commercial application than the longer treatments (6–12 min) reported by other authors (Jemric et al., 2011; Liu et al., 2012); however, there are few data on the effects of HW treatment on fruit naturally infected and in commercial trials. Spadoni et al. (2013) showed a significant reduction (up to 88.2%) of infected fruits after an HW treatment carried out with a dipping machine with a working capacity of 40 bins per hour. Although HW treatment requires specialized equipment, it can be considered a sustainable method, with substantial economic and energy savings, since in industrial packing houses the heating of the water can be obtained from refrigerant gas (ammonia, carbon dioxide, etc.) coming from the cooling store plants, considering that the heat of this gas is usually dispersed in the environment through the condensers. In addition, HW treatment appears particularly attractive to the fruit industry since it could be immediately utilized and incorporated into handling practices before storage without extensive technical modifications. Curing fruit with hot wet air (50°C for 2 h at 95–99% RH) completely inhibited *M. laxa* and *M. fructicola* in four peach and nectarine varieties, without affecting internal and external fruit appearance (Casals et al., 2010). Although the beneficial effects of curing in reducing decay and chilling injury in various fruit species are well documented, curing is not widely utilized on a commercial scale; several factors, such as the duration of treatment and the high cost of heating a large volume of fruit in a confined space, limit its application. However, heat treatments have a greater advantage than other non-conventional control methods of fruit disease (such as BCA or plant bioactive compounds), since they do not require any registration from public agencies like the European Commission or the Food and Drug Administration while, for example, the registration process of a biofungicide in Europe is more difficult than elsewhere, in terms of time and costs (Ehlers, 2011). In addition, more studies should be carried out to determine the effect of heat treatments (HW and curing) on *Monilinia* spp. latent infections. In fact, while the conidia on the fruit surface can be easily inactivated by a dip treatment, the suppression of viable conidia or hyphae localized below the epidermal cell appears more difficult with certain methods that act on the surface of fruit, such as BCAs and volatile compounds, although the heat can stimulate certain host defense mechanisms active also against latent infections. Actually, the activity of heat depends on at least two components: the first is a direct and lethal action on fungal inoculum such as spores and mycelium present on the fruit surface or in the first layers of peel. The second component is an indirect action of heat on the host, mediated by a stress-induced response of fruit including disease resistance (Liu et al., 2012).

Irradiation

The need to achieve fast and effective thermal treatment has resulted in the increased use of RF energy not only to heat foods, but also to disinfect commodities and to control postharvest diseases such as brown rot. Casals et al. (2010) were the first to study RF heating to control postharvest brown rot, demonstrating the efficacy of RF treatment for 18 min of exposure time for brown rot control in peaches, not in nectarines. The effectiveness of treatment was improved by dipping the fruits in water during the RF treatment (Sisquella et al., 2013a). The activity of UV-C, assayed in *in vitro* trials, showed an inhibition of mycelial growth of *Monilinia* spp. after long-wave (320–380 nm) UV exposure (De Cal and Melgarejo, 1999), but UV-C light had no effect against brown rot in cherries (Marquenie et al., 2002) and in peaches (Bassetto et al., 2007).

GRAS Compounds

Common food additives generally recognized as safe (GRAS) include salts such as sodium carbonate, sodium bicarbonate, potassium sorbate, sodium propionate, etc.; they are allowed with no restrictions for many applications in European and North American regulations. They have been tested in numerous trials and appear to be interesting tools to manage postharvest decays including brown rot (Gregori et al., 2008, Palou et al., 2012). In addition to their consistent antimicrobial activity, they are inexpensive, readily available and suitable for postharvest handling practices that use water to float fruit out of field bins and to remove field heat from fruit by hydrocooling. However, the compounds like potassium sorbate and sodium benzoate that showed a good activity in preliminary trials, lacked effectiveness and persistence when tested against *M. fructicola* in small-scale trials after 60 s dips in aqueous solutions at ambient temperatures, showing a reduction of brown rot incidence of less than 40% (Palou et al., 2012). Adversely, a treatment based on potassium sorbate showed good control of *Monilinia* spp.

(over 80% in four of five trials) on peaches and nectarines naturally infected by *Monilinia* spp. (Gregori et al., 2008). However, the GRAS additives also present some drawbacks; they are fungistatic, not very persistent and involve a minimal risk of injury to the fruit and for these reasons are not routinely applied.

Sanitizers

To prevent fruit infections during the postharvest stage, a novel technology has recently been proposed: the use of ozonated or electrolyzed water (EO) for fruit washing. Ozone properties and applications were recently reviewed (Miller et al., 2013); exposure for 1 min in ozonated water ($1.5\,\text{mg g}^{-1}$) significantly reduced brown rot incidence, although the conidia placed in peach wounds before treatment were not affected by ozone (Smilanick et al., 2002). A dip treatment in EO water was used to evaluate its ability to prevent or delay the onset of *Monilinia* rots on fruit during postharvest packaging and in retail shelf environments. The results suggest that this solution may prove to be effective for postharvest sanitation of fruit surfaces prior to packaging and may increase the shelf-life of fruit in commercial settings (Guentzel et al., 2010).

Integrated Methods

All these alternative methods used individually did not always achieve an acceptable level of brown rot reduction and some of these have a poor effect against future infection, after treatment or *vice versa* against established infections before treatment (Droby et al., 2009). To overcome this drawback, integrated strategies have been proposed and widely investigated. The combination of HW and ethanol improved the control of *M. fructicola* in peaches and nectarines compared to HW and ethanol alone (Margosan et al., 1997). Similarly, a combination of peracetic acid and HW showed a greater reduction of brown rot in 'Mountain Gold' and 'Rome Star' peaches inoculated with *M. fructicola* and treated with 40°C-heated peracetic acid ($200\,\text{mgL}^{-1}$) than that observed in fruit treated only with HW or peracetic acid (Sisquella et al., 2013b). When used alone, an isolate of *A. pullulans* (PL5) reduced the *M. laxa* infections on peaches to 25.8%, compared with 74.2% for the untreated control, but when applied together with HW (55°C per 50 s) the brown rot incidence decreased to 17.5%, while when HW was used alone the disease incidence was 30.0%. These results suggest that HW greatly increased the effectiveness of the antagonist (Zhang et al., 2010). The application of sodium bicarbonate (GRAS) improved the efficacy of the antagonist yeast *Cryptococcus laurentii* against brown rot on sweet cherry. It is postulated that the enhancement of disease control is directly caused by the inhibitory effects of the additive on pathogen growth, and indirectly because of the relatively small influence of the additive on the growth of BCA (Qin et al., 2006). A possible synergistic effect between *D. hansenii*, an antagonist yeast, and UV-C irradiation in controlling brown rot incidence on both artificially inoculated and naturally infected peaches was observed by Stevens et al. (1997). The superiority of the combined treatment of UV-C and yeast was probably due to the ability of UV-C to control deep-seated infections such as a latent infection, whereas the yeast controlled only the superficial infection originating in recent wounds. In addition, the control of latent infection by UV-C but not by the antagonist resulted from the induction of fungitoxic substances within tissues of UV-C treated peaches (Wilson et al., 1994). Heat treatment was effective in reducing *M. fructicola* infections that had become established in the field and it did not provide protection for further infections after the treatment, before cool storage while the combination of curing with BCA (*B. subtilis*) showed the best brown rot control on 'Baby Gold' and 'Andros' peaches, first cured (50°C for 2 h), treated with BCA than experimentally inoculated with *M. fructicola* (Casals et al., 2012).

CONCLUDING REMARKS

The importance of brown rot caused by *Monilinia* spp. in stone fruits is evident worldwide, as are the difficulties for controlling the disease development (www.monilinia.org). Although the infections occur in the field, fruit losses are prevalent in the postharvest phase and the control means, proposed until now, are insufficient in different ways. Disease control obtained with fungicide treatments in the pre- and postharvest phases involves a considerable impact on the environment and on human health. The appearance of fungicide resistance in *Monilinia* spp. isolates notably reduces their effectiveness and no fungicide applications are allowed after harvest in Europe. As reviewed, even though significant progress has been achieved in the reduction of pesticide use with the different alternative methods, it seems to be unrealistic to assume that these methods have the same fungicidal activity as fungicides. In this scenario, a multidisciplinary approach that integrates sanitary and alternative strategies has to be investigated more fully. This new strategy for managing brown rot in a sustainable model will start from an adequate prevention of pathogen spread by agronomic interventions such as orchard sanitization to reduce inoculum sources. New tools could also be very useful in the reduction of disease, such as the selection of cultivars less tolerant to *Monilinia* spp. if not resistant, or the development and validation of mathematical models that can be used to predict the incidence of visible brown rot and latent infection at harvest and the subsequent development during the postharvest phase. In the future, the use of postharvest biocontrol and biological products in general will probably continue to increase slowly but will complement or be combined with low-risk chemical fungicides, natural antimicrobial substances and other physical means for an integrated approach to brown rot control (Droby et al., 2009). The tailoring of a complete specific strategy for each situation (species: cherry, peach, apricot, nectarine, plum; climatic and seasonal conditions; conventional or organic production; destination market, etc.) should be implemented. In particular, since stone fruits are characterized by a relatively short postharvest life, the destination of fruits after harvest (time between harvest and sale, distance between the site of production to the point of sale, etc.) must drive the implementation of programs for brown rot management.

REFERENCES

Adaskaveg, J.E., Feliciano, A.J., Ogawa, J.M., 1989. Comparative studies of resistance in peach genotypes to *Monilinia fructicola*. Phytopatology 79, 1183–1184.

Adaskaveg, J.E., Feliciano, A.J., Ogawa, J.M., 1991. Evaluation of the cuticle as a barrier to penetration by *Monilinia fructicola* in peach fruits. Phytopathology 81, 1150.

Adaskaveg, J.E., Schnabel, G., Forster, H., 2008. Disease of peach caused by fungi and fungal-like organisms: biology, epidemiology and management. In: Desmond, R.L., Bassi, D. (Eds.), The Peach: Botany, Production and Uses. CAB International, Oxfordshire, pp. 352–406.

Adikaram, N.K.B., Joyce, D.C., Terry, L., 2002. Biocontrol activity and induced resistance as a possible mode of action for *Aureobasidium pullulans* against grey mould of strawberry fruit. Australas. Plant Pathol. 31, 223–229.

Amiri, A., Brannen, P.M., Schnabel, G., 2010. Induced sensitivity in *Monilinia fructicola* field isolates from South Carolina and Georgia in respiration inhibitor fungicides. Plant Dis. 94, 737–743.

Ark, P.A., Thompson, J.P., 1959. Control of certain plants with antibiotics from garlic (*Alium sativum* L.). Plant Dis. Rep. 43, 276–282.

Bassetto, E., Amorim, L., Benato, E.A., Gonçalves, F.P., Lourenço, S.A., 2007. Effect of UV-C irradiation on postharvest control of brown rot (*Monilinia fructicola*) and soft rot (*Rhizopus stolonifer*) of peaches. Fitopatologia Brasileira 32, 393–399.

Batra, L.R., 1991. World Species of Monilinia (Fungi): their Ecology, Biosystematics and Control. . [Mycologia Memoir No. 16.], J. Cramer, Berlin.

Bautista-Baños, S., Hernandez-Lauzardo, A.N., Velázquez-delValle, M.G., et al., 2006. Chitosan as a potential natural compound to control pre and postharvest diseases of horticultural commodities. Crop Prot. 25, 108–118.

Ben Arfa, A., Combes, S., Preziosi-Belloy, L., Gontard, N., Chalier, P., 2006. Antimicrobial activity of carvacrol related to its chemical structure. Lett. Appl. Microbiol. 43, 149–154.

Bonaterra, A., Mari, M., Casalini, L., Montesinos, E., 2003. Biological control of *Monilinia laxa* and *Rhizopus stolonifer* in postharvest of stonefruit by *Pantoea agglomerans* EPS 125 and putative mechanisms of antagonism. Int. J. Food Microbiol. 84, 93–104.

Byrde, R.J.W., Willetts, H.J., 1977. The Brown Rot Fungi of Fruit. Their Biology and Control. Pergamon Press, New York.

Cao, S., Zheng, Y., Tang, S., Jin, P., Wang, K., Wang, X., 2008. Effect of methyl jasmonate on the inhibition of *Colletotrichum acutatum* infection in loquat fruit and possible mechanisms. Postharvest Biol. Technol. 49, 301–307.

Casals, C., Teixido, N., Vinas, I., Lamarca, N., Usall, J., 2010. Combination of hot water, *Bacillus subtilis* CPA-8 and sodium bicarbonate treatments to control postharvest brown rot on peaches and nectarines. Eur. J. Plant Pathol. 128, 51–63.

Casals, C., Elmer, P.A.G., Vinas, I., Teixido, N., Sisquella, M., Usall, J., 2012. The combination of curing with either chitosan or *Bacillus subtilis* CPA-8 to control brown rot on infections caused by *Monilinia fructicola*. Postharvest Biol. Technol. 64, 126–132.

Ceredi, G., Mari, M., 2011. Moniliosi delle drupacee, prevenzione e difesa per una corretta gestione. Informatore Agrario 23, 57–64.

Chalutz, E., Droby, S., 1997. Biological control of postharvest diseases. In: Boland, G.J., Kuykendall, L.D. (Eds.), Plant-Microbe Interactions and Biological Control. New York, Dekker.

Coté, M.J., Tardiff, M.-C., Meldrum, A.J., 2004. Identification of *Monilinia fructigena, M. fructicola, M. laxa*, and *Monilia polystroma* on inoculated and naturally infected fruit using multiplex PCR. Plant Dis. 88, 1219–1225.

Cox, K.D., Villani, S.M., Raes, J.J., Freier, J., 2011. First report of brown rot on sweet cherry (*Prunus avium*) and plum (*P. domestica*) and shoot blight on apricot (*P. armeniaca*), Kwanzan cherry (*P. serrulata*), and sweet cherry (*P. avium*) caused by *Monilinia laxa* in New York, Rhode Island and Massachusetts. Plant Dis. 92, 1484.

Crisosto, C.H., Machell, F.G., Scott, J., 1995. Factors in fresh market stone fruit quality. Postharvest News Inf. 6, 17N–21N.

De Cal, A., Gell, I., 2009. First report of brown rot caused by *Monilinia fructicola* in peach orchards in Ebro Valley, Spain. Plant Dis. 93, 763.

De Cal, A., Melgarejo, P., 1999. Effects of long-wave UV light on *Monilinia* growth and identification of species. Plant Dis. 83, 62–65.

Ding, C.-K., Wang, C.Y., Gross, K.C., Smith, D.L., 2002. Jasmonate and salicylate induce the expression of pathogenesis-related-protein genes and increase resistance to chilling injury in tomato fruit. Planta 214, 895–901.

Droby, S., Chalutz, E., 1994. Mode of action of biocontrol agents of postharvest diseases. In: Wilson, C.L., Wisniewski, M.E. (Eds.), Biological Control of Postharvest Diseases. Theory and Practice, CRC Press Inc., pp. 63–76.

Droby, S., Wisniewski, M., Macarisin, D., Wilson, C., 2009. Twenty years of postharvest biocontrol research: is it time for new paradigm? Postharvest Biol. Technol. 52, 137–145.

Duchaslavovà, J., Širučková, I., Zapletalová, E., 2007. First report of brown rot caused by *Monilinia fructicola* on various stone and pome fruits in the Czech Republic. Plant Dis. 91, 907.

EFSA, 2011. Pest risk assessment of *Monilinia fructicola* for the EU territory and identification and evaluation of risk management option. EFSA J. 9, 2119.

Ehlers, R.-U., 2011. Regulation of Biological Control Agents. Springer, New York.

Elmer, P.A.G., Gaunt, R.E., 1994. The biological characteristic of dicarbossimide resistant isolates of *Monilinia fructicola* from New Zealand stone fruit orchards. Plant Pathol. 43, 130–137.

Elmer, P.A.G., Spiers, T.M., Wood, P.N., 2007. Effects of pre-harvest foliar calcium sprays on fruit calcium levels and brown rot of peaches. Crop Prot. 26, 11–18.

Emery, K.M., Michailides, T.J., Scherm, H., 2000. Incidence of latent infection of immature peach fruit by *Monilinia fructicola* and relationship to brown rot in Georgia. Plant Dis. 84, 853–857.

EPPO, 2003. *Monilinia fructicola*. Bulletin OEPP/EPPO Bulletin 33, 281–288.
EPPO, 2009. *Monilinia fructicola*. Bulletin OEPP/EPPO Bulletin 39, 337–343.
EPPO 2011. Available on line: http://archives.eppo.org/EPPOReporting/2011/Rse-1106.pdf
EPPO 2012. Available on line: http://faostat.fao.org/site/339/default.aspx
Feliziani, E., Santini, M., Landi, L., Romanazzi, G., 2013. Pre- and postharvest treatment with alternatives to synthetic fungicides to control postharvest decay of sweet cherry. Postharvest Biol. Technol. 78, 133–138.
Forsters, H., Adaskaveg, J.E., 2000. Early brown rot infections in sweet cherry fruit are detected by *Monilinia*-specific DNA primers. Phytopathology 90, 171–178.
Fourie, P.H., Holz, G., 2003. Germination of dry, airborne conidia of *Monilinia laxa* and disease expression on nectarine fruit. Aus. Plant Pathol. 32, 9–18.
Fulton, C.E., Brown, A.E., 1997. Use of SSU rDNA group I intron to distinguish *Monilinia fructicola* from *M. laxa* and *M. fructigena*. FEMS. Microbiol. Lett. 157, 307–312.
Gatto, M.A., Ippolito, A., Linsalataa, V., et al., 2011. Activity of extracts from wild edible herbs against postharvest fungal diseases of fruit and vegetables. Postharvest Biol. Technol. 61, 72–81.
Gell, I., De Cal, A., Torres, R., Usall, J., Melgarejo, P., 2008. Relationship between the incidence of latent infections caused by *Monilinia* spp. and the incidence of brown rot of peach fruit: factor effecting latent infection. Eur. J. Plant Pathol. 121, 487–498.
Gradziel, T.M., Thorpe, M.A., Bostock, R.M., Wilcox, S., 1998. Breeding for brown rot (*Monilinia fructicola*) resistance in clingstone peach with emphasis on the role of fruit phenolics. Acta. Hortic. 465, 161–170.
Gregori, R., Borsetti, F., Neri, F., Mari, M., Bertolini, P., 2008. Effects of potassium sorbate on postharvest brown rot of stone fruit. J. Food. Prot. 71, 1626–1631.
Gueldner, R.C., Reilly, C.C., Pusey, P.L., et al., 1988. Isolation and identification of iturins as antifungal peptides in biological control of peach brown rot with *Bacillus subtilis*. J. Agric. Food. Chem. 36, 366–370.
Guentzel, J.L., Lam, K.L., Callan, M.A., Emmons, S.A., Dunham, V.L., 2010. Postharvest management of gray mold and brown rot on surfaces of peaches and grapes using electrolyzed oxidizing water. Int. J. Food. Microbiol. 143, 54–60.
Guijarro, B., Melgarejo, P., Torres, R., Lamarca, N., Usall, J., De Cal, A., 2007. Effects of different biological formulations of *Penicillium frequentans* on brown rot of peaches. Biol. Control 42, 86–96.
Helmerhorst, E.J., Troxler, R.F., Oppenheim, F.G., 2001. The human salivary peptide histatin 5 exerts its antifungal activity through formation of reactive oxygen species. Proc. Nat. Acad. Sci. 98, 14637–14642.
Hewitt, W.B., Leach, L.D., 1939. Brown-rot Sclerotinias occuring in California and their distribution on stone fruits. Phytopathology 29, 337–351.
Holtz, B.A., Michailides, T.J., Hong, C.X., 1998. Development of apothecia from stone fruit infected and stromatized by *Monilinia fructicola* in California. Plant Dis. 82, 1375–1380.
Hong, C., Michailides, T.J., 1999. Mycelial growth, sporulation, and survival of *Monilinia fructicola* in relation to osmotic potential and temperature. Mycologia 91, 871–876.
Hong, C., Holtz, B.A., Morgan, D.P., Michailides, T.J., 1997. Significance of thinned fruit as a source of the secondary inoculum of *Monilinia fructicola* in California nectarine orchards. Plant Dis. 81, 519–524.
Horka, M., Kubesova, A., Salplachta, J., Zapletalova, E., Horky, J., Slais, K., 2012. Capillary and gel electromigration techniques and MALDI-TOF MS-suitable tools for identification of filamentous fungi. Anal. Chim. Acta. 716, 155–162.
Hughes, K.J.D., Fulton, C.E., McReynold, D., Lane, C.R., 2000. Development of new PCR primers for identification of *Monilinia* species. Bulletin OEPP/EPPO Bulletin 30, 507–511.
Ioos, R., Frey, P., 2000. Genomic variation within *Monilinia laxa, M. fructigena* and *M. fructicola*, and applications to species identifications by PCR. Eur. J. Plant Pathol. 106, 373–378.

Janisiewicz, W.J., Pimeta, R.S., Jurick, W.M., 2011. A novel method for selecting antagonists against postharvest fruit decays originating from latent infections. Biol. Control 59, 384–389.

Jemric, T., Ivic, D., Fruk, G., et al., 2011. Reduction of postharvest decay of peach and nectarine caused by *Monilinia laxa* using hot water dipping. Food and Bioprocess Technol. 4, 149–154.

Jones, A.L., Ehret, G.R., 1976. Isolation and characterization of benomyl-tolerant strains of *Monilinia fructicola*. Plant Dis. Rep. 60, 765–769.

Kable, P.F., 1971. Significance of short-term latent infections in the control of brown rot in peach fruits. Phytopathology 70, 173–176.

Kader, A., Mitchell, F.G., 1989. Maturity and quality in peaches, plums and nectarines: growing and handling for fresh market. In: LaRue, J.H., Johnson, R.S. (Eds.), University California Division of Agriculture and Natural Resource 331, pp. 191–196.

Karabulut, O.A., Cohen, L., Wiess, B., Daus, A., Lurie, S., Droby, S., 2002. Control of brown rot and blue mold of peach and nectarine by short hot water brushing and yeast antagonists. Postharvest Biol. Technol. 24, 103–111.

Keske, C., Amorim, L., May-De Mio, L.L., 2011. Peach brown rot incidence related to pathogen infection at different stages of fruit development in an organic peach production system. Crop Prot. 30, 802–806.

Koenraadt, H., Somerville, S., Jones, A., 1992. Characterization of mutations in the beta-tubulin gene of benomyl-resistant field strains of *Venturia inaequalis* and other plant pathogenic fungi. Phytopathology 82, 1348–1354.

Landgraf, F.A., Zehr, E.I., 1982. Inoculum sources for *Monilinia fructicola* in South Carolina peach orchards. Phytopathology 72, 185–190.

Larena, I., Torres, R., De Cal, A., Linan, M., Melgarejo, P., Domenichini, P., 2005. Biological control of postharvest brown rot (*Monilinia* spp.) of peaches by field applications of *Epicoccum nigrum*. Biol. Control 32, 305–310.

Lee, M.-H., Bostock, R.M., 2007. Fruit exocarp phenols in relation to quiescence and development of *Monilinia fructicola* infections in *Prunus* spp.: A role for cellular Redox? Phytopathology 97, 269–277.

Li, H., Yu, T., 2000. Effect of chitosan on incidence of brown rot, quality and physiological attributes of postharvest peach fruit. J. Sci. Food. Agric. 81, 269–274.

Lichou, J., Myrin, J.F., Breniaux, D., et al., 2002. Une nouvelle moniliose. Phytoma 547, 22–25.

Liu, W.T., Chu, C.L., Zhou, T., 2002. Thymol and acetic acid vapors-phase reduce postharvest brown rot of apricots and plums. HortScience 37, 151–159.

Liu, J., Sui, Y., Wisniewski, M., et al., 2012. Effect of heat treatment on inhibition of *Monilinia fructicola* and induction of disease resistance in peach fruit. Postharvest Biol. Technol. 65, 61–68.

Luo, Y., Ma, Z., Reyes, H.C., Morgan, D., Michailides, T.J., 2007. Quantification of airborne spores of *Monilinia fructicola* in stone fruit orchards of California using real-time PCR. Eur. J. Plant Pathol. 118, 145–154.

Luo, Y., Cox, K.D., Amiri, A., Schnabel, G., 2008. Occurence and detection of the DMI resistance associated with a generic element 'Mona' in *Monilinia fructicola*. Plant Dis. 92, 1099–1103.

Lurie, S., 2002. Temperature management. In: Knee, M. (Ed.), Fruit Quality and its Biological Basis. Academic Press, Sheffield, pp. 107–121.

Ma, Z., Luo, Y., Michailides, T.J., 2003. Nested PCR assays for detection of *Monilinia fructicola* in stone fruit orchards and *Botryosphaeria dothidea* from pistachios in California. J. Phytopathol. 151, 312–322.

Ma, Z., Yoshimura, M., Holtz, B., Michailides, T.J., 2005. Characterization and PCR-based detection of benzimidazole-resistant isolates of *Monilinia laxa* in California. Pest. Manag. Sci. 61, 449–452.

Malandrakis, A.A., Markoglou, A.N., Ziogas, B.N., 2012. PCR-RFLP detection of the E198A mutation conferring resistance to benzimidazole in field isolates of *Monilinia laxa* from Greece. Crop Protect. 39, 11–17.

Mamoci, D.A., Cavoski, I., Simeone,V., Mondelli, D., Al-Bitar, L., Caboni, P., 2011. Chemical composition and *in vitro* activity of plant extracts from *Ferula communis* and *Dittrichia viscosa* against postharvest fungi. Molecules 16, 2609–2625.

Margosan, D.A., Smilanick, J.L., Simmons, G.F., Henson, D.J., 1997. Combination of hot water and ethanol to control postharvest decay of peaches and nectarines. Plant Dis. 81, 1405–1409.

Mari, M., Neri, F., Bertolini, P., 2009. Management of important diseases in Mediterranean high value crops. Stewart Postharvest Rev. 2, 2.

Mari, M., Casalini, L., Baraldi, E., Bertolini, P., Pratella, G.C., 2003. Susceptibility of apricot and peach fruit to *Monilinia laxa* during phenological stages. Postharvest. Biol. Technol. 30, 105–109.

Mari, M., Torres, R., Casalini, L., Lamarca, N., et al., 2007. Control of postharvest brown rot on nectarine by *Epicoccum nigrum* and physico-chemical treatments. J. Sci. Food Agric. Res. 87, 1271–1277.

Mari, M., Leoni, O., Bernardi, R., Neri, F., Palmieri, S., 2008. Control of brown rot on stone-fruit by synthetic and glucosinolate-derived isothiocyanates. Postharvest. Biol. Technol. 47, 61–67.

Mari, M., Martini, C., Guidarelli, M., Neri, F., 2012a. Postharvest biocontrol of *Monilinia laxa, Monilinia fructicola* and *Monilinia fructigena* on stone fruit by two *Aureobasidium pullulans* strains. Biol. Control 60, 132–140.

Mari, M., Martini, C., Spadoni, A., Rouissi, W., Bertolini, P., 2012b. Biocontrol of apple post-harvest decay by *Aureobasidium pullulans*. Postharvest Biol. Technol. 73, 56–62.

Marquenie, D., Michiels, C.W., Geeraerd, A.H., et al., 2002. Using survival analysis to investigate the effect of UVC and heat treatment on storage rot of strawberry and sweet cherry. Int. J. Food. Microbiol. 73, 187–196.

Martini, C., 2012. Phenotype and genotype characterization of *Monilinia* spp. isolates and preformed antifungal compounds in peach peel fruit at different developmental stages. University of Bologna. PhD thesis.

Michailidis, T.J., Johnson, R.S., 1992. Effect of nitrogen fertilization on brown rot (*Monilinia fructicola*) susceptibility in nectarines. Phytopathology 10, 1064.

Michailides, T.J., Ogawa, J.M., Opgenorth, D.C., 1987. Shift of *Monilinia* spp. and distribution of isolates sensitive and resistant to benomyl in California prune and apricot orchards. Plant Dis. 71, 893–896.

Michailides, T.J., Morgan, D.P., Felts, D., 2000. Detection and significance of symptomless latent infection of *Monilinia fructicola* in California stone fruits. Phytopathology 90, S53.

Michailides, T.J., Luo, Y., Ma, Z., Morgan, D.P., 2007. Brown rot of dried plum in California: new insights on an old disease. http://dx.doi.org/10.1094/APSnetFeature-2007-0307. On line APSnet Features.

Miller, F.A., Silva, L.M.C., Brandao, T.R.S., 2013. A review on ozone-based treatments for fruit and vegetables preservation. Food Eng. Rev. 5, 77–106.

Neri, F., 2003. Influence of maturity stage and ripening on sensory quality of peaches, nectarines and plums. EUFRIN Workshop on fruit quality, Bologna, Italy. June 11-14, pp. 141-142.

Neri, F., Mari, M., Brigati, S., Bertolini, P., 2007. Fungicidal activity of plant volatile compounds for controlling *Monilinia laxa* in stone fruit. Plant Dis. 91, 30–35.

Northover, J., Biggs, A.R., 1990. Susceptibility of immature and mature sweet and sour cherries to *Monilinia fructicola*. Plant Dis. 74, 280–284.

Northover, J., Cerkauskas, R.F., 1994. Detection and significance of symptomless infections of *Monilinia fructicola* in plums. Can. Plant Pathol. 16, 30–36.

Nunes, C., 2012. Biological control of postharvest diseases of fruit. Eur. J. Plant Pathol. 133, 181–196.
Ogawa, J.M., Manji, B.I., Rough, D., Sonoda, R.M., 1981. Monitoring and control of benomyl-resistant *Monilinia fructicola*. Phytopathology 71, 246.
Palou, L., Crisosto, C.H., Smilanick, J.L., 2012. Evaluation of food additives and low toxicity compounds as nonpolluting means to control the main postharvest diseases of California peaches. Acta. Hortic. 962, 539–548.
Pellegrino, C., Gullino, M.L., Garibaldi, A., Spadaro, D., 2009. First report of brown rot of stone fruit caused by *Monilinia fructicola* in Italy. Plant Dis. 93, 668.
Penrose, L., 1990. Prolonged field persistence of resistance to benomyl in *Monilinia fructicola*. Crop Prot. 9, 190–192.
Persoon, C.H., 1796. Descriptiones tam novorum quam notabilium fungorum. Observationes Mycologicae. P.P. Wolf, Lypsia.
Petroczy, M., Palkovics, L., 2009. First report of *Monilia polystroma* on apple in Hungary. Eur. J. Plant Pathol. 125, 343–347.
Pimenta, R.S., da Silva, J.F., Bujer, J.S., Janisiewicz, W.J., 2012. Endophytic fungi from plums (*Prunus domestica*) and their antifungal activity against *Monilinia fructicola*. J. Food. Prot. 75, 1883–1889.
Prusky, D., 1996. Pathogen quiescence in postharvest diseases. Annu. Rev. Phytopathol. 34, 413–434.
Pusey, P.L., Wilson, C.L., 1984. Postharvest biological control of stone fruit brown rot by *Bacillus subtilis*. Plant Dis. 68, 753–756.
Qin, G.Z., Tian, S.P., Xu, Y., 2006. Integrated control of brown rot of sweet cherry fruit with food additives in combination with biological control agents. J. Appl. Microbiol. 100, 508–515.
Rouissi, W., Ugolini, L., Martini, C., Lazzari, L., Mari, M., 2013. Control of postharvest fungal pathogens by antifungal compounds from *Penicillium expansum*. J. Food. Prot., (in press).
Sanzani, S.M., Nigro, F., Mari, M., Ippolito, A., 2009. Innovations in the control of postharvest disease of fresh fruits and vegetables. Arab J. Plant Prot. 27, 240–244.
Schena, L., Nigro, F., Pentimone, I., Ligorio, A., Ippolito, A., 2003. Control of postharvest rots of sweet cherry and table grapes with endophytic isolates of *Aureobasidium pullulans*. Postharvest. Biol. Technol. 30, 209–220.
Schlagbauer, H.E., Holz, G., 1989. Occurrence of latent *Monilinia laxa* infections on plums, peaches, and apricots. Phytophylactica 21, 35–38.
Schnabel, G., Bryson, P.K., Bridges, W.C., Brannen, P.M., 2004. Reduced sensitivity in *Monilinia fructicola* to propiconazole in Georgia and implication for disease management. Plant Dis. 88, 1000–1004.
Sellamuthu, P.S., Sivakumar, D., Soundy, P., 2013. Antifungal activity and chemical composition of thyme, peppermint and citronella oils in vapor phase against avocado and peach postharvest pathogens. J. Food Saf. 33, 86–93.
Sharma, R.R., Singh, D., Singh, R., 2009. Biological control of postharvest diseases of fruits and vegetables by microbial antagonists: A review. Biol. Control 50, 205–221.
Sisquella, M., Casals, C., Picouetc, P., Vinas, I., Torres, R., Usall, J., 2013a. Immersion of fruit in water to improve radio frequency treatment to control brown rot in stone fruit. Postharvest Biol. Technol. 80, 31–36.
Sisquella, M., Casals, C., Vinas, I., Teixidó, N., Usall, J., 2013b. Combination of peracetic acid and hot water treatment to control postharvestbrown rot on peaches and nectarines. Postharvest Biol. Technol. 83, 1–8.
Smilanick, J.L., Margosan, D.M., Mlikota Gabler, F., 2002. Impact of ozonated water on the quality and shelf-life of fresh citrus fruit, stone fruit, and table grapes. Ozone: Sci. Eng. 24, 343–356.

Snyder, C.L., Jones, A.L., 1999. Genetic variation between strains of *Monilinia fructicola* and *Monilinia laxa* isolated from cherries in Michigan. Can. J. Plant Pathol. 21, 70–77.
Spadoni, A., Neri, F., Bertolini, P., Mari, M., 2013. Control of Monilinia rots on fruit naturally infected by hot water treatment in commercial trials. Postharvest Biol. Technol. 86, 280–284.
Stevens, C., Khan, V.A., Lu, J.Y., et al., 1997. Integration of ultraviolet (UV-C) light with yeast treatment for control of postharvest storage rots of fruits and vegetables. Biol. Control 10, 98–103.
Svircev, A.M., Smith, R.J., Zhou, T., Hernade, M., Liu, W., Chud, C.L., 2007. Effects of thymol fumigation on survival and ultrastracture of *Monilinia fructicola*. Postharvest Biol. Technol. 45, 228–233.
Szkonik, M., Gilpatrick, J.D., 1977. Tolerance of *Monilinia fructicola* to benomyl in western New York. Plant Dis. Rep. 61, 651–657.
Terry, L.A., Joyce, D.C., 2004. Elicitors of induced disease resistance in postharvest horticultural crops: A brief review. Postharvest Biol. Technol. 32, 1–13.
Thomidis, T., Michailides, T.J., 2010. Development and implementation of cost effective strategies to manage brown rot of peach trees in Imathia, Greece. Eur. J. Plant Pathol. 126, 575–582.
Thomidis, T., Michailides, T.J., Exadaktylou, E., 2009. Contribution of pathogens to peach rot in Northern Greece and their sensitivity to iprodione, carbendazim, thiophanate-methyl and tebuconazole fungicides. J. Phytopathol. 157, 194–200.
Treutter, D., 2005. Significance of flavonoids in plant resistance and enhancement of their biosynthesis. Plant Biol. 7, 581–591.
Tsao, R., Zhou, T., 2000. Interaction of monoterpenoids, methyl jasmonate, and Ca^{2+} in controlling postharvest brown rot of sweet cherries. HortScience 35, 1304–1307.
van Brouwershaven, I.R., Bruil, M.L., van Leeuwen, G.C.M., Kox, L.F.F., 2010. A real-time (TaqMan) PCR assay to differentiate *Monilinia fructicola* from other three brown rot fungi of fruit crops. Plant Pathol. 59. 548–510.
van Leeuwen, G.L.M., van Kesteren, H.A., 1998. Delineation of the three brown rot fungi of fruit crops (*Monilinia* spp.) on the basis of quantitative characteristics. Can. J. Bot. 76, 2042–2050.
Villarino, M., Melgarejo, P., Usall, J., Segarra, J., De Cal, A., 2010. Primary inoculum sources of *Monilinia* spp. in Spanish peach orchards and their relative importance in brown rot. Plant Dis. 94, 1048–1054.
Villarino, M., Sandín-Espana, P., Melgarejo, P., De Cal, A., 2011. High chlorogenic and neochlorogenic acid levels in immature peaches reduce *Monilinia laxa*infection by interfering with fungalmelanin biosynthesis. J. Agric. Food. Chem. 59, 3205–3213.
Whan, J.H., 1976. Tolerance of *Sclerotinia fructicola* to benomyl. Plant Dis. Rep. 60, 200–201.
Willetts, H.J., Byrde, R.J.W., Fielding, A.H., 1977. The taxonomy of the brown rot fungi (*Monilinia* spp.) related to their extracellular cell wall-degrading enzymes. J. Gen. Microbiol. 103, 77–83.
Wilson, C.L., Chalutz, E., 1989. Postharvest biocontrol of Penicillium rots of citrus with antagonistic yeasts and bacteria. Sci. Hortoculturae 40, 105–112.
Wilson, C.L., Wisniewski, M.E., 1989. Biological control of postharvest diseases in fruit and vegetables: an emerging technology. Annu. Rev. Phytopathol. 27, 425–441.
Wilson, C.L., El Ghouth, A., Chalutz, E., et al., 1994. Potential of induced resistance to control postharvest diseases of fruits and vegetables. Plant Dis. 78, 837–844.
Witting, H.P.P., Johnson, K.B., Pscheidt, J.W., 1997. Effect of epiphytic fungi on brown rot blossom blight and infections in sweet cherry. Plant Dis. 81, 383–387.
Xu, X.M., Bertone, C., Berrie, A., 2007. Effects of wounding, fruit age and wetness duration on the development of cherry brown rot in the UK. Plant Pathol. 56, 114–119.

Xu, X.M., Guerin, L., Robinson, J.D., 2001. Effects of temperature and relative humidity on conidial germination and viability, colonization and sporulation of *Monilinia fructigena*. Plant Pathol. 50, 561–568.

Yañez-Mendizibal, V., Usall, J., Vinas, I., et al., 2011. Potential of a new strain of Bacillus subtilis CPA-8 to control the major postharvest diseases of fruit. Biocontrol Sci. Technol. 27, 409–426.

Yao, H.J., Tian, S.P., 2005. Effects of a biocontrol agent and methyl jasmonate on postharvest diseases of peach fruit and the possible mechanisms involved. J. Appl. Microbiol. 98, 941–950.

Zehr, E.I., Luszcz, L.A., Olein, W.C., Newall, W.C., Toler, J.E., 1999. Reduced sensitivity in *Monilinia fructicola* to propiconazole following prolonged exposure in peach orchards. Plant Dis. 83, 913–916.

Zhang, D., Spadaro, D., Garibaldi, A., Gullino, M.L., 2010. Selection and evaluation of new antagonists for their efficacy against postharvest brown rot of peaches. Postharvest Biol. Technol. 55, 174–181.

Zubini, P., 2008. Variation in peach fruit susceptibility to *Monilinia laxa* and gene expression. University of Bologna. PhD thesis.

CHAPTER 8

Aspergillus spp. (Black Mold)

Maribel Plascencia-Jatomea, María Susana Yépiz-Gómez, John Martin Velez-Haro
Departamento de Investigación y Posgrado en Alimentos, Universidad de Sonora, Hermosillo, Sonora, México

Contents

OVERALL IMPORTANCE OF FUNGI

Aspergillus species are widely distributed in nature, being isolated from a variety of substrates or environments such as humans, air, soil, plants and decaying organic matter. This is thanks to the easy dispersion of conidia and their relative size, being suspended in the medium for a long period of time (Warnock et al., 1999; Abarca, 2000). The genus *Aspergillus* consists of a large number of species including opportunistic pathogens, toxin producing and industrially important species, accounting for the majority of agricultural pollution (Palumbo et al., 2008). Some species are drought tolerant, responsible for the deterioration and contamination of food in the world (Radojević et al., 2011).

The first report of this genus is in Micheli's *Nova Plantarum Genera*, published in 1729 then, in the 19th century, 11 groups of this genus were described in more detail (de Vries and Visser, 2001).

Postharvest Decay
http://dx.doi.org/10.1016/B978-0-12-411552-1.00008-9

Thriving in, or at least tolerant of, elevated temperatures and reduced water levels, species of *Aspergillus* and its teleomorph *Eurotium* are the epitome of spoilage fungi. There are few kinds of foods, commodities and raw materials from which aspergilli cannot be isolated consistently. A small number of *Aspergillus* species are associated with plants, and here *Aspergillus* is more directly in competition with *Fusarium* than with *Penicillium* (Pitt and Hocking, 1999).

The genus *Aspergillus* and its species are producers of several mycotoxins, which have a large agricultural, epidemiological and economic impact. It is one of the main agents responsible for mycotoxin agricultural pollution, as well as being associated with common microbial flora of many agronomic crops (Palumbo et al., 2008; Bluma and Etcheverry, 2008). When the spores find the necessary nutrients and favorable environmental conditions, the fungus colonizes quickly and manages to produce several toxins (Abbas et al., 2009). *Aspergillus flavus* and *A. parasiticus* are the main aflatoxin-producing species (Palumbo et al., 2008). Other species, such as *A. peudotamarii*, *A. bonbysis,* and *A. ochraceoroseus* are capable of producing these metabolites in small amounts (Rasooli, 2007). Production of ochratoxin A is mainly associated with *A. carbonarius* and *A. niger* or *nigri*s section species (De Curtis et al. 2012), which has also been reported to produce fumonisin (Chiotta et al., 2011), sterigmatocystin, cyclopiazonic acid and patulin (Passone et al., 2008).

At the industry level, *Aspergillus* species are of great importance because of their wide production of enzymes, mainly in the area of food, coupled with the production of heterologous proteins (de Vries and Visser, 2001). Like some species of pathogenic yeasts, filamentous fungi and molds, under certain circumstances, they have the ability to be pathogenic in humans, being one of the most important opportunistic molds. Approximately 20 species of *Aspergillus* are associated with human infections that are mainly characterized by lung diseases whose occurrence is common in horticulturists (Ziani et al., 2009). However, the vast majority of cases of aspergillosis are caused by only a small number of species, particularly *A. fumigatus*, *A. niger*, *A. terreus*, and *A. flavus*, which have the ability to adapt to growth *in vivo* when the immune response is compromised. In addition, they can cause other diseases, often in immunocompromised individuals (Sullivan et al., 2011). *Aspergillus fumigatus* is responsible for almost 90% of infections in humans (Wang et al., 2009).

Among the fungi most commonly reported from foods, *A. niger* is one of the most common species of this genus, with a broad global distribution. This species is more prevalent in warmer climates, both in field situations and stored foods. It is also recognized as the producing agent of black molds on plant surfaces and is a major cause of the deterioration of seeds (Ziani et al., 2009). The black spores apparently provide conditions that protect from sunlight and UV light, providing a competitive advantage in such habitats. In food, it is one of the most important pollutant species in the world, especially in postharvest fruit products (fresh or dried), some vegetables, and several crops. This fungus is very frequently isolated from sun-dried products such as vine fruits, dried, smoked and cured fish, cocoa beans and spices, and is the most common fungi isolated from nuts (peanuts, pecans, pistachios, hazelnuts, walnuts, kola nuts, coconut, and copra). Among *Aspergillus* species, *A. niger* is by far the most common species responsible for postharvest decay of fresh fruit, including apples, pears, peaches, citrus, grapes, figs, strawberries, mangoes and melons; most of these diseases are sporadic and of minor significance. It also causes serious losses in tomatoes, breadfruit, onions, garlic and yams. Cereals and oilseeds are frequent sources, especially maize, corn snacks, barley, soybeans, rapeseed, rape oil, sorghum, stored and parboiled rice, blackgram, sunflower, seeds, chickpeas and pigeon peas (Pitt and Hocking, 1999).

Although capable of producing toxins such as ochratoxin A and fumonisin B_2 and B_4 (Nielsen et al., 2009), *A. niger* has been considered by the Food and Drug Administration (FDA) as generally recognized as safe (GRAS) and non-toxic under industrial conditions; this represents an advantage in industrial fermentation applications, where it is used for the production of citric acid, industrial enzymes (α-amylase, cellulase, and pectinase) and other types of protein expression (Nielsen et al., 2009). It has been reported that *A. niger* has a cell wall similar to that of *Saccharomyces cerevisiae* and *Candida albicans* (de Groot, 2009), with a production level of homologous proteins of 10–50 g L^{-1} and heterologous proteins in a much smaller range (Jacobs et al., 2009).

Other species of interest are *A. terreus*, an important fungus due to its lovastatin source (cholesterol lowering) and *A. nidulans*, one of the most used for basic genetic research and a common model microorganism for studying eukaryotic cell functions. As well as its involvement in the production of food, *Aspergillus* has lipids that could be used as fuel, because fatty acids can be converted into biodiesel. This process is another example of biocatalysis conducted by lipases from *Aspergillus* and other microorganisms (Lubertozzi and Keasling, 2009).

Aspergillus aculeatus is a plant pathogen and used to identify new extracellular enzymes. *Aspergillus clavatus* is an allergenic fungus and occasionally pathogenic. *Aspergillus* (*Neosartorya*) *fischeri* is non-pathogenic and a close relative of *A. fumigatus* (Andersen and Nielsen, 2009), while *A. oryzae* is a microorganism of great commercial value in Asia due to its use in the production of fermented foods (Wang et al., 2009).

TAXONOMY AND MORPHOLOGY

Aspergillus is a diverse group of fungi which comprises no more than 837 species, among which are classified approximately 10 different teleomorphic genera, such as *A. nidulans*, *A. fumigatus* and *A. flavus*, whose teleomorphic forms belong to the genera *Emericella*, *Neosartorya* and *Petromyces*, respectively. Differences between species of the same genus have been proposed according to their genomic sequence, which is associated with an accelerated evolution rate, restricting their genus diversification to about 200 million years (Krijgsheld et al., 2013). In this genus, several genomes have been published such as *A. fumigatus*, *A. nidulans*, *A. oryzae*, *A. niger*, *A. fumigatus (Neosartorya fischeri)*, and *A. clavatus*; the study of the genomes of *A. flavus*, *A. parasiticus*, and *A. terreus* is in various stages of development, with both, public and private efforts. However, at present, there are few options available to *Aspergillus* (Wang et al., 2009; Lubertozzi and Keasling, 2009). These studies have found genetic relationships between the same species, for example, 20 known genes of *A. niger* are related to protein secretion pathways in *A. nidulans*, including antifungal resistance similarity between the same species of this genus (Andersen and Nielsen, 2009). Some become saprophytic fungi, playing an important role in the environment as recyclers of carbon and nitrogen (Latgé, 1999).

The discovery of penicillin and aflatoxin biosynthesis in various *Aspergillus* species has led to a great interest in the study of secondary metabolism, revealing relationships between various processes in the development stage, as well as nutritional requirements or with the host response, generally in plants (Lubertozzi and Keasling, 2009).

Traditional methods to differentiate *Aspergillus* species rely mainly on their morphological and genetic characteristics (Fig. 8.1). Among recently developed molecular techniques for the identification and classification of these fungi is the comparison of ribosomal DNA (rDNA), and polymerase chain reaction (PCR) amplification and sequencing, as well as comparison of the biosynthetic production of aflatoxin, which is unique among the species of this genus (Abbas et al., 2009).

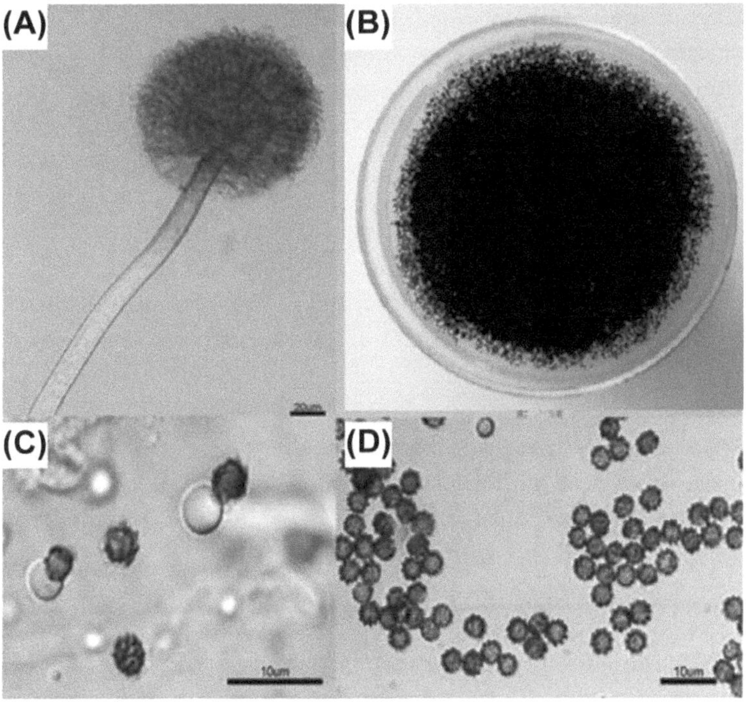

Figure 8.1 (A) Typical *Aspergillus* conidiophores with a radial head; (B) dark colony of *Aspergillus niger*; (C and D) roughened, round conidia with regular low ridges and bars. *(Adapted from Quistorp-Santos et al., 2013.)*

AGRONOMIC AND ENVIRONMENTAL FACTORS FAVORING *ASPERGILLUS* INFECTION

Both fungal diversity and dominance of species in a food are dependent on several abiotic and biotic factors, while the coexistence of microorganisms on the surface of the plant can be determined by food resources (Arroyo et al., 2008; Abbas et al., 2009).

Aspergillus has a wide range of growth. It was found that the growing temperature and pH ranges are between 10–50°C and 2.0–11.0, respectively and it is osmotolerant at salt concentrations of up to 34%. Its filamentous form has developed an enzyme secretion capacity for its growth habitat, growing with a water activity of 0.40. Despite its wide growth range and vast potential, the number of studies regarding this genus is lower in comparison with *Saccharomyces* (Lubertozzi and Keasling, 2009).

Among the conditions that favor fungal infection and mycotoxin production in preharvest foods are intrinsic factors (water activity, the attributes of the plant, the nature of these substrates and nutrient composition), extrinsic (climate change) and implicit (type of microorganism, the load or concentration of spores in food and microbiological ecosystem). To this are added the processing factors, such as type of agronomic practices (Magan and Aldred, 2007). Some of these spoilage conditions are initiated preharvest and others postharvest. The largest number of market fruit and vegetables spoilage conditions occurs after harvesting and although the fungi most often invade bruised and damaged products, some enter specific areas. For example, *A. niger* invades peaches, apricots, onions, and cabbages to cause black mold rot ('smut') of these products; this condition produces a black mycelium and the spoilage is favored by high humidity and warm temperatures (Jay et al., 2005).

It is vital to remember that environmental factors are determinants for coexistence or dominance of fungal species in a particular place, as well as for their mycotoxin production (Lee and Magan, 1999). A change in the environment creates conditions that can become stressful for microorganisms, however, those who manage to adapt create physiological costs and may change the composition of the microbial community, causing changes in the ecosystem such as the energy source and the nutrient fluxes. Despite the long history and intensive nature of *Aspergillus* research, only a few cases of toxin formation by *A. niger* have been reported, however, in no case has *A. niger* been proven to produce aflatoxins or trichotecenes (Schuster et al., 2002; Schimel et al., 2007).

With respect to safety aspects, *A. niger* is generally regarded as a safe organism. This is documented in lists of the organizations responsible for occupational health and safety. In rare cases when persons are exposed to intense spore dust, hypersensitivity reactions have been observed. This species is generally regarded as a non-pathogenic fungus widely distributed in nature. Humans are exposed to its spores every day without disease becoming apparent. Only in a few cases has *A. niger* been able to colonize the human body as an opportunistic invader and, in almost all these cases, the patients have a history of severe illness or immunosuppressive treatment (Schuster et al., 2002).

Speaking of a harvest crop, this refers to a community of plants that are or are not symbiotic with other organisms, both bacteria and fungi. In agriculture, the loss of soil organic carbon due to crops and agricultural practices can support the development of communities dominated by fungi (Singh et al., 2010). *Aspergillus* species are abundant in almost all ecosystems and have a very flexible metabolism, which allows them to consume a wide variety of carbon and nitrogen sources in the environment around them (Flipphi et al., 2009). Saprophytic fungi such as *A. nidulans*, *A. niger* and *A. oryzae* grow readily in nature and dominate in decaying plant material, where they degrade the polysaccharides present (Coutinho et al., 2009).

The 'black fungi' of the genus, mainly *A. niger*, *A. tubingensis*, *A. brasiliensis*, *A. acidus*, *A. carbonarius*, and *A. ibericus*, are capable of damaging food due to their rapid growth, broad pH range and abundance in many environments (Nielsen et al., 2009). Regarding aflatoxin-producing species, factors that can increase its presence include soil composition, the presence of other species in the plant, the harvest handling and environmental conditions, including both rain and change in temperature. Each of these factors can increase or decrease the presence of *A. flavus*, for example, in uncultivated fields that are close to cultivated land (Abbas et al., 2009).

Drought is another factor affecting the presence of this genus in crops. *Aspergillus flavus* can become a very resistant fungus because it can survive in soil after several years of drought. Its conidia are inactive in the soil, however, they germinate when nutritional and environmental conditions are favorable (Abbas et al., 2009). For example, growth and survival of fungi belonging to the genus *Flavus* section are affected by water availability, a limiting factor in the functional ecosystems (Bluma et al., 2008).

The presence of insects is another factor favoring the infection process in the preharvest crop. *Aspergillus* in general is not a fungus that attacks crops at this stage. The association between insect damage and fungal infection occurs only if the insect population is high, increasing the production of mycotoxins at the same time (Abbas et al., 2009). The gaseous atmosphere may also regulate the growth and development of fungi, particularly pathogenic species. Morphological changes in response to increased CO_2 concentration showed an increase in growth, probably as an adaptation to the gas present (Eastwood et al., 2013).

Another factor is competition between pathogenic species. For example, *A. flavus* competes with some *Fusarium* species if there are certain favorable conditions in the medium, such as water activity and temperature, increasing food spoilage and mycotoxin production. To this is added the coexistence of various microorganisms on the plant surface (Giorni et al., 2009).

POSTHARVEST FACTORS FAVORING *ASPERGILLUS* INFECTION

In general, food or stored products are adversely affected by intrinsic factors, such as the water activity of the food and nutritional composition, extrinsic, such as temperature and oxygen levels, processing factors, such as hygiene conditions and type of preservative used and implicit factors such as crop susceptibility and interaction with insects and mites (Magan and Aldred, 2007; Kotan et al., 2009; Kuorwel, et al., 2011). *A. flavus* is a pathogen that causes invasive postharvest decay in citrus fruits during storage (Kotan et al., 2009). Meanwhile, sporulation of *A. nidulans* is strongly dependent on specific environmental conditions such as nutrient availability, light–dark photoperiod, the presence of salts such as KCl and the temperature (Han et al., 2003).

Aspergillus flavus is a weak opportunistic pathogen of many agricultural crops and causes diseases in crops such as maize (corn), cotton, groundnuts (peanuts), as well as tree nuts (Brazil nuts, pecans, pistachio nuts and walnuts). Its ability to attack seeds of both monocots and dicots, and to infect seeds produced both above and below the ground, demonstrates that this fungus has evolved a battery of mechanisms to breach the resistance of the host. Few plant pathogenic fungi have such a broad host range. Compared with *A. fumigatus* and *A. nidulans*, *A. flavus* lacks host specificity (St Leger et al., 2000). It infects corn ears, cotton balls and peanut pods after insect or mechanical damage has occurred. Under weather conditions favorable for its growth, *A. flavus* can cause a significant ear rot on maize; also, because of its ability to grow at low water activity, this species is also capable of colonizing seeds of grains and oil crops. In general, high ambient temperature and plant stress are the two environmental parameters most closely correlated with *A. flavus* infections in plants (Yu et al., 2005).

CONTROL

Conventional Alternatives

Fungicides

The use of fungicides and other agrochemicals, timely irrigation and crop rotation have shown limited success in preventing these pathogens as well as the production of their toxins (Abbas et al., 2009). Traditionally, postharvest food diseases have been controlled by spraying synthetic fungicides such as thiabendazole, imazalil and sodium o-phenylphenate (Kotan, et al., 2009), the consequences of which have led to a search for natural alternative methods that are friendly to the environment, to minimize environmental impact. At present, besides the synthetic chemical formulations, antifungal compounds of natural origin have been used (Raybaudi et al., 2009), the effectiveness of which varies depending on the dose and the physical state of the compound at the time of application. However, an increasing number of chemical fungicides are no longer effective on various fungi, mainly due to development of microbial resistance after several years of exposure to the product, significantly hindering preventive and corrective treatment of disease (Agrios, 1995; Bluma et al., 2008). Added to this there is the rejection by consumers as a result of environmental pollution and damage to health due to the toxicity, carcinogenicity, teratogenicity, hormonal imbalance, among others (Radojević et al., 2011).

In many countries, azoles are the main group of compounds used to control fungi in plants, because of their stability to environmental factors, persistence of its antimicrobial residues and long lifetime. They may be associated or form part of formulations of some fungicides such as benzimidazoles or triazoles. Nowadays, of the total chemical control systems used for fungi control, 25% are derivatives of azoles (Table 8.1) (Quintana-Obregón et al., 2012). Another chemical compound used is didecyl dimethyl ammonium chloride, a broad-spectrum fungicidal active against bacteria and fungus (mold).

For example, in fungicide-treated grapes inoculated with *Aspergillus carbonarius* and *A. niger* aggregate, in different combinations, then dehydrated by reducing a_w for 20 days, the percentages of colonized grapes treated with fungicides were lower, but no differences were observed among fungicides (0.75 kg ha^{-1} of cyprodinil at one and two applications, and 1 kg ha^{-1} of 37.5% cyprodinil and 25% fludioxonil at one and two applications).

Table 8.1 Allowed Concentrations of Azoles in Foods, According to the *Codex Alimentarius*

Azole	Reference	Maximum allowed concentration (mg kg^{-1})	Food
Difenoconazole	Codex alimentarius, 2008	0.02	Wheat, sunflower seeds
Fenbuconazole	Codex alimentarius, 2001	0.05	Cows' milk, beef
	Codex alimentarius, 1999	0.10	Wheat
Penzonazole	Codex alimentarius, 1997	0.2	Tomatoes
	Codex alimentarius, 1997	0.05	Chicken meat
	Codex alimentarius, 1995	0.01	Cows' milk
Tebuconazole	Codex alimentarius, 1997	0.05	Wheat

Source: Quintana-Obregón et al. (2012).

The untreated grapes always showed higher concentrations of ochratoxin A, regardless of the inoculum applied. The four preharvest fungicides reduced the *Aspergillus* section Nigri growth and production of ochratoxin A by *A. carbonarius* during dehydration of grapes, cyprodinil being the most effective (Valero et al., 2007).

Despite the foregoing, application of either partial or total chemicals is still the main practice for controlling plant diseases caused by fungi. In vineyards, most of the treatments involve the application of copper compounds, either alone or in combination with other chemicals. This represents a health risk due to the retention and/or recalcitrance of fungicide residues in the soil, and even grape berries after harvest can be transferred to wine (De Curtis et al., 2012).

Other Alternatives

Plant Breeding

Seed breeding, the use of resistant cultivars or hybrids and biological control are methods that have been used successfully for the control of these fungi (Mannion and Morse, 2012). For example, the use of non-aflatoxigenic *A. flavus* strains reduces *Aspergillus* competitiveness; in the same way, other antagonists of bacterial and fungal origin have been studied and tested as biocontrol method for reduction of mycotoxins (Palumbo et al., 2008). Among the management strategies that may reduce or eliminate aflatoxin contamination of corn, the pre-eminent strategy is to develop preharvest host resistance to aflatoxin accumulation. This strategy has gained prominence due to the natural resistance in corn that can be exploited in plant-breeding strategies. The ability to identify resistant corn genotypes has been enhanced by the development of a laboratory kernel-screening assay and by a strain of *A. flavus* genetically engineered to produce β-glucuronidase, whose activity can be monitored to assess the degree of fungal infection in kernels (Brown et al., 1999).

Antagonists

Monilinia fructicula is one of the antimicrobial agents used for the biological control of *Aspergillus* in strawberries (Droby et al., 2009), while the species *Aureobasidium pullulans* LS30 and *Metschnikowia pulcherrimaes* are capable of reducing infection by *Aspergillus carbonarius* and ochratoxin A production in healthy wine grape bunches (cv. 'Montepulciano') (De Curtis, et al. 2012). Kotan et al. (2009) reported that *Pantoea agglomerans, Bacillus subtilis* and *Burkholderia cepacia* are antagonists of *Aspergillus flavus* on lemon fruits cvs.

'Meyer' and 'Interdonato' under storage conditions, and also noted that *Alcaligenes piechaudii*, *Erwinia chrysanthemi* and *E. rhapontici* are antagonists of *Aspergillus flavus* under the same conditions.

A leaf spot confirmation test indicated that *Trichoderma harzianum* and *A. niger* were able to grow in rice seedlings; this shows that despite the presence of the antagonist *T. harzianum*, *A. niger* was still able to infect the rice seedlings. *Trichoderma harzianum*, though able to contribute to promoting the growth of the rice seedlings, was not successful in suppressing the growth of *A. niger* in *Oryza sativa* (Venus et al., 2013). In a similar study, *Penicillium* species isolated from the rhizosphere of different plants showed very high antagonistic effects on the growth of the mycelium of the onion black rot pathogen *A. niger* in dual culture agar plate assays. *Penicillium roqueforti* and *P. viridicatum* greatly inhibited the growth of *A. niger* by 66% and 60%, respectively, followed by *P. bilaii* (57%) and *P. olsonii* (53%). However, it was also observed that the *Penicillium* species completely overgrew the *A. niger* colony (Khokhar et al., 2012).

For biological control of aflatoxigenic species of *Aspergillus*, fungi which are non-aflatoxigenic can be used. However, to improve this methodology and prevent multiple colonization harvests due to the development of related species, it is vital to understand fully the ecology of these fungi, as their behavior in agricultural soils has yet to be fully elucidated (Abbas et al., 2009). Dorner et al. (2003) determined the efficacy of different formulations of non-toxigenic strains of *A. flavus* and *A. parasiticus* in reducing aflatoxin contamination of peanuts. Although none of the numerical reductions seen in 1996 was statistically significant, the average reduction by all formulations of 86% coupled with the statistically significant average reduction of 92% in 1997 indicates that all formulations were effective in reducing aflatoxin contamination. According to the authors, the technology of incorporating conidia into granules such as Pesta and pregelatinized corn flour are viable options to consider for commercial production of aflatoxin biocontrol formulations.

Genomic Approach

Recently, pathogen control has focused on a genomic approach, which allows understanding of the molecular mechanisms of infection, obtaining a comparison of the processes involved (Karl-Heinz, 2008). Through the molecular biology approach, our knowledge of the biology of *Aspergillus* species and toxin production mechanism continues to improve. We have come to understand the genes and gene expression of these fungi and fungus–plant interaction, improving the control mechanisms (McIntyre et al. 2001; Abbas et al., 2009). The application and development of prevention strategies based on the use of the HACCP approach identified critical control points in the food chain pre- and postharvest (Magan and Aldred, 2007). However, before new control strategies can become agricultural practice, they need to be available to the farmers and several restrictions on availability may occur. First, the technology or machinery is not yet developed for commercial use (i.e. weed control, where intelligent weeders and equipment for precision spraying are still under research and development); second, non-chemical methods, new pesticide application techniques, and reduced doses are all methods that need to be adapted to regional conditions. In some cases, local adaptation is difficult or may even be impossible, even though the method is practiced successfully in other regions. Field sizes and climatic conditions may be important factors; third, a pest control method may work against one pest, but might not be transferable to another (Meissle et al., 2010).

Chitinous Antimicrobial Polymers

Among the naturally occurring antifungal compounds are chitin and chitosan. Chitin is a linear polymer of linked β-(1→4) 2-acetamido-2-deoxy-β-D-glucopyranose and 2-amino-2-deoxy-β-D-glucopyranose, insoluble in water and organic solvents (Kumirska et al, 2011), whereas chitosan is a heterogeneous group of polymers, resulting from the deacetylation of chitin, which mainly differ in molecular weight and degree of deacetylation (Raafat et al., 2008). Because of its versatility, the latter is the most studied and used as a natural antifungal agent *in vitro* and *in vivo* (Ziani et al., 2009).

Few studies have reported chitin antimicrobial activity against *Aspergillus* species. Arthur et al. (2004) observed that high concentrations of chitin affect the growth of hyphae of *A. niger*, producing hyphal swelling. Contrary to the above, chitosan is recognized as the best candidate among the naturally occurring antimicrobial preservatives (Xia et al., 2011).

It has been observed that the genus *Aspergillus* is the best example of fungi that demonstrate resistance or reduced sensitivity to chitosan (Velmurugan et al., 2009). Among the most studied species in evaluating the antimicrobial activity of chitosan and its derivatives are *A. flavus* (Zhang et al., 2009, de Oliveira et al., 2013, Kaur et al., 2012; Jagadish et al., 2012), *A. parasiticus* (Cota-Arriola et al., 2011; Maghsoudlou et al., 2012), *A. fumigatus* (Sabaa et al., 2010), *A. oryzae* (Wang et al., 2011), *A. ochraceus* (Sébastien et al., 2006), and *A. niger* (Ragab et al. 2001; Alfredsen et al., 2004, Jin-Ian et al., 2005; Sebti et al., 2005, 2007 Xiao-Fang et al. 2008; Ziani et al., 2009,

Martínez-Camacho et al., 2010, Wang et al., 2012; Giner et al., 2012, Kumar et al., 2012). These studies have shown that chitosan shows a fungistatic effect mostly on *A. niger* due to direct interaction with the spore cell wall, inducing delayed germination of spores (Plascencia-Jatomea et al., 2003). Most of these studies have been reported *in vitro* using agar media amended with chitosan (from crustaceans) solutions, as well as casting chitosan films.

Recently, Aloui et al. (2014) reported the efficacy of the combined application of chitosan and locust bean gum in combination with citrus essential oils to inhibit *A. flavus in vitro* and on artificially infected dates for a storage period of 12 days. Bergamot oil was the most effective in reducing mycelial growth, followed by bitter orange oil. Both bergamot and bitter orange oils significantly reduced conidial germination and a complete inhibition was obtained at concentrations >2%. Mixtures based on chitosan-2% bergamot oil or chitosan-2% bitter orange oil were the most effective coatings to reduce conidial germination (87–90% inhibition compared with controls). In fruit decay assays, coatings based on chitosan with citrus oils reduced fungal decay (52–62% at day 12). In a similar way, significant inhibition of *A. carbonarius* was observed *in vitro* and *in vivo* in table grapes treated with *Bacillus subtilis* liquid cultures, especially those subjected to cell-free culture treatment. No inhibition was observed for *A. carbonarius* treated with the volatile products of *B. subtilis* and the inhibition was the most significant in red earth grapes, followed by Kyoho then Thompson seedless varieties, when the same fraction of *B. subtilis* culture was used (Jiang et al., 2014). Chitosan-based nanomaterials offer a novel alternative to control the growth of food spoilage fungi such as *Aspergillus* (Cota-Arriola et al., 2013).

Plant Extracts and Essential Oils

Most plants produce secondary metabolites with antimicrobial activity, either as part of their normal growth and development, or in response to pathogen attack or stress, so they can be used as natural additives in different foods (Bluma and Etcheverry, 2008). At present, the use of crude extracts of different spices, herbs and other plant materials is increasingly important in the food area because of their antifungal activity and anti-aflatoxigenic and antioxidant action (Radojević et al., 2011).

Extracts and essential oils of vegetable origin can be used safely as substitute fungicides for control of *Aspergillus* section *Flavi* (Bluma et al., 2008). Essential oils, aromatic and volatile oily liquids have been obtained mainly from leaves, bark, stems or fruits (Gutierrez et al., 2008). It has been shown that the use of natural oils of thyme, lemongrass and other herbs suppresses the production of aflatoxin in harvested crops (Abbas et al., 2009). However, it was observed that the fungi have a higher resistance to natural products compared with bacteria. In vegetables, the antimicrobial activity of these compounds is increased due to a decrease in storage temperature (Rasooli, 2007). In addition, it has been found that ferulic acid and other natural phenolic compounds are effective against the growth of species of *Aspergillus*, mainly *A. flavus* and *A. parasiticus* (Passone et al., 2009).

In general, the antimicrobial activity of essential oils in vegetable dishes is benefited by a decrease in storage temperature and/or a decrease in the pH of the food. All essential oils and their components that have been tested on vegetables appear effective against the natural spoilage flora and food-borne pathogens at levels of 0.1–10 μL g^{-1} in washing water. Fat, protein, and the pH of food systems are important factors affecting the activity of oils. At low pH, the hydrophobicity of some essential oils (i.e. thyme oil and phenolic oleuropein) increases and while they may tend to partition in the lipid phase of the food, they can also dissolve more easily in the lipid phase of the bacterial membrane and have enhanced antimicrobial action (Rasooli, 2007).

Recently, the strong inhibitory effect of essential oil of *Cinnamomum cassia* in different real food systems (table grapes) has been demonstrated against *A.flavus* and *A. oryzae*. The authors also tested the antifungal activity of *Allium tuberosum* and *Pogostemon cablin* (Patchouli, P) essential oils and reported that the main components of tested essential oils were allyltrisulfide 40.05%, cinnamaldehyde 87.2%, and patchouli alcohol 44.5%. The minimal inhibitory concentration of the plant essential oils against *A. flavus* and *A. oryzae* was 250 ppm (*A. tuberosum* and *C. cassia*, respectively), whereas Patchouli essential oil inhibited fungi at a concentration of >1500 ppm. The essential oils exhibited a suppression effect on colony growth at 100–250 ppm for *A. tuberosum*, 25–75 for *C. cassia* and 100–500 for *P. cablin* essential oil (Kocevski et al., 2013).

The antifungal action of *Cuminum cyminum* (cumin) seed essential oil on the ergosterol content in the plasma membrane of *Aspergillus flavus* in stored wheat and chickpeas was determined and a strong antioxidant potential (IC_{50} 0.092 μL mL^{-1}) was observed. The minimum inhibitory concentration and minimum aflatoxin inhibitory concentration were 0.6 and 0.5 μL mL^{-1} respectively, and cymene (47.08%) was found as the major component (Kedia et al., 2014).

In *in vitro* assays, Bluma et al. (2008) evaluated forty traditional plant species in the region of Argentina, extracting essential oils by hydrodistillation and solvent extraction (ethanol), to test the effectiveness against *A. flavus* and *A. parasiticus*; they noted that clove, mountain thyme and poleo essential oils showed the most antifungal effect under all growth parameters analyzed as well as aflatoxin B1 accumulation. Similarly, the *in vitro* antifungal activity of methanol, acetone and ethyl-acetate native plant extracts from *Equisetum telmateia* Ehrh., *Allium flavum* L., *Sedum acre* L., *Sideritismontana* L., *Marubium peregrinum* L. and *Xeranthemum anuum* L. (concentration range of 40–0.156 mg mL^{-1}) grown in Serbia, was tested against *Aspergillus* species, finding a significant antifungal activity against *A. restrictus* and moderate activity against other *Aspergillus* spp. The best results showed ethyl acetate extract of *S. montana*, *X. anuum* and *S. acre* (Radojević et al., 2011).

Various

Xu and Ting (2009) evaluated the effect of fly ash on biomass production of *A. niger*, observing a reduction in biomass with decreasing pH of the medium. Likewise, the effect of glycerol-plasticized wheat gluten films containing potassium sorbate was evaluated on the growth of *Aspergillus niger* by the agar diffusion assay, noting that films containing >10% of potassium sorbate showed antimicrobial activity against the fungus; in addition to antimicrobial activity and biodegradability, films have barrier properties against oxygen and aromatic compounds (Türe et al. 2012).

CONCLUDING REMARKS

The genus *Aspergillus*, widely studied, has come to be of vital importance in our daily life not only as a producer of compounds and enzymes of industrial and medical importance, but also as a pathogen in food and even humans. Naturally occurring chemicals in foods of animal and plant origin often play an important role in controlling the growth of *Aspergillus* species. Combined application of chitosan and plant essential oils is an effective and promising alternative to synthetic antifungal agents for controlling postharvest illness caused by *Aspergillus*. Another alternative is the application of biological control and efforts have been consistently made to promote it in the management of *Aspergillus* species in several commodities around the world, and substantial achievements have been made in most cases. Due to various technical, physiological, and socioeconomic factors, the implementation of natural alternatives has been slow and difficult in many cases. While these challenges will continue to exist, we see a rapidly increasing market demand of natural enemy products and acceptance of conservation biological control in the field. This trend will be mainly driven by the rapidly increasing demand for high quality and pesticide residue-free agricultural goods, the changing of farming systems to form larger farms, and rapid advances in new technologies. Natural alternatives to fungal control will continue to play more critical roles in agricultural production around the world in the years to come.

REFERENCES

Abarca, M.L., 2000. Taxonomía e identificación de especies implicadas en la aspergilosis nosocomial. Rev. Iberoam. de Micología 17, S79–S84.

Abbas, H.K., Wilkinson, J.R., Zablotowicz, R.M., et al., 2009. Ecology of *Aspergillus flavus*, regulation of aflatoxin production, and management strategies to reduce aflatoxin contamination of corn. Toxin Rev. 28, 142–153.

Agrios, G.N., 1995. Fitopatología, second ed.. Limusa, México.

Alfredsen, G., Eikenes, M., Militz, H., Solheim, H., 2004. Screening of chitosan against wood-deteriorating fungi. Scand. J. Forest Res. 19, 4–13.

Aloui, H., Khwaldia, K., Licciardello, F., et al., 2014. Efficacy of the combined application of chitosan and Locust Bean Gum with different citrus essential oils to control postharvest spoilage caused by *Aspergillus flavus* in dates. Int. J. Food Microbiol. 170, 21–28.

Andersen, M.R., Nielsen, J., 2009. Current status of systems biology in *Aspergilli*. Fungal Genet. Biol. 46, S180–S190.

Arroyo, M., Aldred, D., Magan, N., 2008. Environmental factors and preservatives affect carbon utilization patterns and niche overlap of food spoilage fungi. Fungal Ecol. 1, 24–32.

Arthur, F.J.R., Arentshorst, M., Damveld, R.A., Vankuyk, P.A., Klis, F.M., Hondel, C.A.M.J.J., 2004. The cell wall stress response in *Aspergillus niger* involves increased expression of the glutamine: fructose-6-phosphate amidotransferase-encoding gene (gfaa) and increased deposition of chitin in the cell wall. Microbiology 150, 3315–3326.

Bluma, R.V., Etcheverry, M.G., 2008. Application of essential oils in maize grain: impact on *Aspergillus* section *Flavi* growth parameters and aflatoxinaccumulation. Food Microbiol. 25, 324–334.

Bluma, R., Amaiden, M.R., Etcheverry, M., 2008. Screening of Argentine plant extracts: Impact on growth parameters and aflatoxin B1 accumulation by *Aspergillus* section *Flavi*. Int. J. Food Microbiol. 122, 114–125.

Brown, R.L., Chen, Z.-Y., Cleveland, T.E., Russin, J.S., 1999. Advances in the development of host resistance in corn to aflatoxin contamination by *Aspergillus flavus*. Phytopathology 89, 113–117.

Chiotta, M.L., Susca, A., Stea, G., et al., 2011. Phylogenetic characterization and ochratoxin A – fumonisin profile of black *Aspergillus* isolated from grapes in Argentina. Int. J. Food Microbiol. 149, 171–176.

Cota-Arriola, O., Cortez-Rocha, M.O., Rosas-Burgos, E.C., Burgos-Hernández, A., López-Franco, Y.L., Plascencia-Jatomea, M., 2011. Antifungal effect of chitosan on the growth of *Aspergillus parasiticus* and production of aflatoxin B1. Polym. Int. 60, 937–944.

Cota-Arriola, O., Cortez-Rocha, M.O., Burgos-Hernández, A., Ezquerra-Brauer, J.M., Plascencia-Jatomea, M., 2013. Controlled release matrices and micro/nanoparticles of chitosan with antimicrobial potential: development of new strategies for microbial control in agriculture. J. Sci. Food Agric. 93, 1525–1536.

Coutinho, P.M., Andersen, M.R., Kolenova, K., et al., 2009. Post-genomic insights into the plant polysaccharide degradation potential of *Aspergillus nidulans* and comparison to *Aspergillus niger* and *Aspergillus oryzae*. Fungal Genet. Biol. 46, S161–S169.

De Curtis, F., de Felice, D.V., Ianiri, G., De Cicco, V., Castoria, R., 2012. Environmental factors affect the activity of biocontrol agents against ochratoxigenic *Aspergillus carbonarius* on wine grape. Int. J. Food Microbiol. 159, 17–24.
de Groot, P.W.J., Brandt, B.W., Horiuchi, H., Ram, A.F.J., de Koster, C.G., Klis, F.M., 2009. Comprehensive genomic analysis of cell wall genes in *Aspergillus nidulans*. Fungal Genet. Biol. 46, S72–S81.
de Oliveira, P.R., Takaki, M., Gorayeb, T.C., et al., 2013. Synthesis, characterization and antifungal activity of quaternary derivatives of chitosan on *Aspergillus flavus*. Microbiol. Res. 168, 50–55.
deVries, R.P., Visser, J., 2001. *Aspergillus* enzymes involved in degradation of plant cell wall polysaccharides. Microbiol. Mol. Biol. Rev. 65, 497–522.
Dorner, J.W., Colea, R.J., Connick, W.J., Daigle, D.J., McGuire, M.R., Shasha, B.S., 2003. Evaluation of biological control formulations to reduce aflatoxin contamination in peanuts. Biol. Control 26, 318–324.
Droby, S., Wisniewski, M., Macarisin, D., Wilson, C., 2009. Twenty years of postharvest biocontrol research: Is it time for a new paradigm? Postharvest Biol. Technol. 52, 137–145.
Eastwood, D.C., Herman, B., Noble, R., Dobrovin-Pennington, A., Sreenivasaprasad, S., Burton, K.S., 2013. Environmental regulation of reproductive phase change in *Agaricus bisporus* by 1-octen-3-ol, temperature and CO_2. Fungal Genet. Biol. 55, 54–66.
Flipphi, M., Sun, J., Robellet, X., et al., 2009. Biodiversity and evolution of primary carbon metabolism in *Aspergillus nidulans* and other *Aspergillus* spp. Fungal Genet. Biol. 46, S19–S44.
Giner, M.J., Vegara, S., Funes, L., et al., 2012. Antimicrobial activity of food-compatible plant extracts and chitosan against naturally occurring micro-organisms in tomato juice. J. Sci. Food Agric. 92, 1917–1923.
Giorni, P., Magan, N., Battilani, P., 2009. Environmental factors modify carbon nutritional patterns and niche overlap between *Aspergillus flavus* and *Fusarium verticillioides* strains from maize. Int. J. Food Microbiol. 130, 213–218.
Gutierrez, J., Barry-Ryan, C., Bourke, P., 2008. The antimicrobial efficacy of plant essential oil combinations and interactions with food ingredients. Int. J. Food Microbiol. 124, 91–97.
Han, K.H., Lee, D.B., Kim, J.H., et al., 2003. Environmental factors affecting development of *Aspergillus nidulans*. J. Microbiol. 41, 34–40.
Jacobs, D.I., Olsthoorn, M.M.A., et al., 2009. Effective lead selection for improved protein production in *Aspergillus niger* based on integrated genomics. Fungal Genet. Biol. 46, S141–S152.
Jagadish, R.S., Divyashree, K.N., Viswanath, P.P., Srinivas, B.R., 2012. Preparation of N-vanillyl chitosan and 4-hydroxybenzyl chitosan and their physico-mechanical, optical, barrier, and antimicrobial properties. Carbohyd. Polym. 87, 110–116.
Jay, J.M., Loessner, M.J., Golden, D.A., 2005. Modern Food Microbiology, seventh ed. Springer.
Jiang, C., Shi, J., Liu, Y., Zhu, C., 2014. Inhibition of *Aspergillus carbonarius* and fungal contamination in table grapes using *Bacillus subtilis*. Food Control 35, 41–48.
Jin-Ian, X., Chun, W., Zhen-yuan, N., An-an, P., Xin, G., 2005. Structure, properties and application to water-soluble coatings of complex antimicrobial agent Ag-carboxymethyl chitosan-thiabendazole. J. Cent. South Univ. Technol. 12, 526–530.
Karl-Heinz, K., 2008. Compatible host–microbe interactions: Mechanistic studies enabling future agronomical solutions. J. Plant Physiol. 165, 1–4.
Kaur, P., Thakur, R., Ashok, C., 2012. An *in vitro* study of the antifungal activity of silver/chitosan nanoformulations against important seed borne pathogens. Int. J. Sci. Technol. Res. 1, 83–86.

Kedia, A., Prakash, B., Mishra, P.K., Dubey, N.K., 2014. Antifungal and antiaflatoxigenic properties of *Cuminum cyminum* (L.) seed essential oil and its efficacy as a preservative in stored commodities. Int. J. Food Microbiol. 168–169, 1-7.

Khokhar, I., Haider, M.S., Mukhtar, I., Mushtaq, S., 2012. Biological control of *Aspergillus niger*, the cause of black-rot disease of *Allium cepa* L. (onion), by *Penicillium* species. J. Agrobiol. 29, 23–28.

Kocevski, D., Du, M., Kan, J., Jing, C., Lačanin, I., Pavlović, H., 2013. Antifungal effect of *Allium tuberosum, Cinnamomum cassia*, and *Pogostemon cablin* essential oils and their components against population of *Aspergillus* species. J. Food Sci. 78, M731–M737.

Kotan, R., Dikbas, N., Bostan, H., 2009. Biological control of post harvest disease caused by *Aspergillus flavus* on stored lemon fruits. Afri. J. Biotechnol. 8, 209–214.

Krijgsheld, P., Bleichrodt, R., van Veluw, G.J., et al., 2013. Development in *Aspergillus*. Stud. Mycol. 74, 1–29.

Kumar, S., Koh, J., Kim, H., Gupta, M.K., Dutta, P.K., 2012. A new chitosan-thymine conjugate: synthesis, characterization and biological activity. Int. J. Biol. Macromol. 50, 493–502.

Kumirska, J., Weinhold, M.X., Thöming, J., Stepnowski, P., 2011. Biomedical activity of chitin/chitosan based materials-influence of physicochemical properties apart from molecular weight and degree of N-acetylation. Polymers 3, 1875–1901.

Kuorwel, K.K., Cran, M.J., Sonneveld, K., Miltz, J., Bigger, S.W., 2011. Essential oils and their principal constituents as antimicrobial agents for synthetic packaging films. J. Food Sci. 76, R164–R177.

Latgé, J.P., 1999. *Aspergillus fumigatus* and *Aspergillosis*. Clin. Microbiol. Rev. 12, 310–350.

Lee, H.B., Magan, N., 1999. Enviromental factors and nutritional utilization patterns affect niche overlap indices between *Aspergillus ochraceus* and other spoilage fungi. Lett. Appl. Microbiol. 28, 300–304.

Lubertozzi, D., Keasling, J.D., 2009. Developing *Aspergillus* as a host for heterologous expression. Biotechnol. Adv. 27, 53–75.

Magan, N., Aldred, D., 2007. Post-harvest control strategies: minimizing mycotoxins in the food chain. Int. J. Food Microbiol. 119, 131–139.

Maghsoudlou, A., Yahya, M., Morteza, K., Mohamad, G., 2012. Evaluation of anti-fungal activity of chitosan and its effect on the moisture absorption and organoleptic characteristics of pistachio nuts. International Journal on Advanced Science. Eng. Inf. Technol. 2, 65–69.

Mannion, A.M., Morse, S., 2012. Biotechnology in agriculture: Agronomic and environmental considerations and reflections based on 15 years of GM crops. Prog. Phys. Geography. DOI: 10.1177/0309133312457109.

Martínez-Camacho, A.P., Cortez-Rocha, M.O., Ezquerra-Brauer, J.M., et al., 2010. Chitosan composite films: thermal, structural, mechanical and antifungal properties. Carbohydr. Polym. 82, 305–315.

McIntyre, M., Müller, C., Dynesen, J., Nielsen, J., 2001. Metabolic engineering of the morphology of Aspergillus. In: Scheper, T. (Ed.), Advances in Biochemical Engineering/biotechnology. Metabolic Engineering, vol. 73. Springer-Verlag, Berlin, pp. 103–128.

Meissle, M., Mouron, P., Musa, T., et al., 2010. Pests, pesticide use and alternative options in European maize production: current status and future prospects. J. Appl. Entomol. 134, 357–375.

Nielsen, K.F., Mogensen, J.M., Johansen, M., Larsen, T.O., Frisvad, J.C., 2009. Review of secondary metabolites and mycotoxins from the *Aspergillusniger* group. Anal. Bioanal. Chem. 395, 1225–1242.

Palumbo, J.D., O'Keeffe, T.L., Abbas, H.K., 2008. Microbial interactions with mycotoxigenic fungi and mycotoxins. Toxins Rev. 27, 261–285.

Passone, M.A., Resnik, S., Etcheverry, M.G., 2008. The potential of food grade antioxidants in the control of *Aspergillus* section *Flavi*, interrelated mycoflora and aflatoxin B1 accumulation on peanut grains. Food Control 19, 364–371.

Passone, M.A., Ruffino, M., Ponzio,V., Resnik, S., Etcheverry, M.G., 2009. Postharvest control of peanut *Aspergillus* section *Flavi* populations by a formulation of food-grade antioxidants. Int. J. Food Microbiol. 131, 211–217.
Pitt, J.I., Hocking, A.D., 1999. Fungi and food spoilage, second ed. Editorial Gaithersburg, Aspen publication.
Plascencia-Jatomea, M.,Viniegra, G., Olayo, R., Castillo-Ortega, M.M., Shirai, K., 2003. Effect of chitosan and temperature on spore germination of *Aspergillus niger*. Macromol. Biosci. 3, 582–586.
Quintana-Obregón, E.A., Plascencia-Jatomea, M., Cortez-Rocha, M.O. 2012. Impacto del uso de los azoles: implicaciones agrícolas y daños a la salud. In: Ramírez de León, J.A., Uresti-Marín, R.M., Aldana-Madrid, M.L., Loarca-Piña, M.G.F. (Eds.), Avances de ciencia y tecnología alimentaria en México. Plaza y Valdez, S.A. de C.V. México.
Quistorp-Santos, B.A., Obed-Ntwampe, S.K., Doughari, J.H., 2013. Continuous biotechnological treatment of cyanide contaminated waters by using a cyanide resistant species of Aspergillus awamori. In: Petre, M. (Ed.), Biochemistry, Genetics and Molecular Biology. Environmental Biotechnology – New Approaches and Prospective Applications. Intech Open Access Company.
Raafat, D., Bargen, K.V., Haas, A., Hans-Georg, S., 2008. Insights into the mode of action of chitosan as an antibacterial compound. Appl. Environ. Microbiol. 74, 3764–3773.
Radojević, I.D., Stanković, M.S., Stefanović, O.D., Topuzović, M.D., Čomić, L.R., Ostojić, A.M., 2011. Anti-*Aspergillus* properties of different extracts from selected plants. Afri. J. Microbiol. Res. 5, 3986–3990.
Ragab, M.M., Ragab, M.M.M., El-Nagar, M.A., Farrag, E.S., 2001. Effect of chitosan and its derivatives as an antifungal and preservative agent on storage of tomato fruits. Egypt. J. Phytopathol. 29, 107–116.
Rasooli, I., 2007. Food preservation – A biopreservative approach. Food 1, 111–136.
Raybaudi, M.R., Mosqueda, M.J., Soliva, F., Martin, B.O., 2009. Control of pathogenic and spoilage microorganisms in fresh/cut fruits and fruit juices by traditional and alternative natural antimicrobials. Compr. Rev. Food Sci. Food Saf. 8, 157–180.
Sabaa, M.W., Nadia, A.M., Riham, R.M., Neveen, M.K., Soliman, M.A.E.L., 2010. Synthesis, characterization and antimicrobial activity of poly (N-vinyl imidazole) grafted carboxymethyl chitosan. Carbohydr. Polym. 79, 998–1005.
Schimel, J., Balser, T.C., Wallenstein, M., 2007. Microbial stress-response physiology and its implications for ecosystem function. Ecology 88, 1386–1394.
Schuster, E., Dunn-Coleman, N., Frisvad, J.C., van Dijck, P.W.M., 2002. On the safety of *Aspergillus niger* – a review. Appl. Microbiol. Biotechnol. 59, 426–435.
Sébastien, F., Stéphane, G., Copinet, A., Coma, V., 2006. Novel biodegradable films made from chitosan and poly(lactic acid) with antifungal properties against mycotoxinogen strains. Carbohydr. Polym. 65, 185–193.
Sebti, I., Martial-Gros, A., Carnet-Pantiez, A., Grelier, S., Coma, V., 2005. Chitosan polymer as bioactive coating and film against *Aspergillus niger* contamination. J. Food Sci. 70, M100–M104.
Sebti, I., Chollet, E., Degraeve, P., Noel, C., Peyrol, E., 2007. Water sensitivity, antimicrobial, and physicochemical analyses of edible films based on HPMC and/or chitosan. J. Agric. Food Chem. 55, 693–699.
Singh, B.K., Bardgett, R.D., Smith, P., Reay, D.S., 2010. Microorganisms and climate change: terrestrial feedbacks and mitigation options. Nat. Rev. Microbiol. 8, 779–790.
Sullivan, D.J., Moran, G.P., Coleman, D.C., 2011. Fungal infections of humans. In: Kavanagh, K. (Ed.), Fungi: Biology and Applications. John Wiley & Sons, Ltd, pp. 257–278.
St Leger, R.J., Screen, S.E., Shams-Pirzadeh, B., 2000. Lack of host specialization in *Aspergillus flavus*. Appl. Environ. Microbiol. 66, 320–324.

Türe, H., Gällstedt, M., Hedenqvist, M.S., 2012. Antimicrobial compression-moulded wheat gluten films containing potassium sorbate. Food Res. Int. 45, 109–115.

Valero, A., Marín, S., Ramos, A.J., Sanchis, V., 2007. Effect of preharvest fungicides and interacting fungion *Aspergillus carbonarius* growth and ochratoxin A synthesis in dehydrating grapes. Lett. Appl. Microbiol. 45, 194–199.

Velmurugan, N., Kumar, G.G., Sang, S.H., Kee, S.N., Yang, S.L., 2009. Synthesis and characterization of potential fungicidal silver nano-sized particles and chitosan membrane containing silver particles. Iran. Polym. J. 18, 383–392.

Venus, B.H., Kintanar, I.M.P., Reyes, J.G.M., 2013. Biocontrol potential of *Aspergillus niger* on *Oryza sativa* (Rice) by *Trichoderma harzianum* under experimental conditions. A Research Paper. Philippine Science High School–Central Visayas Campus, Talaytay, Argao, Cebu.

Wang, J., Lian, Z., Wang, H., Jin, X., Liu, Y., 2012. Synthesis and antimicrobial activity of Schiff base of chitosan and acylated chitosan. J. Appl. Polym. Sci. 123, 3242–3247.

Wang, K., Ussery, D.W., Brunak, S., 2009. Analysis and prediction of gene splice sites in four *Aspergillus* genomes. Fungal Genet. Biol. 46, S14–S18.

Wang, L., Liu, F., Jiang, Y., et al., 2011. Synergistic antimicrobial activities of natural essential oils with chitosan films. J. Agric. Food Chem. 59, 12411–12419.

Warnock, D.W., Arthington-Skaggs, B.A., Li, R.K., 1999. Antifungal drug susceptibility testing and resistance in *Aspergillus*. Drug Resist. Updat. 2, 326–334.

Xia, W., Liu, P., Zhang, J., Chen, J., 2011. Biological activities of chitosan and chitooligosaccharides. Food Hydrocolloids 25, 170–179.

Xiao-Fang, L., Xiao-Qiang, F., Sheng, Y., Ting-Pu, W., Zhong-Xing, S., 2008. Effects of molecular weight and concentration of chitosan on antifungal activity against *Aspergillus niger*. Iran. Polym. J. 17, 843–852.

Xu, T.J., Ting, Y.P., 2009. Fungal bioleaching of incineration fly ash: Metal extraction and modeling growth kinetics. Enzyme Microbial Technol. 44, 323–328.

Yu, J., Cleveland, T.E., Nierman, W.C., Bennett, J.W., 2005. *Aspergillus flavus* genomics: gateway to human and animal health, food safety, and crop resistance to diseases. Rev. Iberoam. de Micol. 22, 194–202.

Zhang, B., Dong-Feng, W., Hai-Yan, L., Ying, X., Li, Z., 2009. Preparation and properties of chitosan–soybean trypsin inhibitor blend film with anti-*Aspergillusflavus* activity. Ind. Crops Prod. 29, 541–548.

Ziani, K., Fernández-Pan, I., Royo, M., Maté, J.I., 2009. Antifungal activity of films and solutions based on chitosan against typical seed fungi. Food Hydrocolloids 23, 2309–2314.

CHAPTER 9

Thielaviopsis paradoxa, *Thielaviopsis basicola* (Black Rot, Black Root Rot)

I.G.N. Hewajulige[1] and R.L.C. Wijesundera[2]

[1]Senior Deputy Director/Principal Research Scientist, Food Technology Section, Industrial Technology Institute, Colombo, Sri Lanka; [2]Senior Professor of Plant Pathology, Department of Plant Sciences, University of Colombo, Colombo, Sri Lanka

Contents

INTRODUCTION

Considerable increases in production and distribution of agricultural products to local, national and international markets have been achieved through modern agricultural developments introduced in the last few

Postharvest Decay
http://dx.doi.org/10.1016/B978-0-12-411552-1.00009-0

decades. Despite these production improvements, high postharvest losses of horticultural commodities are prevalent due to improper postharvest handling practices and postharvest spoilage through diseases caused by several fungi and also bacteria. The fungi include *Thielaviopsis paradoxa* (De Seynes) Hohn and *Thielaviopsis basicola* (Berk and Broome) Ferraris, which belong to the family Ceratocystidaceae. These two fungi are plant pathogenic in nature and cause diseases in agronomic crops, fruit trees, and timber trees and also in some ornamental foliage plants. The host range of *T. paradoxa* includes economically important crops such as pineapple, coconut, sugarcane, date palm, banana, sorghum, cocoa, sweet potato and corn. The fungus *T. basicola* infects a wide range of greenhouse plants, herbaceous ornamentals and several important vegetable and field crops, including carrot, bean, tobacco and peanut. The most common postharvest disease caused by *T. paradoxa* is black rot disease in pineapple. *T. basicola* is responsible for the development of postharvest diseases such as black rot disease in carrots.

Preharvest orchard sprays and postharvest fungicides applied as sprays or dips with food grade waxes have been shown to be effective in controlling postharvest diseases. However, the use of fungicides for extended periods can lead to the emergence of fungicide-resistant strains. Furthermore, the residues of fungicides present on surfaces of perishable commodities may be harmful to consumers. The growing concerns of the consumer have generated an interest in the development of alternative approaches to postharvest disease control. Hot water treatment, irradiation, use of GRAS (generally regarded as safe) compounds and biological control are identified as successful alternatives to replace fungicide usage.

This chapter describes the overall importance of *T. paradoxa* and *T. basicola*, their taxonomical and morphological features, infection process, agronomic, environmental and postharvest factors which favor the two fungi and the potential control strategies.

OVERALL IMPORTANCE OF FUNGI

Thielaviopis paradoxa

The fungus *Thielaviopis paradoxa* was first reported from pineapple (*Ananas comosus* (L.) Merr.) in 1886 in France by De Seynes. Since then, the fungus has been reported from a range of crops including coconut palm (*Cocos nucifera* L.) (Dade, 1928), oil palm (*Elaeis guineesis*), date palm (*Arecu catheru* L.) (Kile, 1993) and sugar cane (*Saccharum officinarum* L.) (Mc Martin, 1937; Chang and Jensen, 1974). The disease mostly occurs in the tropics

(Ploetz et al., 2003) and has been reported from more than 40 countries (Wismer, 1961).

Thielaviopsis paradoxa is a soil-borne wound parasite that can affect all parts of the plant (Py et al., 1987). In pineapple, *T. paradoxa* causes three distinct diseases, leaf spot, butt rot (basal rot of the asexual propagative part), and fruit rot (Fig. 9.1A). Butt rot and fruit rot cause serious economic damage to the crop compared to leaf spots (Cho et al., 1977). The disease levels of leaf spot are low in most of the commercial cultivars of pineapple. Pineapple fruit rot, which is also known as black rot, soft rot, stem-end-rot or water blister is the most common and well-known postharvest disease of pineapple and causes serious losses (Ploetz, 2003). This economically important disease is present in all pineapple producing countries in the world (Broadely et al., 1993). In coconut (*Cocos nucifera*), the causal organism of the stem bleeding disease reported in growing countries such as India, Jamaica, Sri Lanka, Malaysia, Mauritius, Philippines, Puerto Rico and the USA is also *T. paradoxa*. The fungus produces not only stem bleeding disease but a range of other diseases including bud rot, leaf spot, heart rot and root decay (Fig. 9.1B). One of the most economically important diseases of sugarcane (*Saccharum officinarum*), pineapple disease of sugarcane, is also caused by *T. paradoxa*. The disease derives its name as the symptoms caused by the disease are similar to the symptoms of pineapple black rot disease. This disease is prevalent in most of the sugarcane growing countries

Figure 9.1 Disease symptoms of black rot in (A) pineapple fruit and (B) coconut stem.

in the world (Martin et al., 1961). The fungus also induces seedling decay following planting, causing poor initial stands of sugarcane (Raid, 2012). Black scorch, an economically important disease of date palm (*Phoenix dactylifera*), which is reported in all date palm growing areas of the world, is also caused by *T. paradoxa*.

The symptoms of soft rot include soft, watery rot in the fruit flesh, which liquefies rapidly at ambient temperature (25°C) and gives off a characteristic, sweetish, ethereal odor (Py et al., 1987). Subsequently, the diseased tissue turns dark as the dark colored mycelium of the causal fungus forms conidia on the wounded surface and and results in oozing of juice from infected tissue.

Thielaviopis basicola

Thielaviopsis basicola is a soil inhabiting fungus which attacks many plant species in the families, Fabaceae, Solanaceae and Cucurbitaceae. It causes a widespread root rot problem on carrot and several greenhouse ornamental plants such as poinsettia, geranium, pansy, petunia and Japanese holly (Wills and Lambe 1978). *Thielaviopsis basicola* invades the root system of the host plants and causes the disease named black root rot (King and Presley, 1942). This disease is mostly prevalent in seedlings although mature plants can also be infected. The disease reduces the nutrient uptake of infected plants and subsequently weakens the root systems. It has also been reported from a variety of commercial crops such as cotton, beans, lettuce, lupin, peas, tobacco and soybean (Allen, 1990; O'Brien and Davis, 1994; Mondal et al., 2004).

Black root rot disease symptoms in cotton seedlings initiate as brown to purplish-black lesions on roots which finally turn black and appear shrunken. This make the roots very fragile and the plant can be easily pulled out from the ground. In mature cotton plants, symptoms start with abnormal swelling of the root and subsequent disintegration of tissues forming brown to purplish-black colored cavities. The disease symptoms in the aerial part of the cotton plants are characterized by leaf discoloration, reduction in foliage size and withering of leaves. Swelling of the stem above the soil surface is also evident in some instances (King and Presley, 1942).

In tobacco and beans, it is reported that foliar tissues are also susceptible to the black root rot disease and the plant tissue exhibits necrosis (Hecht and Bateman, 1964; Punja, 2004). Walker et al. (2000) indicated that the death of seedlings of mature plants will not occur although seedlings with extensive root damage may perish. Many researchers have reported that host specificity and the degree of pathogenicity of *T. basicola* are highly variable. The pathogenic isolates of *T. basicola* isolated from pointsettia, an ornamental plant, have been found to be moderately pathogenic on beans, but non-pathogenic on tobacco whereas tobacco isolates have been found to be non-pathogenic on both beans and pointsettia (Keller and Shank 1955; Lloyd and Lockwood 1963). The cotton root rot disease-causing isolate of *T. basicola* was not only pathogenic to cotton but also found to be highly pathogenic to lupin and soybean but non-pathogenic towards lettuce (Mondal et al., 2004). These interactions suggest that the differences in genotype between different *T. basicola* isolates result in differences in host specificity related to pathogenicity.

TAXONOMY AND MORPHOLOGY

Taxonomy and Morphology of *Thielaviopsis paradoxa*

Thielaviopsis paradoxa belongs to the filamentous Ascomycetes group (Sánchez et al., 2007). A detailed study of phylogenetic and taxonomic evaluation of the pathogen *T. paradoxa* (De Seyen.) Hohn, *Chalara paradoxa* (synonym), teleomorph: *Ceratocytis paradoxa* (Dade) *C. Moreau* has been conducted by Paulin-Mahady et al. (2002).

Thielaviopsis paradoxa forms thick walled resting spores, chlamydospores, and produces infective asexual spores or conidia. *T. paradoxa* produces two types of conidia; the first type is hyaline, cylindrical and small microspores of 8.5–12.7 × 2.5–4.7 μm in size (Fig. 9.2A). The second type is noticeably bigger, brown colored, oval in shape (macrospores) and mainly responsible for the black color in the final stage of development of disease.

Taxonomy and Morphology of *T. basicola*

Thielaviopsis basicola (synamorph *Chalara elegans*) can be categorized as an imperfect fungus due to the lack of apparent sexual or teleomorphic stage. *Thielaviopsis basicola* is also classified according to its conidia type and other

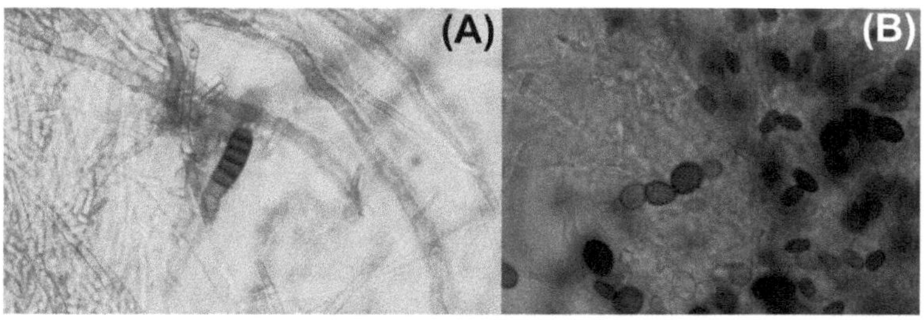

Figure 9.2 Conidia of (A) *Thielaviopsis paradoxa* and (B) *T. brasicola*.

general features of morphology such as shape and color of the colony and conidia (Ranjan, 2002). It is now established that *T. basicola* belongs to Subdivision Ascomycetes, Class Hyphomecetes, Order Moniales/Hyphales and Family Dematiciae (Ranjan, 2002; Agrios, 2005).

Variation in morphology of *T. basicola* colonies has been observed, and two distinct variants are recognized as 'gray' and 'brown' (Punja and Sun, 1999). Other variants include 'albino' and 'mycelial' aberrant phenotypes (Punja and Sun, 1999). Coumans et al. (2011) have also observed variation in colony morphology, color and size of *T. basicola* cultures in the presence of different plant root extracts. *Thielaviopsis basicola* produces two distinct types of spores, conidia and chlamydospores. The conidia are also known as phialospores or endoconidia. *T. basicola* endoconidia are asexual, non-motile spores produced from phialides in a basipetal fashion and are hyaline, cylindrical with rounded ends and variable in size (Delvecchio et al., 1969). In cultures, endoconidia are produced after 24 h of incubation (Shew and Meyer, 1992). Generally, less than 1% of endoconidia survive in soil for 15 months (Schippers, 1970). The chlamydospores, which are also known as aleuriospores or amylospores, are also asexual, non-motile spores produced at the tip of hyphae. A chlamydospore (see Fig. 9.2B) chain of individual spores is composed of thick-walled compartments surrounded by a distinct outer wall (Delvecchio et al., 1969). The number and size of individual spores in the chain vary with the *T. basicola* isolate (Punja and Sun, 1999). The chlamydospores are pigmented and contain melanin. Their thick walls and the presence of melanin protect them from adverse conditions. It is these features that allow them to function generally as resting spores (Tsao and Bricker, 1966). Chlamydospores can survive in soil for several years. This provides *T. basicola* with a long-term survival strategy. In cultures, chlamydospores are produced after 3 days of incubation (Shew and Meyer, 1992). Differences in colony appearance, color, zonation, growth rate, production of spores, length of chains of chlamydospores and pathogenicity have been observed among cultures originating from a single chlamydospore (Huang and Patrick, 1971).

BIOLOGY OF FUNGAL INFECTION PROCESS

The main point of entry of *T. paradoxa* into the pineapple fruit is through the broken fruit stalk or peduncle. Infection of the peduncle occurs soon after harvest. As a result, characteristic cone-shaped symptoms develop, with the pineapple core as the axis and the base of the fruit as the base (Frossard, 1978). Any bruising on the fruit surface that occurs during harvesting can also be invaded by the pathogen. The rot spreads rapidly throughout the area round the wound. In both cases, the disease can be identified by its characteristic odor and by pressing the skin. The severity of this disease depends on the degree of bruising or wounding that occurs to the outer surface of the fruit during harvesting, handling and packaging. The level of fungal inoculum on the fruit and the storage temperature during transport and marketing also contribute to the enhancement of severity. Damage is particularly likely to occur in fruits that have been stored, and losses from 20 to 80% have been observed in batches of fresh fruit stored for the export market.

In coconuts, it shows symptoms of soft, yellow decaying patches, which become dark and finally blacken with age. With the decay of the trunk pith tissues, reddish brown or rust-colored liquid flows from the point of entry of the pathogen. Affected palms defoliate and die imparting a high economic loss of this long-lasting perennial crop. The symptoms of pineapple disease of sugarcane are detected 2–3 weeks after planting and the symptoms include internal tissue breakdown turning tissues a reddish brown color and eventually black upon establishment of the fungus. Nodes of the seedlings act as partial barriers to the spread of the rot but, in susceptible varieties, the entire seedling may become colonized by the fungus (Raid, 2012). This disease imparts severe economic losses by retarding bud and shoot development, preventing root initiation and development (Byther and Moore, 1974) and affecting early shoot vigor. Occasionally, this fungus also causes considerable damage in standing crops especially when they have rot or borer damage, or any other similar injury (Mazano, 1975; Natarajan and Subba Raja, 1976). Severe drought conditions also accelerate the damage by spreading the pathogen throughout the cane. Due to the production of ethyl acetate, the affected tissues emit a pineapple odor, which has a synergistic influence. In such cases, the fungus spreads rapidly throughout the cane leading to leaf yellowing and ultimately withering of the plant (Martin et al., 1961). In date palm, symptoms are usually expressed in four distinct forms such as black scorch on the leaves, inflorescence blight, heart or trunk rot and bud rot on palms of all ages. Decay is more serious when the fungus attacks the terminal bud and heart which leads to the death of the palm. By development of lateral buds from uninjured portions of meristamatic tissues, some palms recover from this disease (Zaid et al., 2002).

AGRONOMIC AND ENVIRONMENTAL FACTORS FAVORING *T. PARADOXA* AND *T. BASICOLA* INFECTION

Thielaviopsis paradoxa

The conidia produced by *T. paradoxa* remain viable in the soil and plant debris and are also freely found in all pineapple and sugarcane growing areas. Plant pathogens require a certain minimum temperature to carry out their activities. *Thielaviopsis paradoxa* grows more actively at temperatures between 21 and 22°C and is almost completely inactivated below 10°C. The moisture level in the soil or environment is another important factor for the germination of conidia and penetration of the host by the germ tube (Agrios, 2005). The infective asexual conidia disseminate by splashing on the fruit during rain. Infection occurs mainly before harvest, via insect punctures, growth cracks or directly via the natural crevices between individual fruitlets or peduncle. Being a wound pathogen, the pineapple black rot disease-causing fungus gains entry through the cut stem of the peduncle or through wounds sustained during handling of the fruits.

Thielaviopsis basicola

Being a soil-borne fungus, the soil environment influences radial growth, conidia production and germination of this fungus. Conidia disseminate via contaminated soil, plant and soil amendment or directly by workers and machinery. Irrigation water has also been observed as a mode of dissemination (Nehl et al., 2004), thus allowing the fungus to spread within and among adjacent fields. Recent studies have revealed that it also can spread through insect vectors (El-Hamalawi, 2008a, b) such as shore flies. Soil temperature and water-holding capacity of the soil are two other important factors that favor infection and severity of disease development in host plants. The optimum growth temperature for *T. basicola* in culture is between 20 and 30°C (Lucas, 1955), whereas soil temperatures below 24–26°C increase the root rot disease severity caused by the fungus in a variety of hosts (Lloyd and Lockwood, 1963). However, survival rates of the fungus are greater at temperatures between 10 and 18°C (Papavizas and Lewis, 1971).

Root rot disease caused by *T. basicola* is more prevalent in cool, humid and wet weather as such environmental conditions favor this fungus. Papavizas and Lewis (1971) reported that the survival of *T. basicola* is lower in soils with water-holding capacities of 45% or greater than 15%. Hence, severity of this disease is less with well-drained soils with a lower water-holding capacity.

Soil pH is another important factor which indirectly affects the distribution and growth of soil microorganisms by altering the solubility of ions in the soil (Kaufman and Williams, 1964). Soils with a pH greater than 5.6 promote black root rot disease while a pH less than 5.2 suppresses the disease (Bateman, 1962; Harrison and Shew, 2001).

Damage of carrot crops due to *T. basicola* in the field is slight or absent except for damping-off of seedlings, because the fungus cannot penetrate intact plant tissues (Punja et al., 1992). Extensive damage to stored carrots may result as the fungus infects the wounds caused by mechanized harvesting procedures. Infections are not prevalent at 4°C, but the inoculum remains alive and can cause disease even weeks later when the carrots are placed at room temperature for sale.

POSTHARVEST FACTORS FAVORING *T. PARADOXA* AND *T. BASICOLA* INFECTION

Black rot disease of pineapple starts in the field when the causal organism enters the fruit through broken or cut peduncle and other wounds such as bruising damage or through natural openings on the shell. Usually, the infection occurs 8–12 h after wounding (Cho et al., 1977). The high temperature and relative humidity prevalent in the tropical climates where pineapple is predominantly grown aggravate the development of the disease. The severity of this disease depends on some postharvest factors such as degree of wounding or bruising during harvesting and packing, the level of inocula on the fruit and storage during transportation and marketing (Swroop Kumar, 2007).

Black rot is not common in the field unless the fruit is wounded or overripe. Under high humidity conditions, the conidia may readily be produced on pineapple residues and be disseminated by wind to healthy pineapple fruits attached to plants. Inoculum levels on fruit at harvest may vary according to the environment conditions prior to harvest (Rohrbach and Schmitt, 1994). The high correlation between moisture or rainfall prior to harvest and disease incidence following harvest has resulted in the name 'water core rot'. The spores of the fungus could also be disseminated through mechanical devices such as knives and other tools used for harvesting the crop. Therefore, it is advisable to clean the tools using a suitable disinfectant at specified time intervals during harvesting. Washing of harvested fruits has also been shown to aggravate black rot disease and it has been observed by the industry that pineapple black rot occurs more commonly on washed fruit rather than unwashed fruit (Swroop Kumar, 2007).

The most important factors that affect disease development during storage are temperature, humidity and gases in the environment. Storage at low temperature is the most effective way of shelf-life extension of many fruits and vegetables as low temperatures tend to reduce decay due to fungi. This is due to the retardation of the maturation process, thereby prolonging the disease resistance associated with immaturity or by inhibiting pathogen growth and development (Dennis, 1987). Relative humidity in the storage environment affects the activity of fungus. Controlled atmosphere (CA) storage achieved via increasing the CO_2 level and decreasing the O_2 level have been used to slow down respiration of both pathogen and the host and thereby suppressing the development of postharvest diseases. Even though disease development could be delayed by keeping the fruit in cold storage, disease may progress rapidly once the fruit is taken out of storage and put on the display shelves (Rohrbach and Phillips, 1990).

Infection of carrot by *T. basicola* is probably related to initial inoculum density in the field and cross-contamination especially during hydrocooling (Punja et al., 1992; Weber and Tribe, 2004). Eshel et al. (2009) found that the tissue wounds caused by postharvest brushing of carrots increased disease incidence, whereas avoiding the brush process eliminated the development of the disease during storage and post-storage shelf-life compared to unbrushed carrots. Postharvest tissue brushing of carrots has been introduced in most Israeli carrot packing houses in the last few years, aimed at improving produce marketing. Experiments conducted with carrot cultivars 'Dordogne' and 'Nairobi' revealed that brushed carrots developed black root rot disease symptoms during the shelf-life simulation period following the process of simulated breakdown in the cool chain and cold storage. The disease caused superficial symptoms in up to 80% of the carrot root area, although no soft rot developed (Eshel et al., 2009). Correlation of symptoms to the brushing process was high and non-brushed carrots developed no symptoms in most cases.

Short exposure (2–3 s) of brushed carrots to steam following brushing reduced the incidence of carrot black root rot decay by 50–75%. Increasing the exposure time to up to 4 s resulted in an 80% reduction in disease development with no significant difference relative to the 3 s exposure. The results further revealed that 4 s exposure to steam caused tissue-burn damage and color change in more than 50% of the carrots, (Eshel et al., 2009).

CONTROL

Conventional Control

Agronomic and Postharvest Handling Practices

The control of black rot in pineapple intended for canning or for the local fresh fruit market is basically a matter of good farm management and packing house hygiene. Proper management practices include soil preparation, crop rotation, selection of suitable varieties, especially pest- and disease-free planting materials, removal of plant debris and infected plant parts, timely weed and fertilizer management and selection of an efficient irrigation system. The number of *T. paradoxa* chlamydospores present in the field depends upon the amount of plant residue left from the previous crop which serves as a substrate for growth and reproduction of the fungus. Spores are produced in large numbers on diseased material during warm, wet weather. It is therefore essential that all pineapple refuse or discarded parts be continually removed from the packing shed and its vicinity. The packing shed, boxes and all other areas that the fruit comes into contact with should be sprayed with a suitable sanitizing agent such as sodium hypochloride, ozonated water etc. and culled fruits promptly destroyed. Good hygienic practices in the packing shed are beneficial in reducing losses due to this disease. It is advisable to sort fruits to avoid packing of injured, sunburned or wet fruits. Extreme care should be taken in picking and handling of fruit to avoid bruising.

Crop rotation with non-host crops, especially monocots, has been shown to reduce the inoculum density of *T. basicola* in soil and the incidence of black root rot (Holtz and Weinhold, 1994). Rotation with other host plants may contribute to the cumulative increase of inocula from year to year, such as the use of soybean in cotton farming systems (Mondal et al., 2004). Addition of organic matter to the soil plays an important role in reducing the incidence of several soil-borne plant pathogens including *T. basicola* and *T. paradoxa*. Nambiar and Ramanujam (1992) observed that the survival of chlamydospores of *T. paradoxa* causing stem bleeding of coconut was greatly reduced by Neem cake amendment. Lowering the soil pH from 6.5 to 5.5 or lower was also beneficial in controlling black root rot disease but may not be as effective as the use of fungicides (Graham, 1994). However, in addition to pH, exchangeable ions such as calcium, aluminum, nitrogen, and phosphates available in the soil in combination with pH contribute more to increasing or suppressing disease development depending on the circumstances. The suppressive effects on *T. basicola* are directly attributed to the increased aluminum solubility, which appears to be fungistatic at low levels and fungitoxic at high levels (Meyer et al., 1994). A similar correlation has been reported between soil pH and ammonia, with antagonistic effects being produced on growth of *T. basicola* indirectly by low soil pH and directly by free ammonia. Ammonia induces plants to release putrescine, a fungistatic compound (Harrison and Shew, 2001) and also acts directly as a fungitoxic compound by penetrating and disrupting the fungal cell membrane (Candole and Rothrock, 1998). Chalamydospore production is also inhibited by high levels of aluminum at pH less than 5.0 (Fichtner et al., 2006). Alkaline soils containing high levels of calcium have also been reported to promote the disease (Meyer and Shew, 1991). Acidic soils that suppress the disease generally contain high levels of aluminum, phosphate or nitrogen, which have been reported to inhibit spore germination and hyphal growth (Meyer et al., 1994; Delgado et al., 2006).

Common biotic factors which influence the growth of *T. basicola* include microorganisms that are found to coexist in the soil and the organic content of the soil. A number of bacteria such as *Psudomonas*, *Agrobacterium*, *Bacillus*, *Alcaligenes*, *Streptomyces*, and *Trichoderma*, are antagonistic to many soil-borne pathogenic fungi including *T. basicola*.

Chemical Control

With the expansion of domestic and international markets, the use of chemical pesticides and fungicides on horticultural commodities has increased. With the application of chemical treatments postharvest losses could be minimized. Therefore, to reduce postharvest damage due to fungi, fungicides are extensively used by present day farmers. Further to meet the stringent quality requirements of many markets, improved phytosanitary procedures and effective quarantine treatments are also necessary.

The timing of application and type of fungicide used depend primarily on the target pathogen and the time infection occurs. Field application of fungicides is often recommended for postharvest pathogens which infect produce before harvest. Repeated application of protectant fungicides during the growing season, and/or strategic application of systemic fungicides are used for this purpose. In the postharvest phase, fungicides are often applied to control infections already established in the surface tissues of produce or to protect against infections which may occur during handling and storage. In the case of occurring during and after harvest, fungicides can be used to interrupt pathogen development. Fungicides for the control of wound-invading pathogens such as *T. paradoxa* should be applied as soon as possible after harvest. If infection is well advanced at the time of postharvest treatment, control will be difficult to achieve.

The method of application of fungicides involves total immersion or the application of a spray or paste to the cut stem. It is important to treat the fruit with a fungicide within 6–12 h after harvest and before packing and shipping. Refrigeration serves to suppress the development of symptoms, since at 8°C the fungus grows hardly at all. However, fruit known to be infected should not be shipped, because significant decay could occur before arrival of the required fruit to the packing house (Salunkhe and Desai, 1984).

To control postharvest diseases of horticultural commodities, the chemical fumigant, methyl bromide, fungicides such as benzimidazole (benlate and benomyl) and carbendazim formulations (bevistine) are used (Py et al., 1987; Rohrbach and Phillips, 1990). Methyl bromide has been used as a preliminary pre-plant soil fumigant (often for high value crops such as strawberries and tomatoes) as well as a postharvest treatment since the early 1960s (Snowdon, 1990). Phosphene is another fumigant used in tropical countries because it is relatively cheap, effective and easy to apply. Conventional soil fumigation with fungicides can be used to control black root rot though it does not eradicate the fungal pathogen completely, but reduces or suppresses *T. basicola* present in the soil (Matthiessen and Kirkegaard, 2006). However, application of soil fumigation is limited when soils are rich in clay as such soil has poor dispersion and penetration of fumigants. Seeds pretreated with fungicides, or host resistance-inducing chemicals can reduce the incidence of black root disease.

Benomyl, sold as benlate, is a broad-spectrum fungicide, effective against a large number of important fungal pathogens. It is highly active and suppresses infections caused by *Rhizoctonia*, *Thielaviopsis*, *Ceratocystis*, *Fusarium* and *Verticillium*. Rot of vegetative materials (suckers) used for propagation of pineapple was controlled by dipping in benomyl, carbendazim or thiabendazole followed by an insecticide dip prior to planting. The pretreatment of suckers with 1–2% of fungicide solution is highly effective in controlling *T. paradoxa*. However, the use of benomyl is now banned or restricted in many countries. Dipping the roots in iprodione (Rovral™, Rhone-Poulenc Agro-chemie, Lyon, France) before storage is used commercially to control black root rot disease caused by *T. basicola* (Eshel et al., 2009). The black rot disease caused by *T. paradoxa* is controlled by application of 1% sodium salicylanilide to the cut surface of the stem of pineapple within 5 h of trimming (Salunkhe and Desai, 1984). It was reported that Dowicide™ A and benomyl (0.3 kg active ingredient per acre) are also used to control the fungus (Cook, 1975). Postharvest wax application in pineapple is a common practice especially for fruits intended for distant markets. Waxing suppresses the symptoms of the disorder, internal browning and reduces the water loss from the fruit. Commercial grade waxes such as Sta-Fresh 7055, 2952 could be effectively used to alleviate the symptoms of internal browning of pineapple (Hewajulige et al., 2006; Hu et al., 2011). Thiabendozole or benomyl may be added to the wax formulations but are reported to be less effective when mixed with water (Cho et al., 1977).

Although these fungicides are widely used on horticultural commodities, high residue levels observed on the flesh limit their use. The residual effect is very high in freshly eaten horticultural commodities; hence pre- and postharvest fungicide treatments can be hazardous to human life. Intensive use of these compounds has often led to the proliferation of tolerant strains of fungi, resulting in the efficacy being lost within one or two seasons. Although the fungicide can be applied to roots or to leaves, the residues may still be found in the fruit. Therefore, the residual effects of these fungicides used on horticultural commodities are causing serious health problems for consumers. On the other hand, these fungicides also cause environmental pollution. For example, methyl bromide is now clearly identified as a significant stratospheric ozone-depleting chemical, and agricultural use of this material contributes to environmental degradation. Due to these problems, methyl bromide is to be phased out internationally and is banned presently in many countries. Other chemicals also cause environmental problems when they enter water bodies and destroy beneficial macroorganisms in the environment. Therefore, there is an increasing lobby against the use of these chemicals on freshly eaten commodities. The need has arisen to seek other alternative methods to increase the storage life and quality of horticultural commodities without using chemicals.

Alternative Control Methods

Physical and chemical treatments have been used for control of black rot. The physical treatments include low-temperature storage, heat therapy and ionizing radiation. Among these, heat therapy and low temperature have been used with some success. However, most of the physical treatments have their limitations and it is reported that low temperature may arrest the growth of the pathogen but has adverse effects on the ripening of some fruits (Sharma and Alam, 1998).

Hot water treatment, irradiation treatment, the use of GRAS (generally regarded as safe) compounds such as benzoic acid, salicylic acid and acetic acid and biological control are identified as successful alternatives to fungicides.

Hot Water Treatment

Postharvest heat treatment of fruits is used for insect disinfestations, disease control and to modify fruit responses to other stresses and maintain fruit quality during storage (Paull and Chen 2000). Hot water treatment was first reported in 1922 to control decay on citrus fruits, and has been extended for use in insect disinfestations (Fallik, 2004). Then, in 1929, Baker and co-workers developed vapor heat treatments against the Mediterranean fruit fly (Couey, 1989). The specific temperature at which the germination and growth of conidia and mycelia of a pathogen are affected is known as the thermal inactivation point (Sivakumar, 2001). Thermal death time is the minimum temperature fatal to all microorganisms after exposure for 10 minutes. The thermal death point of *T. paradoxa* is between 52.5° C and 53°C (Ames, 1915).

Research conducted in Sri Lanka indicated that a 3 min hot water dip treatment at 54°C, applied to the trimmed peduncle of pineapple was effective in controlling the incidence of black rot (Wilson Wijeratnam et al., 2005). As the thermal death point of *T. paradoxa* is between 52.5°C and 53°C (Ames, 1915), treatment at 54°C for 3 min was selected to protect the quality of the fruit and allow for marginal fluctuations in temperature during commercial operations. The specially designed metal frame on which the fruit were placed protected them from exposure to hot water but permitted the stems to be immersed was used for this study (Wilson Wijeratnam et al., 2005). Pineapple crown and shell tissues discolored when the entire pineapple fruit was immersed in hot water at 54°C for 3 min. A low level of *in vitro* growth with mean colony count of 1±0.7 of *T. paradoxa* conidia survival was observed at 58°C. This emphasizes the importance of preventing spore build up in hot water treatment systems in commercial operations. Hence, care must be taken to change the water in treatment tanks before the spore count of the pathogen reaches critical levels (Wilson Wijeratnam et al., 2005).

GRAS Compounds

GRAS compounds namely, potassium metabisulfite, calcium chloride, sodium carbonate, tartaric acid, salicylic acid, acetic acid etc. offer much promise in postharvest disease management of fruits and vegetables. These compounds may be used as either a postharvest dip treatment or a preharvest spray treatment in orchards. Complete inhibition of spore germination of *T. paradoxa,* was observed at a concentration of 2% acetic acid and radial mycelial growth of the organism was also inhibited at a level of 3% acetic acid. Black rot disease of pineapple was minimal after a 7-day storage period at 28±2°C when cut ends of the stem of fruits were subjected to a 3 min acetic acid dip at either 4% or 5% compared to untreated inoculated controls (Wilson Wijeratnam et al., 2006). The effect of GRAS compounds, sodium bicarbonate, calcium chloride and acetic acid in different concentrations (25–125 mM) on the growth of mycelium and spore germination of *T. paradoxa* on artificial media was evaluated. It was found that 100 mM of sodium bicarbonate completely inhibited the mycelial growth and spore germination of the fungus (Rathnayake and Thilakarathne, 2012). Alvindia et al. (2004) also revealed that spore germination of *T. paradoxa* isolated from bananas was completely inhibited by sodium carbonate, 4 g L^{-1}, sodium hypochlorite, 5 g L^{-1}, sodium bicarbonate, calcium chloride and sodium chloride, 6 g L^{-1}. Copes et al. (2003) observed that chlorine dioxide doses ranging from 15.0 to 45.5 mg L^{-1} resulted in 50% mortality for conidia of *T. basicola* and for aleuriospores of *T. basicola.*

Plant Extracts

Reduction of postharvest decay was also obtained through plant extracts, a variety of essential oils and volatile compounds. Tamarind seed polysaccharides cause effective storability of pineapple fruits. Citrox®, a formulation of soluble bioflavonoid extracted from citrus fruit peel was successfully used to reduce the incidence of black rot of pineapple in most packing houses in Sri Lanka. Citrox® at a concentration of 0.5 % is effective in controlling the symptoms of *T. paradoxa* of pineapple (author's unpublished data).

These alternatives are part of an overall integrated pest management system and must be combined with other pest control tools to achieve an economically viable level of management.

Antagonistic Microorganisms

Biological control through the use of antagonistic microorganisms has recently emerged as a viable disease management strategy (Wisniewski et al., 1989). Antagonistic microorganisms have been found for major pathogens of banana (Mortuza and Ilag, 1999), pineapple (Lim and Rohrbach, 1980), citrus fruits (Singh, 1980) and apples (Heye and Andrews, 1983; Janisiewcz 1988a, b). The most viable approach in using antagonistic microorganisms to control postharvest diseases is the promotion and management of natural antagonists that already exist on plant extracts and are well adapted to the target plant surfaces. The potential for a resident microorganism to be an effective biological control agent depends on its ability to colonize food sources in the plant without damaging the cells (Alvinda and Natsuaki, 2008).

The potential use of *Trichoderma* species as a biocontrol agent of plant diseases was first recognized in the early 1930s and, in subsequent years, has been used in the control of many plant diseases (Elad et al., 1980). Various species of *Trichoderma* were reported to be effective in the growth promotion and yield promotion in various crops including cucumber and cotton seedlings (Ugur and Altintas, 2008). The biological activity of different species of *Trichoderma*, which are commercially used as biocontrol agents against plant fungal pathogens, are based on different mechanisms, such as the production of antifungal metabolites, mycoparasitism and competition for space and nutrients (Sivakumar 2001). *Trichoderma asperellum*, a local isolate from pineapple fields in Sri Lanka, has shown superior inhibitory action against *T. paradoxa* compared to other strains of the fungus. A talc formulation of this fungus completely inhibited radial mycelia growth of *T. paradoxa* at concentrations of 1% and above (Wijesinghe et al., 2011). In this study, the antagonistic activity of *T. asperellum* was observed only in the presence of actively growing mycelia of *T. asperellum*. Therefore, it appears that the antagonistic activity in this instance is due to the direct action of *T. asperellum* on pathogen hyphae and not due to any extracellular metabolites produced by the fungus. The authors confirmed that *T. asperellum* has a fungicidal effect on the pathogen and microscopic examinations revealed direct parasitism of *T. paradoxa* by forming coils around the pathogen hyphae (Wijesinghe et al., 2010, 2011). The two important criteria in the implementation of a biofungicidal mode of biocontrol program are the ability to develop a method for large-scale production of the candidate organism and the formulation of its biomass into a commercially acceptable product (Lewis and Larkin, 1998).

Reyes et al. (2004) revealed that epiphytic antagonist microorganisms present on pineapple fruit also contributed to the reduction of the severity of the black rot of pineapple. The most frequently isolated yeasts from the fruit surface were reported to inhibit *in vitro* growth of *T. paradoxa* and *Pichia guilliermondii* was identified as the most inhibitory. A 50% reduction of black rot severity was observed with a treatment of a mixture of five selected yeast isolates. *P guilliermondii* and the yeast mixture reduced spore germination, germ tube length, and dry weight of *T. paradoxa*. According to the authors, the use of *Pichia* or the yeast mixture was comparable with current industry practice of holding fruit at a low temperature (8–10°C) and fungicide.

CONCLUDING REMARKS

This chapter has described possible control strategies of postharvest diseases caused by *T. paradoxa* and *T. basicola*. However, proper preharvest management to reduce the inoculum potential, harvesting at the correct stage of maturity, the method of harvesting and proper handling of perishable crops throughout the supply chain, selection of proper packaging materials and suitable environment management during storage are also equally important in reducing the diseases. The effects of use of hot water treatment, irradiation, use of GRAS compounds and use of biocontrol agents which offer much promise to extend the storage life while maintaining the overall quality should be investigated further. More effort has to be focused on improving or inducing defense mechanisms of the fruit or vegetable and thereby reducing the incidence and severity of these diseases. Combined treatments of the above promising technologies would result in better disease control of these two fungi while maintaining the postharvest quality of perishables.

REFERENCES

Agrios, G.N., 2005. Plant Pathology, fifth ed. Elsevier Academic Press Inc.

Allen, S.J., 1990. *Thielaviopsis basicola*, a new record on cotton in Australia. Aust. Plant Pathol. 19, 2425.

Alvindia, D.G., Natsuaki, K.T., 2008. Evaluation of fungal epiphytes isolated from banana fruit surfaces for biocontrol of banana crown rot disease. Crop Prot. 27, 1200–1207.

Alvindia, D.G., Kobayashi, T., Natsuaki, K.T., Tanda, S., 2004. Inhibitory influence of inorganic salts on banana postharvest pathogens and preliminary application to control crown rot. J. Gen. Plant Pathol. 70, 61–65.

Ames, A., 1915. The temperature relations of some fungi causing storage rots. Phytopathology 5, 10–19.

Bateman, D.F., 1962. Relation of soil pH to development of poinsettia root rots. Phytopathology 52, 559–566.

Broadley, R.H., Wassman, R.C., Sindair, E., 1993. Pineapple Pests and Disorders, Department of Primary Industries. Queensland, pp. 12–14.

Byther, S.R., Moor, P.H., 1974. Inhibition of sugarcane rooting by *Ceratocystis paradoxa*. Can. J. Bot. 52, 761–766.

Candole, B.L., Rothrock, C.S., 1998. Using marked strains to assess the effect of hairy fetch amendment on the inoculum densities of *Thielaviopsis basicola, Pythium ultimum* and *Rhizoctonia solani*. Soil Biol. Biochem. 30, 443–448.

Chang, V.C.S., Jensen, L., 1974. Transmission of pineapple disease organism of sugarcane by nitidulid beetles in Hawaii. J. Econ. Entomol. 67, 190–192.

Cho, J.J., Rohrbach, K.G., Apt, W.J., 1977. Induction and chemical control of rot caused by *Ceratocystis paradoxa* on pineapples. Phytopathology 67, 700–703.

Cook, A.A., 1975. Diseases of Tropical and Subtropical Fruits and Nuts. Hafner Press, New York.

Copes, W.E., Chastaganer, G.A., Hummel, R.L., 2003. Activity of chlorine dioxide in a solution of ions and pH against *Thielaviopsis basicola* and *Fusarium oxysporum*. Plant Dis. 88, 188–194.

Couey, H.M., 1989. Heat treatment for control of postharvest disease and insect pest fruits. HortScience 24, 198–202.

Coumans, J.V.F., Harvey, J., Backhouse, D., et al., 2011. Proteomic assessment of host-associated microevolution in the fungus *Thielaviopsis basicola*. Environ. Microbiol. 13, 576–588.

Dade, H., 1928. *Ceratostomella paradoxa*, the perfect stage of *Thielaviopsis paradoxa* (De Seynes). Van Honnel Br. Mycol. Soc. Transcripts 13, 184–194.

Delgado, A., Franco, G.M., Paez, J.I., Vega, J.M., Carmona, E., Aviles, M., 2006. Incidence of cotton seedling diseases caused by *Rhizoctonia solani* and *Thielaviopsis basicola* in relation to previous crop, residue management and nutrients availability in soils in SW Spain. J. Phytopathol. 154, 710–714.

Delvecchio, V.G., Corbaz, R., Turian, G., 1969. An ultrastructural study of the hyphae, endoconidia and chlamydospores of *Thielaviopsis basicola*. J. Gen. Microbiol. 58, 23–27.

Dennis, C., 1987. Fungi. In: Weichmann, J. (Ed.), Postharvest Physiology of Vegetables. Marcel Dekker, New York, pp. 157–177.

De Seynes, J., 1886. Recherches Veg. Infer. III. In: Martin, J.P., Abbert, E.V., Huges, C.G. (Eds.). Pineapple Diseases, Sugarcane Diseases of the World, vol. 1. Ire Partie, Paris, pp. 28–43.

Elad, Y., Chat, I., Katan, Y., 1980. *Trichoderma harzianum* and a biocontrol agent effective against *Sclerotium rolfsii* and *Rhizoctonia solani*. Phytopathology 70, 119–121.

El-Hamalawi, Z.A., 2008a. Attraction, acquisition, retention and spatiotemporal distribution of soilborne plant pathogenic fungi by shore flies. Ann. Appl. Biol. 152, 169–177.

El-Hamalawi, Z.A., 2008b. Acquisition, retention and dispersal of soilborne plant pathogenic fungi by fungus gnats and moth flies. Ann. Appl. Biol. 153, 195–203.

Eshel, D., Regev, R., Orenstein, J., Droby, S., Gan-Mor, S., 2009. Combining physical, chemical and biological methods for synergistic control of postharvest diseases: A case study of black root rot of carrot. Postharvest Biol. Technol. 54, 48–52.

Fallik, E., 2004. Review prestorage hot water treatments (immersion, rinsing and brushing). Postharvest Biol. Technol. 32, 125–134.

Fichtner, E.J., Hesterberg, D.L., Smyth, T.J., Shew, H.D., 2006. Differential sensitivity of *Phytophthora parasitica* var. nicotianae and *Thielaviopsis basicola* to monomeric aluminum species. Phytopathology 96, 212–220.

Frossard, P., 1978. Desinfection des ananas contre *Thielaviopsis paradoxa*. Fruits 25, 785–791.

Graham, J.H., 1994. Control of black root rot on citrus seedlings in peat-based media. Proc. Florida State Horticultural Soc. 107, 21–26.

Harrison, U.J., Shew, H.D., 2001. Effects of soil pH and nitrogen fertility on the population dynamics of *T. basicola*. Plant Soil 228, 147–155.

Hecht, E., Bateman, D.F., 1964. Nonspecific acquired resistance to pathogen resulting from localized infections by *Thielaviopsis basicola* or virus in tobacco leaves. Phytopathology 54, 523–530.

Hewajulige, I.G.N., Wilson Wijeratnam, S., Wijesundera, R.L.C., 2006. Pre-harvest application of calcium to control black heart disorder in Mauritius pineapples during low-temperature storage. J. Sci. Food Agric. 86, 420–424.

Heye, C.C., Andrews, J.H., 1983. Antagonism of *Althelia bombacina* and *Chetomium globosum* to apple scab pathogen *Venturia inaequalis*. Phytopathology 73, 650–654.

Holtz, B.A., Weinhold, A.R., 1994. *Thielaviopsis basicola* in San Joaquin Valley soils and the relationship between inoculum density and disease severity of cotton seedlings. Plant Dis. 78, 986–990.
Hu, H., Li, X., Dong, C., Chen, W., 2011. Effect of wax on the quality of pineapple fruit in cold storage. International Conference on Agriculture and Biosystem Engineering. Adv. Biosyst. Eng. 1, 198–200.
Huang, H.C., Patrick, Z.A., 1971.Variability of *Thielaviopsis basicola* in culture. Can. J. Bot. 49, 1041–1047.
Janisiewicz, W.J., 1988a. Biocontrol of postharvest disease of apple with antagonist mixtures. Postharvest Pathol. Mycotoxins 78, 194–198.
Janisiewicz, W.J., 1988b. Postharvest biological control of blue mold on apples. Phytopathology 77, 481–485.
Kaufmann, D.D., Williams, L.E., 1964. Effect of mineral, fertilization and soil reaction on soil fungi. Phytopathology 54, 134–139.
Keller, J.R., Shank, J.B., 1955. Pointsettia root rot. Phytopathology 54, 134–139.
Kile, G.A., 1993. Plant diseases caused by species of *Ceratocytis sensu* stricto and Chalara. In: Wingfield, M.J., Seifert, K.A., Webber, J.F. (Eds.), Ceratocystis and Ophiostoma: Taxonomy, Ecology and Pathogenicity. APS Press, St Paul, Minnesota, pp. 173–183.
King, C.J., Presley, J.T., 1942. A root rot of cotton caused by *Thielaviopsis basicola*. Phytopathology 32, 752–761.
Lewis, J.A., Larkin, R.P., 1998. Formulation of the biocontrol fungus *Cladorrhinum foecundissimum* to reduce damping-off disease caused by *Rhizoctonia solani* and *Phythium ultimum*. Biol. Control 12, 182–190.
Lim, T.K., Rohrbach, K.G., 1980. Role of *Penicillium funiculosum* strains in the development of pineapple fruit diseases. Phytopathology 70, 663–665.
Lloyd, A.B., Lockwood, J.L., 1963. Effect of soil temperature, host variety and fungus strain on *Thielaviopsis* root rot of peas. Phytopathology 53, 329–331.
Lucas, G.B., 1955. The cardinal temperatures and pH response of *Thielaviopsis basicola*. Mycologia 47, 793–798.
Martin, J.P., Abbot, E.V., Hughes, G.C., 1961. Sugarcane Diseases of World, vol. 1, Elsevier Publication Company, New York. p. 223.
Matthiessen, J., Kirkegaard, J.A., 2006. Biofumigation and enhanced biodegradation: opportunity and challenge in soilborne pest and disease management. Crit. Rev. Plant Sci. 25, 235–265.
Mazano, S.K., 1975. Pineapple diseases on standing sugarcane in Nigeria. Sugarcane Pathol. Newslett. 14, 3–4.
McMartin, A., 1937. Pathological conditions affecting growth of sugarcane plant cuttings from Natal. South Afri. Sugar J. 21, 353–359.
Meyer, J.R., Shew, H.D., 1991. Soils suppressive to black root rot of burley tobacco caused by *Thielaviopsis basicola*. Phytopathology 81, 946–954.
Meyer, J.R., Shew, H.D., Harrison, U.J., 1994. Inhibition of germination and growth of *Thielaviopsis basicola* by aluminium. Phytopathology 84, 598–602.
Mondal, A.H., Nehl, D.B., Allen, S.J., 2004. First report of *Thielaviopsis basicola* on soybean in Australia. Australa. Plant Pathol. 33, 451–452.
Mortuza, M.G., Ilag, L.L., 1999. Potential for biocontrol of *Lasiodiploidea theobromae* (Pat.) Griff. and Maubl. In banana fruits by *Trichoderma* species. Biol. Control 15, 235–240.
Nambiar, K.K.N., Ramanujam, B., 1992. Current studies on etiology and management of stem bleeding disease of coconut. Indian Coconut J. 23, 2–4.
Natarajan, S., Subba Raja, K.T., 1976. Infection by *Ceratocystis paradoxa* Moreau on standing sugarcanes of some sugarcane clones. Sugarcane Pathol. Newslett. 17, 25–28.
Nehl, D.B., Allen, S.J., Mondal, A.H., Lonergan, P.A., 2004. Black root rot: a pandemic in Australian cotton. Australa. Plant Pathol. 33, 87–95.

O'Brien, R.G., Davis, R.D., 1994. Lettuce black root rot – a disease caused by *Chalara elegans*. Australa. Plant Pathol. 23, 106–111.

Paulin-Mahady, A.E., Harrington, T.C., MacNew, D., 2002. Phylogenic and taxonomic evaluation of *Chalara, Chalaropdid* and *Thielaviopsis* anamorphs associated with *Ceratocystis*. Mycologia 94, 62–72.

Paull, R.E., Chen, N.J., 2000. Heat treatment and fruit ripening. Postharvest Biol. Technol. 21, 21–37.

Papavizas, G.C., Lewis, J.A., 1971. Survival of endoconidia and chlamydospores of *Thielaviopsis basicola* as effected by soil environmental factors. Phytopathology 61, 108–113.

Ploetz, R.C., Lim, T.K., Menge, J.A., Rohrbach, K.G., Michailides, T.J., 2003. Common pathogens of tropical fruit. In: Ploetz, R.C. (Ed.), Diseases of Tropical Fruit Crops. CABI Publishing, Wallingford, pp. 1–20.

Ploetz, R.C., 2003. Diseases of Tropical Fruit crops. CABI Publishing, Wallingford.

Punja, Z.K., 2004. Virulence of *Chalara elegans* on bean leaves, and host-tissue response to infection. Can. J. Plant Pathol. 26, 52–62.

Punja, Z.K., Sun, L.J., 1999. Morphological and molecular characterization of *Chalara elegans* (*Thielaviopsis basicola*), cause of black root rot on diverse plant species. Can. J. Bot. 77, 1801–1812.

Punja, Z.K., Chittaranjan, S., Gaye, M.M., 1992. Development of black root rot caused by *Chalara elegans* on fresh market carrots. Can. J. Plant Pathol. 14, 299–309.

Py, C., Lacoeuilhe, J.J., Teisson, C., 1987. The Pineapple, Cultivation and Uses. Maisonneuve, G.P. Larose, Vicroria-Cousin, Paris.

Raid, R.N., 2012. Pineapple disease of sugarcane. Agronomy Department Publication SS-AGR-205. IAFS Extension University of Florida, USA (edis.ifas.ufl.edu/sc005, accessed July 2013).

Ranjan, S.S., 2002. Introduction to Fungi. Anmol Publications Pvt Ltd, New Delhi.

Rathnayake, R.M.R.N.K., Thilakarathne, B.M.K.S., 2012. Control of *Thielaviopsis* black rot of pineapple (*Ananas comosus*). Proc. Res. Symp., 503–507. Institute of Postharvest Technology, Sri Lanka.

Reyes, M.E.Q., Rohrbach, K.G., Paull, R.E., 2004. Microbial antagonists control postharvest black rot of pineapple fruit. Postharvest Biol. Technol. 33, 193–203.

Rohrbach, K., Schmitt, D.P., 1994. Pineapple. In: Ploetz, R.C., Zentmyre, G.A., Nishijima, W.T., Robrch, K.., Ohr, H.D. (Eds.), Compendium of Tropical Fruit Diseases. American Phytopathology Society, St Paul MN, pp. 45–55.

Rohrbach, K.G., Phillips, D.J., 1990. Postharvest diseases of pineapple. Acta Horticulturae 269, 503–507.

Salunkhe, D.K., Desai, B.B., 1984. Postharvest Biotechnology of Fruits, vol. 11, CRC press, Inc, Boca Raton, Florida. 3–10.

Sánchez, V., Rebolledo, O., Picaso, M.R., et al., 2007. In vitro antagonism of *Thielaviopsis paradoxa* by *Trichorderma longibrachiatum*. Mycopathologia 63, 49–58.

Schippers, B., 1970. Survival of endoconidia of *Thielaviopsis basicola* in soil. Neth. J. Plant Pathol. 76, 206–211.

Sharma, N., Alam, M., 1998. Postharvest Diseases of Horticultural Perishables. International Book Distributing Co., India. pp. 66–79.

Shew, H.D., Meyer, J.R., 1992. Thielaviopsis. In: Singleton, L.L., Mihail, J.D., Rush, C.M. (Eds.), Methods for Research on Soilborne Pathogenic Fungi. APS Press, St Paul, MN, pp. 171–174.

Singh, V., 1980. Control of *Alternaria* rot in citrus fruit. Aust. Plant Pathol. 9, 12–13.

Sivakumar, D., 2001. Alternative methods to control postharvest diseases of Rambutan. University of Colombo, Sri Lanka. PhD Thesis.

Snowdon, A.L., 1990. A Colour Atlas of Post-harvest Disease and Disorders of Fruits and Vegetables, vol. 1, Wolfe Scientific Ltd, London. pp. 140–157.

Swaroop Kumar, H.M., 2007. Pathological and Biochemical Investigations on Stem End Rot Disease in Pineapple Ananas comosus (L,) Merr. PhD thesis. University of Mysore, India.
Tsao, P.H., Bricker, J.L., 1966. Chlamydospores of *Thielaviopsis basicola* as surviving propagules in natural soils. Phytopathology 56, 1012–1014.
Ugur, B., Altintas, S., 2008. Effects of the commercial product based on *Trichoderma harzianum* on plant, bulb and yield characteristics of onion. Sci. Horticulture 116, 219–222.
Walker, N.R., Kirkpatrick, T.L., Rothrock, C.S., 2000. Influence of *Meloidogyne incognita* and *Theieloaviopsis basicola* on early-season disease development and cotton growth. Plant Dis. 84, 449–453.
Weber, R.W.S., Tribe, H.T., 2004. Moulds that should be better known: *Thielaviopsis basicola* and *T. thielavioides*, two ubiquitous moulds on carrots sold in shops. Mycologist 18, 6–10.
Wijesinghe, C.J., Wilson Wijeratnam, R.S., Samarasekera, J.K.R.R., Wijesundera, R.L.C., 2010. Biological control of *Thielaviopsis paradoxa* on pineapple by an isolate of *Trichoderma asperellum*. Biol. Control 53, 285–290.
Wijesinghe, C.J., Wilson Wijeratnam, R.S., Samarasekera, J.K.R.R., Wijesundera, R.L.C., 2011. Development of a formulation of *Trichoderma asperellum* to control black rot disease on pineapple caused by *Thielaviopsis paradoxa*. Crop Prot. 30, 300–306.
Wills, W.H., Lambe, R.C., 1978. Pathogenicity of *Thielaviopsis basicola* to Japanese holly (*Ilex crenata*). Plant Dis. Rep. 62, 859–863.
Wilson Wijeratnam, R.S., Hewajulige, I.G.N., Abeyratne, N., 2005. Postharvest hot water treatment for the control of *Thielaviopsis* black rot of pineapple. Postharvest Biol. Technol. 36, 323–327.
Wilson Wijeratnam, R.S., Hewajulige, I.G.N., Perera, M.G.D.S., 2006. The control of the black rot disease and the application of the novel wax formulation for extending storage life of pineapple. Acta Horticulturae 702, 407–411.
Wismer, C.A., 1961. Pineapple disease. In: Martin, J.P., Abbett, E.V., Huges, C.G. (Eds.). Sugarcane Diseases of the World, vol. 1. Elsevier Publishing Company, Amsterdam, pp. 223–245.
Wisniewski, M.E., Wilson, C.L., Hersberger, W., 1989. Characterization of inhibition of *Rhizopus stolonifer* germination and growth by *Enterobacter cloacae*. Can. J. Bot. 67, 2317–2323.
Zaid, A., De Wat, P.F., Djerbi, M., Oihabi, A.C., 2002. Diseases and pests of date palm. In: Zaid, A., Arias-Jimenez, E. (Eds.), Date Palm Cultivation. Food and Agriculture Organization of the United Nations, Rome, pp. 227–281.

CHAPTER 10

Lasiodiplodia theobromae in Citrus Fruit (Diplodia Stem-End Rot)

Jiuxu Zhang
Florida Department of Citrus, Lake Alfred, Florida, USA

Contents

OVERALL IMPORTANCE OF THE FUNGUS

Lasiodiplodia theobromae is an economically important fungal pathogen with a worldwide distribution in tropical and subtropical regions (Voorhees, 1942). It has over 500 different kinds of host plants and can cause many diseases such as tree dieback, root rot, fruit rots and leaf spot (Punithalingam, 1980).

This fungus is a key pathogen that causes postharvest decay on many important horticultural fruits such as citrus, avocados, mangos, papayas,

Postharvest Decay
http://dx.doi.org/10.1016/B978-0-12-411552-1.00010-7

bananas and guava in tropical and subtropical regions in the world (Smoot et al., 1971; Mascarenhas et al., 1995; Alvindia and Natsuaki, 2007; Hashem and Alamri, 2009; Abeywickrama et al., 2012). This postharvest decay can lead to significant economic losses, and the fungus can also cause preharvest diseases on some horticultural crops, such as mango tree dieback disease (Khanzada et al., 2004). This chapter will focus on a review of the biology of *L. theobromae* and the commercial control strategies for citrus in Florida, USA.

Citrus is a popular and economically important horticultural crop that is widely cultivated in tropical and subtropical regions in the world. Citrus can rightly be regarded as a universal fruit with production in over 100 countries spanning six continents. In addition, citrus is the most important tree fruit crop in the world, with the world production far exceeding that of other deciduous tree fruits such as apples, pears, peaches and plums (Saunt, 1990).

Production of all types of citrus in the world in the 2009–2010 crop year was at about 82.5 million metric tons, of which 59.8% were oranges, 26.4 % mandarins, 7.2% lemons and limes, and 6.6% grapefruit and pomelos. The production of grapefruit and oranges in the USA was first (about 30%) and second (about 11%) in terms of the total world production in the 2009–2010 crop year, respectively (USDA-NASS, 2011).

The Florida citrus industry is a vital component of the Florida economy. It consists of two segments, with fresh fruit accounting for about 10% and processed citrus products accounting for about 90% in economic value. According to the report of the University of Florida on the economic impact of Florida's citrus industry in the 2007–2008 crop season, the Florida citrus industry generated a total economic impact of about $8.91 billion, including about $1.02 billion from the fresh citrus fruit industry (Rahmani and Hodges, 2009). There are many factors affecting citrus production, citrus quality and marketing values. In recent years, citrus greening disease (also called Huanglongbing or HLB) caused by *Candidatus Liberibacter asiaticus* has become the most important problem causing a significant decline in Florida citrus production (Hodges and Spreen, 2012). However, postharvest decay still remains an important problem affecting the fresh fruit quality, shelf-life and economical values of citrus fruit. Diplodia stem-end rot caused by *L. theobromae* is one of the most important postharvest decays occurring on citrus fruit produced in Florida and in other warm, humid tropical and subtropical regions (Brown and Eckert, 2000; Ismail and Zhang, 2004). In the literature, many of the publications relating to citrus Diplodia stem-end rot described the causal fungal name as *Diplodia natalensis*, a synonym of *L. theobromae*, and the term Diplodia stem-end rot was frequently used to differentiate this decay from the other citrus stem-end rots, such as Phomopsis stem-end rot caused by *Phomopsis citri*, and Alternaria stem-end rot caused by *Alternaria alternata*. In order to be consistent with the existing literature, the term Diplodia stem-end rot is used throughout this chapter.

TAXONOMY AND MORPHOLOGY

Based on CABI databases for fungal classifications online (www.indexfungorum.org/), the current taxonomy status of *L. theobromae* is as follows: Kingdom: fungi; Phylum: Ascomycola; Class: Dothideomycetes; Order: Botryosphaeriales; Family: Botryosphaeriaceae; Genus: *Lasiodiplodia*; Species: *L. theobromae* (Pat.) Griffon & Maubl. Synonyms: *Botryodiplodia theobromae* Pat., and *Diplodia natalensis* Pole-Evans. Teleomorph: *Botryosphaeria rhodina* (Berk. & M.A. Curtis) Arx.; synonym: *Physalospora rhodina* Berk. & M.A. Curtis.

Many other synonyms for *L. theobromae* can also be found from the CABI databases online. These synonyms include: *L. nigra* Griffon & Maubl., *L. triflorae* B.B. Higgins, *L. tubericola* Ellis & Everh., *Diplodia ananassae* Sacc., *D. cacaoicola* Henn., *D. gossypina* Cooke; *D. theobromae* (Pat.) W. Nowell, *D. tubericola* (Ellis & Everh.), *Botryodiplodia ananassae* (Sacc.) Petr., *B. gossyii* Ellis & Barthol., *B. elasticae* Petch, *B. theobromae* Pat, and *B. tubericola* (Ellis & Everh.) Petr. Voorhees (1942) reported that the same fungus had been described under many different names and on many host species at various times, and researchers had reduced many of these names to synonyms of *L. theobromae*. These names are seldom used at present, given the evolution of the taxonomy of this fungus over a long period of time.

Lasiodiplodia theobromae on potato dextrose agar (PDA) medium grows rapidly at 20–30°C, with a white to light or dark gray aerial mycelia on PDA (Fig. 10.1A and B), and later produces abundant black pigmentation,

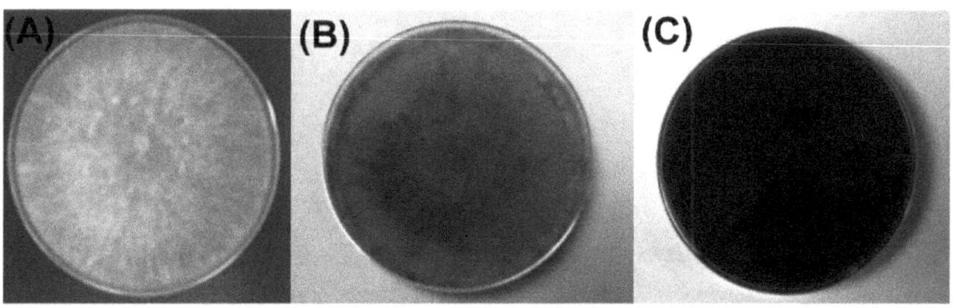

Figure 10.1 *Lasiodiplodia theobromae* grown on potato dextrose agar medium at 24°C for (A) 2 days, (B) for 10 days, and (C) for 10 days (reverse side of the culture).

which can be clearly viewed from the reserve side of the PDA culture (Fig. 10.1C). In the case of PDA as a fungal cultural medium, mycelial growth, pigmentation and pycnidium productions are largely affected by the dextrose concentrations in the medium. The higher the dextrose concentration, the more pigment production and the less pycnidium formation (Alan et al., 2001).

Lasiodiplodia theobromae can produce pycnidia when the fungus is grown on sterilized dead citrus twigs. Pycnidia formed and developed by the fungus on dead citrus twigs in groves are subglobose to globose (300–700 μm in diameter) and single or grouped in a stroma and papillate. Conidiophores are hyaline, simple, sometimes septate, rarely branched cylindrical, and arising from the inner layers of cells lining the pycnidial cavity. Conidia (10–8 × 17–43 μm) are formed in the pycnidia. Young conidial cells are hyaline and granular and have no septum, and mature conidia are striated and contain one septum (Brown and Eckert, 2000). Once the two-celled spores are formed, they readily survive from one season to the next on citrus trees and can cause latent infections on citrus fruit during the winter and spring in groves (Brown, 1971).

Perithecia of the fungus can be formed singly or in groups on citrus deadwood. They are globose (225–300 μm in diameter) and papillate, and could produce ascospores (11–14 × 30–35 μm) that are hyline, one-celled, and elliptical to ellipsoidal. The ascospores in citrus groves become airborne and are responsible for long-distance dispersal to newly planted citrus trees (Brown and Eckert, 2000).

BIOLOGY OF *L. THEOBROMAE* INFECTION PROCESS

Lasiodiplodia theobromae is a facultative parasite that completes its life cycle on dead twigs in groves by growing and sporulating in the deadwood of citrus trees. Water transmits fungal spores from deadwood to the surfaces of immature fruit where the fungus colonizes the dead tissue of the button. The fungal colonization remains latent (quiescent) and does not cause any fruit decay before harvest. Infections develop after harvest, especially under conditions of high temperatures and relative humidity. The optimal temperature for fungal mycelial growth is about 30°C (Zhang and Swingle, 2005). The fungus usually infects citrus fruit from the button (calyx and disk) at the stem-end of the fruit. It proceeds through the core more quickly than the rind, leading to the development of soft brown to black decay symptoms at both ends of the fruit (Fig. 10.2). The pathogen usually

Figure 10.2 Stem-end rot caused by *Lasiodiplodia theobromae* on 'Ambersweet' orange hybrids.

develops unevenly in the rind, forming finger-like projections of black to brown discolorations at the lesion margin between the segments. The fungus may also infect fruit through injuries, giving a soft brown to black lesion that enlarges very rapidly. Decay develops rapidly during and after excessive fruit ethylene degreening treatment and can be observed in fruit at the packing house. It is often observed on arrival at market or shortly thereafter (Ismail and Zhang, 2004). Normally, this fungus does not sporulate on infected fruit, and the decay does not spread from infected fruit to healthy fruit in packed containers.

AGRONOMIC AND ENVIRONMENTAL FACTORS FAVORING *L. THEOBROMAE* INFECTION

Citrus Growing Regions and Climates

Occurrence and distribution of Diplodia stem-end rot are largely related to environments characterized by hot and humid climates. World citrus production is distributed in the tropical and subtropical regions, which are located approximately 40° latitude each side of the equator, where favorable soil and climate conditions occur (Saunt, 1990). In the regions where citrus is produced, the climates also vary greatly; thus, citrus growth, yield and the physiological properties and quality of the fruit are affected differently. Tropical regions are characterized by hot and humid weather and abundant rainfall. The average temperature is greater than 18.3°C, and there are little diurnal or seasonal fluctuations in temperature. Subtropical regions, where a large part of the world's citrus production is located, could be separated into humid and dry (arid or semi-arid) growing areas. Dry subtropical regions including Mediterranean-type climates are characterized by hot, dry days and cool nights, winter-dominated rainfall, and dry summers. Some of the major citrus producers in these regions include Spain and California. Humid subtropical regions, such as Florida and Brazil, are characterized by high average temperatures and high humidity and rainfall. These different climatic environments in various subtropical regions in the world affect not only citrus growth and development but also the distribution and occurrence of Diplodia stem-end rot; however, little to none of this decay on citrus fruit has been reported in dry subtropical regions such as California and Spain (Ismail and Zhang, 2004; Smilanick et al., 2006a; Fischer et al., 2009). This disparity indicates that hot and humid tropical and subtropical weather favors *L. theobromae* survival, proliferation and infection on citrus in groves.

High rainfall during the summer months favors *L. theobromae* growth, sporulation on deadwood, and dispersal and infection on citrus fruit. Diplodia stem-end rot on citrus fruit rarely occurs when the fruit is still on the trees, and the decay shows up on fruit only after harvest when the conditions are conducive for disease development. Diplodia stem-end rot originates from latent infections of *L. theobromae* established in necrotic tissue of the fruit button (calyx and disk) on trees in the grove (Brown and Wilson, 1967, 1968; Brown and Burns, 1998). The fungus grows and sporulates on the deadwood of citrus trees. The fungus does not cause noticeable diseases on citrus stems and leaves, however, it has been reported to cause diseases such as tree decline, gummosis and dieback on other host fruit crops such as mango, peach, and apricot trees (Li et al., 1995; Khanzada et al., 2004; Úrbez-Torres et al., 2008). Fungal pycnidia are formed on deadwood, and spores of the fungus can be discharged from the pycnidia. The spores discharged from pycnidia initially are predominately single-celled spores; the single-celled spores later change to two-celled spores if the single-celled spores are not killed. Climatic factors significantly affect spore discharge from pycnidia and changes of spore forms. Brown (1971) studied the pycnidial release and survival of *L. theobromae* spores on citrus and found that hyaline, single-celled spores were released from pycnidia growing on citrus deadwood when the pycnidia became wet or when they were dried following wetting. This author further observed that after single-celled spores were discharged from pycnidia under 100% relative humidity (RH), the single-celled spores were changed into dark-colored two-celled spores within 5–6 hours. This process was interrupted when single-celled spores were placed at humidity below 100% RH, but continued upon return to 100% RH. The humidity conditions favorable for pycnidial spore release and the changes from single-celled spores to two-celled spores could be easily met in hot, humid tropical and subtropical citrus growing regions such as Florida, as rain occurs frequently during the summer. In addition, two-celled spores survive better than the single-celled spores in citrus groves. Once the two-celled spores are formed, they readily survive from one season to the next and could serve as inoculum sources for fruit infections. Brown and McCornack (1969) also reported that postharvest Diplodia stem-end rot incidence was less prevalent on the fruit harvested during drought than on fruit harvested from the same trees before the drought, and the recovery of *L. theobromae* by isolation was also less during the periods of drought. These findings further indicate that the successful sporulation and infection of *L. theobromae* in groves is closely related to climate, rainfall and humidity.

Conditions of Citrus Trees and Grove Management Practices

Lasiodiplodia theobromae primarily invades weak and dying tissues on citrus trees (Childs, 1952). The fungal spores produced on deadwood can be spread by rain and irrigation water to the citrus fruit surface, where they colonize the necrotic tissue on the button of immature fruit. The fungus remains latent on necrotic tissue caused by injury on the sepals and disk wall of the button and does not grow or infect the healthy button tissue due to the wound periderm formed on the button in response to injury (Brown and Wilson, 1967, 1968). The availability and amount of deadwood on citrus trees directly affect the populations of the fungus in groves. More deadwood available on citrus trees supports more fungal growth, proliferation and sporulation, leading to an increased population on citrus trees in groves. Postharvest Diplodia stem-end rot is usually less prevalent on fruit harvested from healthy trees that are 10 years or younger, largely due to the presence of much less deadwood compared to older trees (Brown and Miller, 1999).

Other cultural practices, such as irrigation methods, can also provide favorable conditions for *L. theobromae* to grow and sporulate on deadwood and spread the spores to immature fruit to infect the necrotic tissue of the fruit button in the grove. Irrigation systems using overhead water spraying methods could be more favorable for spreading the fungal spores from deadwood to fruit surfaces than other systems, such as under tree dripping or microsprinkler methods.

Chemical applications for other citrus diseases and pest control could also affect the population and sporulation of this fungus on citrus trees in groves. Less use of *L. theobromae*-suppressive chemicals in groves could favor its proliferation, sporulation and infection. Fungicides such as azoxystrobin and pyraclostrobin are commercially used in Florida to control some citrus preharvest diseases, such as citrus scab caused by *Elsinoë fawcettii*, Alternaria brown rot caused by *Alternaria alternata*, greasy spot caused by *Mycosphaerella citri*, and melanose caused by *Phomopsis citri* (Dewdney, 2013a, b, c, d). There is little information available regarding the effects of azoxystrobin and pyraclostrobin on *L. theobromae* in groves since it does not cause any noticeable disease on citrus trees. However, Zhang and Timmer (2007) reported that grove spray of azoxystrobin and pyraclostrobin shortly before harvest reduced the postharvest Diplodia stem-end rot incidence on 'Fallglo' tangerine hybrids in one of two tests, especially for fruit harvested 2 days after grove application of the chemicals. Benzimidazole fungicides, such as carbendazim, thiabendazole (TBZ), benomyl and thiophanate methyl, are effective for the control of postharvest Diplodia stem-end rot, Penicillium decays and preharvest postbloom fruit drop (Peres et al., 2004; Smilanick et al., 2006a). However, with the exception of TBZ, which is registered for postharvest citrus fruit treatment, none of the other benzimidazole fungicides are registered for preharvest or postharvest applications on citrus in the USA.

Postharvest Ethylene Degreening

Ethylene degreening treatment of early season harvested citrus fruit is a common commercial practice to stimulate the breakdown of chlorophyll and to reveal the characteristic orange and yellow fruit peel colors used for marketing purposes (McCornack, 1971; Smilanick et al., 2006b). The external color of citrus fruit often relates more to climatic conditions than to internal maturity (Wardowski et al., 2006). Mature fruit with green rind is more common in warm, humid tropical and subtropical regions such as Florida than in more arid climates such as California, where lower temperatures (below 10°C) at night during the growing season promote the degradation of chlorophyll (Brown and Miller, 1999). The commercial ethylene degreening conditions utilized for citrus fruit grown in different regions are quite different, especially regarding temperatures used, largely due to the physiological properties of the fruit formed under different climatic conditions. Each citrus growing region must determine its own optimum degreening conditions regarding chlorophyll degradation speed, desirable carotenoid buildup, and the effects of fruit senescence and decay development. In Florida, temperatures maintained at 28–29°C achieve degreening in the least amount of time because chlorophyll degradation occurs most rapidly at this temperature (Grierson and Newhall, 1953; Brown and Miller, 1999; Ritenour et al., 2012b). In other citrus producing regions, such as California, Israel and Spain, the degreening treatment is more often conducted at 21–25°C to allow for rapid degreening and some carotenoid synthesis along with the chlorophyll degradation (Brown and Miller, 1999; Wardowski et al., 2006). The recommended conditions for ethylene degreening treatment of Florida citrus fruits are: 28–29°C, 90–95% RH, and ethylene gas 3–5 $\mu L\ L^{-1}$ (ppm). The fruit should be treated only as long as necessary to obtain the adequate peel color, depending on fruit variety and degree of color break (Ritenour et al., 2012b). The ethylene degreening treatment of citrus fruit does not improve consumption quality; on the contrary, it contributes to physiological disorders, senescence, and loss from decay such as Diplodia stem-end rot and anthracnose (Brown, 1980).

Numerous studies in the last 70 years have reported that ethylene degreening treatment of citrus fruit significantly increased the incidence and severity of Diplodia stem-end rot (Brooks, 1942; Voorhees, 1944; Loucks and Hopkins, 1946; Barmore and Brown, 1985; Brown, 1986, 1994; Brown and Burns, 1998; Zhang, 2004). Ethylene degreening treatment also enhances citrus postharvest anthracnose caused by *Colletotrichum gloeosporioides* (Brown, 1975; Zhang and Timmer, 2007). The concentration of ethylene and the duration of the degreening treatment are positively correlated with Diplodia stem-end rot incidence and severity (Barmore and Brown, 1985; Brown, 1986; Brown and Lee, 1993; Zhang, 2004). Brown (1986) studied the effects of ethylene on Diplodia stem-end rot by degreening 'Valencia' oranges at 0, 1, 10 and 50 $\mu L\ L^{-1}$ of ethylene for 96 h at 30°C and 94–96% RH and showed a disease incidence of 4, 24, 47 and 65%, respectively, after 4 weeks' storage. In other studies, Zhang (2004) reported that after 'Valencia' oranges were degreened at 0, 5 and 50 $\mu L\ L^{-1}$ concentrations at 30°C and 92–96% RH for 60 h, Diplodia stem-end rot incidence of 3.3, 20.0, and 66.7%, respectively was observed after 2 weeks of storage at 21°C.

It is interesting and important to understand why ethylene treatment of citrus fruit increases Diplodia stem-end rot occurrence and development. The following mechanisms or reasons have been proposed by researchers: (1) ethylene stimulates the formation of abscission zones between fruit and button tissue (Voorhees, 1944; Brown and Wilson, 1968), largely due to enhanced fruit abscission enzymes (Brown and Burns, 1998), to form natural openings which allow the pathogen to infect fruit (Brown and Wilson, 1968; Barmore and Brown, 1985); (2) ethylene directly stimulates the growth of *L. theobromae* (Brown and Lee, 1993); (3) ethylene degreening temperature of 28–29°C in Florida is close to the optimum temperature for *L. theobromae* mycelial growth (Voorhees, 1944; Zhang and Swingle, 2005); and (4) ethylene reduces natural fruit resistance by triggering certain biochemical events in fruit rind that make fruit tissues more susceptible to the fungus (Brown, 1986; Zhang, 2004). It should also be noted that the effects of ethylene degreening of citrus fruit on Diplodia stem-end rot under the Florida packing house system could be attributed to a combination of the factors described above.

Postharvest Fruit Holding Time and Temperature Before Fungicide Treatments

High ambient temperature and longer fruit holding time before postharvest fungicide application favor Diplodia stem-end rot development. Packing houses sometimes hold harvested field fruit (untreated fruit) at ambient temperature for up to a few days if they do not have enough cold storage capacity and if the shipment of packed fruit to market is slow. This delay could create favorable conditions for Diplodia stem-end rot development. If ethylene degreened fruit are held at ambient temperature for a few days before further packing line processing and fungicide treatments, the fruit could be more vulnerable to *L. theobromae* infection and rot development, since the ethylene degreening treatment increases the abscission of button tissues from fruit and leads to the natural openings available for *L. theobromae* infection from necrotic tissues of button (Brown and Wilson, 1968). Brown and Miller (1999) indicated that the efficacy of postharvest fungicide treatments diminished if the application was delayed more than 24 h after harvest.

In Florida, early season citrus fruit are harvested from September to November, and the fruit normally need to be subjected to ethylene degreening treatment to achieve commercial acceptance of fruit color by consumers (Barmore and Brown, 1985; Zhang and Timmer, 2007). At this time of year in Florida, the ambient temperature is also quite high and thus favorable for *L. theobromae* growth and infection (Smilanick et al., 2006a).

Relatively high temperatures favor the growth of *L. theobromae*. It has been reported that the fungus grows well over a range of temperatures from 15 to 35°C, with an optimum temperature at about 30°C, while there is no or little growth at 10°C and at 40°C (Voorhees, 1942; Alan et al., 2001;

Zhang and Swingle, 2005). When 'Valencia' and 'Pineapple' oranges were exposed to temperatures of 21 and 35°C, respectively, and about 95% RH for 48 h, and then stored at 21°C for 2 weeks, Diplodia stem-end rot incidences on fruit treated at 35°C were 1.3 to 11.2 times higher than those on the control fruit treated at 21°C in several separate tests. High temperature may also affect the physiological properties of the harvested fruit, allowing an enhanced formation of the abscission zone between fruit and button, thus promoting *L. theobromae* infection in fruit tissue and making fruit more susceptible to fungal infection (Zhang and Swingle, 2005).

CONTROL

Effective control strategies of postharvest Diplodia stem-end rot should be applied as an integrated decay control system to prevent and maximally reduce fungal inoculum production, fruit infections and decay development and to eradicate existing infections by various approaches during both the preharvest and postharvest stages. The following disease control measures and practices may be applied separately or in a sequence procedure to achieve successful and commercially acceptable control of postharvest Diplodia stem-end rot of citrus fruit.

Conventional Alternatives

Cultural Practices in Groves

Some cultural practices in citrus groves can reduce pathogen inoculum and infection levels, subsequently leading to a reduced level of postharvest Diplodia stem-end rot. Good cultural practices such as nutrition and irrigation management and pest and weed control to maintain vigorous and healthy citrus trees with minimum deadwood could reduce *L. theobromae* population levels in groves. Selective removal of deadwood by pruning can aid in the reduction of *L. theobromae* inoculum levels. However, it is generally not economically feasible in commercial practice to try to remove all small deadwood (Jackson, 1991).

Early season fruit harvests that require extensive postharvest degreening treatment should be made from trees with minimal deadwood since it is likely that these fruit have low levels of *L. theobromae* inoculum and infection. Fruit harvest methods could also affect postharvest Diplodia stem-end rot incidences due to the amount of necrotic button tissues left on fruit, which affect the amount of pathogen inoculum and latent infections of the fungus. Debuttoning of fruit at harvest or after harvest could result in a reduction of Diplodia stem-end rot due to the lesser amounts of button tissues remaining on the fruit (Winston, 1950; Smoot et al., 1971; Brown, 1980, 1994). A comparison of pulling versus clipping oranges for postharvest decay showed that pulled fruit had less Diplodia stem-end rot than clipped ones (Hopkins and Loucks, 1944). Pulling of fruit for most citrus varieties instead of removing them with clippers has been recommended for the same reason and has extensively been practiced in Florida (Smoot et al., 1971). This practice is also popular in South Africa and in some other citrus-producing regions (Eckert and Eaks, 1989).

Harvesting fruit with better natural orange color (color break), either by spot-picking for better color or delaying harvest time to allow for better natural orange/yellow color development, could shorten the required time duration for postharvest degreening treatment and lead to reduced levels of Diplodia stem-end rot (Brown and Miller, 1999).

Preharvest Chemical Spraying in Groves

If early season harvested fruit need ethylene degreening treatment or need to be held in packing houses at ambient temperature for longer than the normal holding time, preharvest application of appropriate fungicides in groves shortly before fruit harvesting could be practiced to help citrus postharvest disease management (Eckert and Eaks, 1989; Ritenour et al., 2004; Smilanick et al., 2006b). The benzimidazole fungicides benomyl and thiophanate methyl have been reported to be effective for the control of Diplodia stem-end rot and other postharvest diseases when these fungicides are sprayed on citrus trees shortly before harvest (Brown and McCornack, 1969; Brown and Albrigo, 1970; Brown, 1974; Ritenour et al., 2004; Salvatore and Ritenour, 2007; Zhang and Timmer, 2007). When 'Sunburst' tangerines were harvested up to 4 weeks after the trees in the grove were sprayed with benomyl or thiophanate methyl at an application rate of 2.2 kg ha^{-1} in both 2003 and 2004 experiments, benomyl and thiophanate methyl reduced Diplodia stem-end rot incidences on the fruit by 77–100% compared to the controls during the 2 weeks of fruit postharvest storage at 21°C and 95% RH (Zhang and Timmer, 2007). The residues of benomyl and thiophanate methyl in citrus fruit peel after spraying are maintained at effective levels for more than a month (Brown and Albrigo, 1972; Smilanick et al., 2006b). A recommended time for fruit harvest after preharvest benomyl application is 2 days to 3 weeks (Wardowski and Brown, 1993). Benomyl used to be the only effective fungicide registered by the US EPA on citrus preharvest application for postharvest decay control. However, US EPA canceled its re-registration due to the registrants' voluntary cancellation request in 2002 (US EPA, 2002). Thiophanate methyl was only temporarily permitted for citrus preharvest application for post-bloom fruit drop and Diplodia stem-end rot control in Florida through the US EPA Section 18 emergency exemption from 2002 to 2009 (US EPA, 2008). At present, there are no registered preharvest fungicides that can be used shortly before harvesting for the control of postharvest Diplodia stem-end rot in Florida. However, benomyl or thiophanate methyl may still be registered for citrus grove application for preharvest citrus disease control in other countries. Research efforts seeking new effective preharvest fungicides for the control of Diplodia stem-end rot are ongoing in Florida.

Postharvest Fungicide Drenching Treatment

Early season harvested Florida citrus cultivars often need degreening treatment to improve their yellow/orange colors for marketing purposes. If a suitable preharvest fungicide spray shortly before harvest in grove is not applied, a postharvest drenching with an appropriate fungicide before degreening treatment would be significantly beneficial in reducing postharvest Diplodia stem-end rot, since this fungus can be significantly enhanced by the ethylene degreening treatment (Barmore and Brown, 1985; Brown and Miller, 1999; Ismail and Zhang, 2004; Zhang and Timmer, 2007). Postharvest fruit drenching with a fungicide, TBZ or imazalil, at 0.1% in drenching suspensions immediately after harvest is a common commercial practice for decay control in Florida (Brown, 1977; Brown and Miller, 1999; Ismail and Zhang, 2004; Ritenour et al., 2012b) and other citrus-producing regions (Smilanick et al., 2006a). Moreover, if fruit do not need degreening treatment but are delayed for more than 24 h before being treated with fungicides in a packing house, they should also be drenched with TBZ or imazalil at a concentration of 0.1% to prevent excessive decays (Brown and Miller, 1999). Drenching fruit with adequate amounts of fungicide suspensions should result in better coverage and decay control efficacies. A commercial bin-drenching system has been a common fungicide application practice in Florida packing houses. Fruit in field bins that are then often loaded onto field trucks are transported through a tunnel and drenched with a fungicide (TBZ or imazalil) suspension before being placed in the degreening room (Brown and Miller, 1999; Smilanick et al., 2006a). Fruit are normally drenched for 3–4 min and then drained for another 4–5 min before being removed from the drenchers. The drenching system is a recovery application of fungicide suspensions. TBZ and imazalil are commonly used in the drenching treatment since both are effective for controlling Diplodia stem-end rot (Table 10.1). The fungicide concentration (the recommended rate for TBZ and imazalil is 0.1% in the suspensions) in the drenching tank should be monitored, and the desired strength should be maintained by adding additional fungicide.

Commercially, chlorine is often added to the TBZ-drenching suspension to control some other postharvest fruit decay pathogens such as *Geotrichum citri-aurantii*, *Phytophthora* species, and TBZ-resistant *Penicillium* isolates. Chlorine must constantly be added to the TBZ-drencher, and the free available chlorine in the drenching suspensions should be maintained at 50–100 mg L^{-1} (Brown and Miller, 1999). The suspension's pH values can significantly affect the forms (molecular chlorine Cl_2, hypochlorous acid HOCl, and hypochlorite OCl^-) of free chlorine available in water. At low pH, especially below pH 4, more Cl_2 can evolve, and Cl_2 is harmful to workers. Above pH 7.5, more hypochlorite (OCl^-) can be formed. Hypochlorous acid (HOCl) is a strong and active sporicidal component, and hypochlorite (OCl^-) is a weak sporicide (Brown and Miller, 1999). A recommended optimum pH for chlorine-added TBZ drenching suspension is 6.5–7.5, which should be adjusted and maintained. TBZ is only slightly soluble in water, therefore, the suspension must be constantly agitated to ensure uniformity of TBZ concentration during applications. Chlorine cannot be added to an imazalil drenching system since they are not compatible. Instead, some heating or other sanitizers are applied in imazalil drenching suspensions for sanitation purposes (Ritenour et al., 2012b).

In the USA, a new postharvest fungicide, fludioxonil, has been registered by the US EPA for citrus postharvest decay control (US EPA, 2004a).

Table 10.1 Effects of Postharvest Fungicides on Diplodia Stem-End Rot Control of Citrus Fruits

Fungicide	Diplodia Stem-end Rot
Thiabendazole	+++
Imazalil	++
Sodium o-phenylphanate	+
Azoxystrobin	+
Fludioxonil	++
Pyrimethanil	+
Guazatine*	++

+: Some control; ++: moderate control; +++: good control.
Brown, 1983, 1988; Brown and Miller, 1999; Smilanick et al., 2006a; Zhang, 2007, 2009, 2012
*Guazatine is not registered in the USA.

Fludioxonil has been reported to be an effective fungicide for Diplodia stem-end rot control in Florida, and it performs similarly to imazalil and slightly less than TBZ for Diplodia stem-end rot control using both simulated drenching and packing-line application systems (Zhang, 2007, 2012). A drenching application of aqueous fludioxonil at 250–1200 mg L^{-1} using 'Ambersweet' oranges reduced Diplodia stem-end rot incidences by 75.7–88.6% compared to the controls (Zhang, 2007). It has been reported that chlorine does not affect the activity of fluodioxonil and is compatible with aqueous fludioxonil suspensions (Adaskaveg et al., 2004; Zhang, 2007; Ritenour et al., 2012b). This indicates that fludioxonil can be used in commercial chlorine-fungicide drenching systems for Diplodia stem-end rot control in Florida. In fact, fludioxonil (600–1000 mg L^{-1}) has been recommended for a commercial Florida fungicide drenching system application for Diplodia stem-end rot and other citrus decay control (Ritenour et al., 2012b).

Another postharvest fungicide, guazatine, provides some control of Diplodia stem-end rot, particularly if applied by a drenching system before degreening, but is not as effective as TBZ (see Table 10.1) (Brown, 1983, 1986). A study showed that guazatine and TBZ reduced Diplodia stem-end rot incidence by 86 and 94%, respectively, when 'Valencia' oranges were dipped in fungicide suspensions at 1000 mg L^{-1} for 15 seconds, and then stored for 3 weeks at 21°C and 85% RH (Brown, 1986). Guazatine has been used commercially on citrus fruit for many years in many countries, but it has not been registered in the USA, where it can neither be used nor can fruit treated with it be imported (Smilanick et al., 2006a).

Citrus fruit fungicide drenching can be used not only for fruit that need to be degreened or delayed for packing house processing but also be employed as a routine citrus postharvest decay control measure to achieve consistent and successful decay control in an overall integrated decay control system. Some citrus packing houses in Florida conduct the fruit fungicide drenching treatment throughout the whole fruit season.

Ethylene Degreening Management

Mature fruit with green rind are common for early season fruit, particularly citrus growing in hot, humid subtropical regions such as Florida. Degreening of citrus fruit is a process by which the green color is removed from the peel by treating the fruit with ethylene gas to degrade chlorophyll and reveal yellow and orange colors for commercial and marketing purposes. However, if fruit degreening is not handled appropriately, one of the consequences is that Diplodia stem-end rot may significantly be enhanced (Barmore and Brown, 1985; Ismail and Zhang, 2004). The incidence and severity of this fungus are positively correlated with the concentration of ethylene used and the duration of the degreening (Brown, 1986; McCornack, 1971; Zhang, 2004). Zhang (2004) reported that Diplodia stem-end rot incidence in 'Valencia' oranges treated with ethylene at 50 μL L^{-1} was 3.3 times greater than in fruit treated with 5 μL L^{-1} for 60 h after the fruit were stored for 2 weeks at 21°C and 92–96% RH. Another study (Barmore and Brown, 1985) showed that Diplodia stem-end rot incidence of 'Valencia' oranges that were treated with ethylene at 50 μL L^{-1} for 96 h was about two times higher than that of the fruit treated with the same concentration of ethylene for 48 h after the treated fruit were stored for 4 weeks at 29–30°C and 94–96% RH. Therefore, to minimize Diplodia stem-end rot, the fruit degreening treatment should be carefully and appropriately handled.

If citrus fruit degreening is needed, the ethylene concentration should be controlled at 3–5 μL L^{-1}, and the duration of degreening should be only as long as necessary to obtain the adequate peel color depending on fruit variety and degree of color break (Ritenour et al., 2012b). To minimize the duration of ethylene degreening treatment, fruit on trees in the grove should be allowed to have sufficient color break before harvest. In Florida, a target degreening duration for mandarin-type varieties, such as 'Fallglo', 'Murcott', 'Rubinson' and 'Sunburst' tangerines or tangerine hybrids, should be within 36 h; and for orange varieties, such as 'Hamlin', 'Ambersweet', 'Navel' oranges, and grapefruit varieties such as 'Star Ruby', 'Flame', 'Marsh' and 'Duncan' grapefruits, the target degreening duration should be within 72 h. Ethylene concentrations in degreening facilities should be accurately controlled and monitored. Using higher ethylene concentrations (such as more than 10 μL L^{-1}) than recommended does not expedite degreening, but instead significantly enhances Diplodia stem-end rot (Brown and Miller, 1999). Ethylene is explosive at concentrations of 3% or higher, and care must be taken to prevent leaks and avoid ignition from any spark or flame (Brown and Miller, 1999). The recommended ventilation and air circulation during degreening of Florida fresh citrus fruits are that fresh air should enter the degreening room at the rate of one air change per h, and air movement in the degreening room should be a minimum of 10 cubic feet per min (CFM) per field box or 100 CFM per pellet box (Ritenour et al., 2012a).

The recommended temperature for degreening of Florida citrus is 28–29°C, since it is the optimum temperature for rapidly degrading the chlorophyll (Brown and Miller, 1999; Ritenour et al., 2012a). However, relatively low temperatures (21–25°C) are recommended and used in other citrus-producing regions such as California and Israel, in which some orange pigments can be synthesized during chlorophyll degradation (Brown and Miller, 1999). Since the temperature (28–29°C) of Florida degreening conditions is close to the optimum temperature (30°C) for the mycelial growth of *L. theobromae*, it also increases Diplodia stem-end rot development (Zhang and Swingle, 2005). Some packing houses in Florida use relatively low temperatures (22°C) to degreen some oranges and have good success (Metcalf, 2001, internal communication). In the degreening rooms, relative humidity (90–95%), air circulation and fresh air exchange also have to be adequate and appropriate to ensure successful degreening and to minimize other adverse effects on the fruit, such as fruit water loss, stem-end rind breakdown and fruit deformation after packing. Carbon dioxide (CO_2) generated by fruit respiration during degreening also affects the degreening efficacy. The rate of degreening is reduced at a CO_2 concentration of 0.1% and is nearly stopped at a concentration of 1.0% or higher (Brown and Miller, 1999).

To reduce maximally the risk of Diplodia stem-end rot development as a result of fruit degreening treatments, fruit should be drenched with a suitable postharvest fungicide such as TBZ, imazalil or fludioxonil as discussed above in addition to the appropriate management of degreening conditions and treatments.

Packing Line Processing

Fungicide application during the fruit packing line operation is another important component to a successful, integrated citrus postharvest decay control system. Fungicides can be applied at different steps and forms based on the properties of different fungicides and target decays during fresh citrus packing line processing in Florida. Three postharvest fungicides, TBZ, imazalil and sodium o-phenylphanate (SOPP), are commercially used for Diplodia stem-end rot control during packing line processing (Ismail and Zhang, 2004). TBZ and imazalil are more effective than SOPP for Diplodia stem-end rot control (see Table 10.1). The efficacies of TBZ, imazalil and SOPP for Diplodia stem-end rot control could be affected by many factors such as fungicide concentrations, formulations, application methods, citrus varieties, time of the harvest, fruit degreening treatment, temperature and moisture. A study (Zhang, 2009) showed that aqueous imazalil and TBZ (1000 mg L^{-1}) applied to ethylene-degreened 'Pineapple' oranges during packing line processing reduced Diplodia stem-end rot incidences by 70.7

and 84.0%, respectively, after the fruit were stored for 3 weeks at 21°C and about 95% RH.

SOPP is still commonly used in Florida packing houses for citrus postharvest decay control and fruit disinfection for citrus canker bacterium although SOPP is less effective than TBZ and imazalil for Diplodia stem-end rot control. SOPP is usually applied in a non-recovery foam or, less often, in the form of a recovery drencher or flood during the time that fruit are washed (Brown and Miller, 1999). SOPP is soluble in water and is hydrolyzed extensively in aqueous solutions. Effective concentrations of the undissociated form, o-phenylphenol (HOPP), and dissociated form, o-phenylphenate (OPP) ion are dependent upon the strength of the solution (the concentration of SOPP) and pH values (Eckert and Eaks, 1989). The undissociated form, HOPP, is lethal to microorganisms and injures the peel of citrus at concentrations of 200–400 mg L^{-1}, depending upon the temperature of the solution and the time of contact. The dissociated form, OPP, is not phytotoxic, and a solution of SOPP containing excess alkali to suppress the hydrolysis of SOPP is generally safe for treatment of citrus fruit (Eckert and Eaks, 1989; Smilanick et al., 2006a). A commercially recommended solution formulation of SOPP used in Florida contains 2% SOPP at pH 11.5–12, 0.2% sodium hydroxide (for pH control), and 1% hexamine to minimize phytotoxicity (Ritenour et al., 2012b). Non-recovery foam application of SOPP is more popular in commercial operations in Florida, in which SOPP solution foam is applied to fruit and brushed for 15–30 seconds and then followed by a water rinse during packing line processing.

Imazalil and TBZ are commonly applied in an aqueous suspension before waxing or in the water emulsion wax to control Diplodia stem-end rot decay during packing line processing (Brown and Miller, 1999; Ismail and Zhang, 2004). A non-recovery drip or spray of fungicide suspensions to washed fruit while fruit are moving on a bed of slowly rotating brushes is commonly used during packing line processing in Florida. Both imazalil and TBZ are effective for Diplodia stem-end rot control when they are applied appropriately. However, TBZ is generally more effective than imazalil for Diplodia stem-end rot control, especially on ethylene degreened fruit (Brown, 1984). Application of aqueous suspensions of imazalil and TBZ, especially imazalil, is more effective than the application of these fungicides with a water-based wax formulation at the same concentration (1000 mg L^{-1}) for Diplodia stem-end rot control due to the adverse effect of the wax on fungicide movement and activity (Brown, 1984; Brown and Miller, 1999). A higher concentration (2000 mg L^{-1}) of imazalil in a wax suspension is usually needed to achieve a similar level of decay control efficacy compared to an aqueous fungicide form (1000 mg L^{-1}) (Brown, 1984). Commercially recommended rates for imazalil and TBZ are 1000 mg L^{-1} for aqueous application and 2000 mg L^{-1} for water-based wax application (Ismail and Zhang, 2004; Ritenour et al., 2012b). TBZ is relatively insoluble in water, and preparations should be constantly agitated during its application. Residues of TBZ on treated fruit largely remain on the fruit surface since it is non-systemic, while imazalil is more systemic, so some of the fungicide moves into the fruit rind.

In the last several years, three new postharvest fungicides, azoxystrobin, fludioxonil and pyrimethanil, have been registered by the US EPA for citrus postharvest decay control (US EPA, 2004a, b, 2009; Kanetis et al., 2007). The US EPA classifies these new compounds as reduced-risk fungicides (Adaskaveg et al., 2004). The new fungicides, azoxystrobin, fludioxonil and pyrimethanil, and existing fungicides, imazalil, TBZ and SOPP, belong to different chemical groups and have different modes of action when compared to one another. The registrations of these new fungicides have provided the citrus industry in the USA with additional chemical tools and modes of action for effectively controlling citrus postharvest decays and for managing fungicide resistance problems, particularly for *Penicillium* pathogens (Adaskaveg et al., 2004). The efficacy of the three new fungicides is quite different for Diplodia stem-end rot control (Zhang, 2009, 2012). Fludioxonil is much more effective for Diplodia stem-end rot control than azoxystrobin and pyrimethanil, and it performs similarly to imazalil and slightly less than TBZ (Zhang, 2012). Therefore, fludioxonil appears to be a new alternative postharvest fungicide to TBZ and imazalil for Diplodia stem-end rot control on citrus fruit in Florida.

The management of pathogen resistance to fungicides could play an important role in citrus postharvest decay control. Fungicide resistance problems have frequently been reported with *Penicillium* pathogens (Holmes and Eckert, 1999; Kinay et al., 2007; Perez et al., 2011). However, fungicide resistance has seldom been reported as a problem for Diplodia stem-end rot control. This may be partly due to the fact that *L. theobromae* does not usually sporulate on infected fruit, and postharvest infection starts from the latent infection which occurs on trees in groves. To prevent and reduce the potential development of fungicide resistance of *L. theobromae*, the use of different fungicides with different modes of action at different stages of the fruit packing process could help minimize and prevent the potential development of Diplodia fungicide resistance.

Low Temperature

Low-temperature storage of fruit is an essential aspect of the postharvest decay control program, since heat removal from fruit slows down metabolic processes that lead to the loss of both fruit quality and disease resistance (Eckert and Eaks, 1989). Temperatures near ambient conditions favor fungal decay, rapid water loss, ascorbic acid (vitamin C) loss, and fruit softening, while excessive cold may cause chilling injury (Brown and Miller, 1999; Grierson and Miller, 2006). Low temperature also directly suppresses pathogen growth and decay development on fruit. Moreover, low temperature is especially effective for the control of fungal pathogens with high temperature requirements for their optimum mycelial growth, such as *L. theobromae*. The development of Diplodia stem-end rot is almost completely inhibited at 10°C (Brown, 1986; Brown and Eckert, 2000). This result is consistent with the temperature requirement for optimum growth of *L. theobromae*. As described previously, this fungus has a temperature range of 10–40°C for its mycelial growth, with the optimum temperature at about 30°C (Zhang and Swingle, 2005). However, for low-temperature storage of citrus fruit, care must also be taken to avoid chilling injury, particularly in grapefruit, which are sensitive to chilling temperatures. The suitable low temperature for citrus fruit storage varies with citrus varieties, producing regions, maturity and postharvest treatments. Recommended optimal storage temperatures are 10–15, 0–1.1, 4.4, and 10°C for Florida grapefruit, oranges, mandarins and lemons/limes, respectively (Ritenour et al., 2012b). Oranges grown in other regions such as Arizona, California, Australia, Israel and South Africa are usually stored at relatively higher temperatures (4.4–10°C) than those recommended for Florida oranges (Brown and Miller, 1999). Generally, low temperature only retards and delays fungal growth and disease development during fruit storage; decays develop after infected fruits are transferred to the ambient environmental conditions. To eradicate the diseases, appropriate fungicide treatments of fruit are still needed.

Pre-cooling can also be used as a practical approach in packing houses to improve fruit quality and reduce decays, particularly for Diplodia stem-end rot. Pre-cooling is a technique that removes field heat from citrus fruits. Common pre-cooling methods for citrus fruit include room cooling and forced-air cooling. Hydro-cooling is seldom used largely due to the increased risk of spreading decay pathogens.

Other Alternatives

Heat Treatment

Heat treatments for Diplodia stem-end rot control are generally less effective compared to the control of *Penicillium* decay, largely due to the latent infection of Diplodia stem-end rot pathogen developed during the preharvest stage and their containment within button tissues, which are more protected from physical and chemical treatments. In addition, *L. theobromae* requires a relatively high temperature for its optimum growth. Thus, quite a high level of heat treatment is required effectively to inhibit its growth. However, if the temperature is too high or heat treatment is applied for too long, fruit damage can occur. Smoot (1963) studied hot water treatment of Florida citrus fruit for postharvest decay control using 'Hamlin' oranges artificially inoculated with *L. theobromae* and temperatures of 47.8–53.9°C and 5 min in a fruit dip method. The results showed that the treatments at 50–53.9°C effectively suppressed Diplodia stem-end rot for only one week when fruit were stored at 21°C after heat treatments (Smoot, 1963). However, at 53.9°C, heat injury (scalding) was observed in all tested fruits. Ritenour et al. (2003) studied hot water treatments for Diplodia stem-end rot control on Florida Ruby Red and 'Marsh' grapefruits by using short duration (10–120 seconds) and high temperatures (50–65°C) combinations and showed that Diplodia stem-end rot was significantly reduced when fruit were dipped for 10–60 sec in hot water at 59°C or higher, or at 56°C for 120 sec. However, all of these effective temperature and duration combinations caused grapefruit peel scalding, with the single exception of treatment at 59°C for 10 sec (Ritenour et al., 2003). Some of our recent studies showed that Diplodia stem-end rot incidences on 'Murcott' tangerine hybrids and 'Valencia' oranges were reduced by 77.3 and 46.7%, respectively, after the fruit were dipped into hot water at 49°C for 30 s, and then stored at 29°C and 95–100% RH for 3 weeks, compared to the control fruit that were dipped into the water at ambient temperature (22°C), and then stored at the same conditions and period of time (Ritenour and Zhang, unpublished data). The currently available information above indicates that hot water treatment of Florida citrus fruit alone appears to have limited potential to be a commercially feasible approach for Diplodia stem-end rot control, and further research is needed in this area.

Heat treatment combined with fungicide treatments may be the better approach for citrus Diplodia stem-end rot control. Heat treatment of citrus fruit could increase the levels of fungicide residues in fruit tissues

(Schirra et al., 1996). Our tests also showed that heated TBZ suspension increased TBZ residue level in heat-treated grapefruit. Based on a test to compare the TBZ residue levels when 'Flame' grapefruit were dipped into the TBZ suspension (600 mg L^{-1}) at 50°C or at 24°C for 1 min, TBZ residue based on whole fruit weight was 0.904 μg g^{-1} from the heated TBZ suspension at 50°C compared to 0.402 μg g^{-1} from the TBZ suspension at 24°C (Zhang and Trama, unpublished data). However, there is little available information on heated fungicide treatments relating to Diplodia stem-end rot control. Presumably, hot water combined with a suitable postharvest fungicide could be a useful approach in achieving effective Diplodia stem-end rot control since fungicide residue levels are increased in citrus fruit tissues when fungicide suspensions are heated. Further study on this subject is needed, and is ongoing in our research lab.

Plant Growth Regulators

Some plant growth regulators, such as 2,4-dichlorophenoxyacetic acid (2,4-D) and gibberellic acid (GA), used on citrus during the preharvest and/or postharvest stages could maintain better fruit peel quality, delay overripe color development, decrease late season fruit drop, prolong both pre- and postharvest life of fruit, and reduce postharvest stem-end rots, including Diplodia stem-end rot (Coggins, 1969; Ferguson et al., 1982). Generally, the percentage of green button preserved on harvested fruit is inversely related to the incidence of Diplodia stem-end rot, which develops during prolonged storage (Stewart, 1949). The effectiveness of 2,4-D in preserving green button and in controlling stem-end rots has been studied and demonstrated by preharvest spraying of citrus trees with aqueous 2,4-D, postharvest dipping of citrus fruit with aqueous solution containing 2,4-D, or postharvest fruit application of waxes containing 2,4-D (Stewart, 1949). Ferguson et al. (1982) reported that pre- and/or postharvest application of GA plus 2,4-D significantly reduced stem-end rot incidences and increased grapefruit storage life. 2,4-D for reducing the Diplodia stem-end rot incidence of citrus fruit has been reported to be related to the suppression of the fruit abscission enzymes such as polygalacturonases and cellulases by 2,4-D (Brown and Burns, 1998). When 'Valencia' fruit were degreened at 55 μL L^{-1} ethylene for 42 h and then inoculated with *L. theobromae* mycelial inoculum after the fruit were dipped with 2,4-D isopropyl ester solution (1 mL L^{-1}) for 15 s, the fruit treatment with 2,4-D reduced decay by 75 and 46%, respectively, at 6 and 14 days of fruit storage compared to the controls (Brown and Burns, 1998). 2,4-D has often been commercially incorporated into wax formulations at 500 mg L^{-1} to treat citrus fruits during packing line processing to control stem-end rots caused by *L. theobromae* and other fungal pathogens during long distance shipments or prolonged storage (Brown and Miller, 1999; Smilanick et al., 2006a).

CONCLUDING REMARKS

Citrus is one of the most important fruit crops in the tropical and subtropical regions in the world. Many factors affect citrus production, quality and economic value from the field to the consumer. Diplodia stem-end rot caused by *L. theobromae* is one of the most significant postharvest decays of citrus fruit produced in the hot and humid regions in the world. Diplodia stem-end rot develops from fungal latent infections that occur on immature fruit on citrus trees in the grove. The occurrence and development of this fungus are affected by many preharvest and postharvest factors, including deadwood on the trees, rainfall, cultural practices, fruit harvesting methods, postharvest ethylene degreening treatment, timing and types of fungicide treatments, and temperatures during fruit transit and storage. Postharvest Diplodia stem-end rot control can be achieved by preventing, inhibiting, reducing and eradicating the fungal growth, sporulation, infections and the reactivation of latent infections and decay development by using both physical and chemical approaches in the pre- and postharvest stages. An integrated disease control system with chemical application as the core component can be employed effectively to control postharvest Diplodia stem-end rot. The Diplodia stem-end rot control system should also be integrated into the whole citrus postharvest decay control system to control all citrus postharvest diseases. The control strategies of citrus Diplodia stem-end rot described above should also provide useful information for the control of this fungus on other horticultural fruits.

REFERENCES

Abeywickrama, K., Wijerathna, C., Rajapaksha, N., Sarananda, K., Kannangara, S., 2012. Disease control strategies for extending storage life of papaya (*Carica papaya*), cultivars 'Red Lady' and 'Rathna'. Ceylon Journal of Science. Biol. Sci. 41, 27–34.

Adaskaveg, J.E., Kanetis, L., Soto-Estrada, A., Förster, H., 2004. A new era of postharvest decay control in citrus with the simultaneous introduction of three new 'reduced-risk' fungicides. Proc. Int. Soc. Citriculture 10 (999), 1004.

Alan, M.S., Begum, M.F., Sarkar, M.R., Islam, M.R., Alam, M.S., 2001. Effect of temperature, light and media on growth, sporulation, formation of pigments and pycnidia of *Botryodiplodia theobromae*. Pak. J. Biol. Sci. 4, 1224–1227.

Alvindia, D., Natsuaki, K.T., 2007. Control of crown rot-causing fungal pathogens of banana by inorganic salts and surfactant. Crop Prot. 26, 1667–1673.

Barmore, C.R., Brown, G.E., 1985. Influence of ethylene on increased susceptibility of oranges to *Diplodia natalensis*. Plant Dis. 69, 228–230.
Brooks, C., 1942. Prevention of stem-end rot. Proc. Florida State Hortic. Soc. 55, 61–69.
Brown, G.E., 1971. Pycnidial release and survival of *Diplodia natalensis* spores. Phytopathology 61, 559–561.
Brown, G.E., 1974. Postharvest citrus decay as affected by benlate applications in the grove. Proc. Florida State Hortic. Soc. 87, 237–240.
Brown, G.E., 1975. Factors affecting postharvest development of *Colletotrichum gloeosporioides* in citrus fruits. Phytopathology 65, 404–409.
Brown, G.E., 1977. Application of benzimidazole fungicides for citrus decay control. Proc. Int. Soc. Citriculture 1, 28273–28277.
Brown, G.E., 1980. Fruit handling and decay control techniques affecting quality. In: Nagy, S., Attaway, J. (Eds.), Citrus Nutrition and Quality, ACS Symposium Series, No. 143. The American Chemical Society, pp. 193–224.
Brown, G.E., 1983. Control of Florida citrus decays with quazatine. Proc. Florida State Hortic. Soc. 96, 335–337.
Brown, G.E., 1984. Efficacy of citrus postharvest fungicides applied in water or resin solution water wax. Plant Dis. 68, 415–418.
Brown, G.E., 1986. Diplodia stem-end rot, a decay of citrus fruit increased by ethylene degreening treatment and its control. Proc. Florida State Hortic. Soc. 99, 105–108.
Brown, G.E., 1994. Responses to ethylene of citrus stem-end rot and its causal agent, *Diplodia natalensis*. Proc. Florida State Hortic. Soc. 99 (107), 298–301.
Brown, G.E., Albrigo, L.G., 1970. Grove application of benlate for control of postharvest citrus decay. Proc. Florida State Hortic. Soc. 83, 222–225.
Brown, G.E., Albrigo, L.G., 1972. Grove application of benomyl and its persistence in orange fruit. Phytopathology 62, 1434–1438.
Brown, G.E., Burns, J.K., 1998. Enhanced activity of abscission enzymes predisposes oranges to invasion by *Diplodia natalensis* during ethylene degreening. Postharvest Biol. Technol. 14, 217–227.
Brown, G.E., Eckert, J.W., 2000. Diplodia stem-end rot. In: Timmer, L.W., Gernsey, S.M., Graham, J.H. (Eds.), Compendium of Citrus Diseases, second ed. APS Press, St Paul, pp. 43–44.
Brown, G.E., Lee, H.S., 1993. Interactions of ethylene with citrus stem-end rot caused by *Diplodia natalensis*. Phytopathology 83, 1204–1208.
Brown, G.E., McCornack, A.A., 1969. Benlate, an experimental preharvest fungicide for control of postharvest citrus fruit decay. Proc. Florida State Hortic. Soc. 82, 39–43.
Brown, G.E., Miller, W.R., 1999. Maintaining fruit health after harvest. In: Timmer, L.W., Duncan, L.W. (Eds.), Citrus Health Management. APS Press, St Paul, pp. 175–188.
Brown, G.E., Wilson, W.C., 1967. Stem-end rot fungi in oranges: Entry and possible use of preharvest fungicides for control. Proc. Florida State Hortic. Soc. 80, 301–305.
Brown, G.E., Wilson, W.C., 1968. Mode of entry of *Diplodia natalensis* and *Phomopsis citri* into Florida oranges. Phytopathology 58, 736–739.
Childs, J.F., 1952. Pruning citrus in relation to disease control. Proc. Florida State Hortic. Soc. 65, 83–88.
Coggins, C.W., 1969. Use of growth regulators to delay maturity and prolong shelf-life of citrus. Acta Hortic. 34, 469–472.
Dewdney, M.M., 2013a. Greasy spot. In: Roger, M.E., Dewdney, M.M. (Eds.), Florida Citrus Pest Management Guide, EDIS –SP43. University of Florida, Gainesville, pp. 77–78.
Dewdney, M.M., 2013b. Melanose. In: Roger, M.E., Dewdney, M.M. (Eds.), Florida Citrus Pest Management Guide, EDIS –SP43. University of Florida, Gainesville, pp. 79–80.
Dewdney, M.M., 2013c. Citrus scab. In: Roger, M.E., Dewdney, M.M. (Eds.), Florida Citrus Pest Management Guide, EDIS –SP43. University of Florida, Gainesville, pp. 87–88.

Dewdney, M.M., 2013d. Alternaria brown rot. In: Roger, M.E., Dewdney, M.M. (Eds.), Florida Citrus Pest Management Guide, EDIS –SP43. University of Florida, Gainesville, pp. 89–91.

Eckert, J.W., Eaks, I.L., 1989. Posthharvest disorders and diseases of citrus fruits. In: Reuther, W., Calavan, E.C., Carman, G.E. (Eds.), The Citrus Industry, vol. 5. University of California Press, Berkeley, pp. 179–260.

Ferguson, L., Ismail, M.A., Davies, F.S., Wheaton, T.A., 1982. Pre- and postharvest gibberellic acid and 2,4-dichlorophenoxyacetic acid applications for increasing storage life of grapefruit. Proc. Florida State Hortic. Soc. 95, 242–245.

Fischer, I.H., Ferreira, M.D., Sposito, M.B., Amorim, L., 2009. Citrus postharvest diseases and injuries related to impact packing lines. Sci. Agricola Piracicaba Braz. 66, 210–217.

Grierson, W., Miller, W., 2006. Storage of citrus fruits. In: Wardowski, W.F., Miller, W.M., Hall, D.J., Grierson, W. (Eds.), Fresh Citrus Fruits, second edn. Florida Science Source, Inc, Longboat Key, pp. 547–581.

Grierson, W., Newhall, W.F., 1953. Degreening conditions for Florida citrus. Proc. Florida State Hortic. Soc. 66, 42–46.

Hashem, M., Alamri, S., 2009. The biocontrol of postharvest disease (*Botryodiplodia theobromae*) of guava (*Psidium guajava* L.) by the application of yeast strains. Postharvest Biol. Technol. 53, 123–130.

Hodges, A.W., Spreen, T.H., 2012. Economic impacts of citrus greening (HLB) in Florida, 2006/07–2010/11. University of Florida, Gainesville, FL. Electronic Data Information Source (EDIS) FE903. http://edis.ifas.ufl.edu/fe903.

Holmes, G.J., Eckert, J.W., 1999. Sensitivity of *Penicillium digitatum* and *P. italicum* to postharvest citrus fungicides in California. Phytopathology 89, 716–721.

Hopkins, E.F., Loucks, K.W., 1944. Pulling versus clipping of oranges in respect of loss from stem-end rot and blue mold. Proc. Florida State Hortic. Soc. 57, 80–86.

Ismail, M.A., Zhang, J., 2004. Post-harvest citrus diseases and their control. Outlooks Pest Manag. 15, 29–35.

Jackson, L.K., 1991. Citrus growing in Florida, third ed. University of Florida Press, Gainesville.

Kanetis, L., Forster, H., Adaskaveg, J., 2007. Comparative efficacy of the new postharvest fungicides azoxystrobin, fludioxonil, and pyrimethanil for managing citrus green mold. Plant Dis. 91, 1502–1511.

Khanzada, M.A., Lodhi, A.M., Shahzad, S., 2004. Pathogenicity of *Lasiodiplodia theobromae* and *Fusarium solani* on mango. Pak. J. Bot. 36, 181–189.

Kinay, P., Mansour, M.F., Gabler, F.M., Margosan, D.A., Smilanick, J.L., 2007. Characterization of fungicide-resistant isolates of *Penicillium digitatum* collected in California. Crop Prot. 26, 647–656.

Li, H., Cao, R., Mu, Y., 1995. In vitro inhibition of *Botryosphaeria dothidea* and *Lasiodiplodia theobromae*, and chemical control of gummosis disease of Japanese apricot and peach trees in Zhejinag province, China. Crop Prot. 14, 187–191.

Loucks, K.W., Hopkins, E.F., 1946. A study of the occurrence of Phomopsis and of Diplodia rots in Florida oranges under various conditions and treatments. Phytopathology 36, 750–757.

Mascarenhas, P., Behere, A., Shama, A., Padwal-Desai, S.R., 1995. Postharvest spoilage of mango (*Mangifera indica*) by *Botryodiplodia theobromae*. Mycol. Res. 100, 27–30.

McCornack, A.A., 1971. Effect of ethylene degreening on decay of Florida citrus fruit. Proc. Florida State Hortic. Soc. 84. 279–272.

Peres, N.A.R., Souza, N.L., Furtado, E.L., Timmer, L.W., 2004. Evaluation of systems for timing of fungicide sprays for control of postbloom fruit drop of citrus in Brazil. Plant Dis. 88, 731–735.

Perez, E., Blanco, O., Berreta, C., Dol, I., Lado, J., 2011. Imazalil concentration for *in vitro* monitoring of imazalil resistant isolates of *Penicillum digitatum* in citrus packinghouses. Postharvest Biol. Technol. 60, 258–262.
Punithalingam, E., 1980. Plant Diseases Attributed to *Botryodiplodia theobromae*. Pat. J. Cramer. Vaduz.
Rahmani, M., Hodges, A., 2009. Economic impacts of the Florida citrus industry in 2007–08. University of Florida, Gainesville. Electronic Data Information Source (EDIS) FE802. http://edis.ifas.ufl.edu/fe802.
Ritenour, M.A., Karruppah, K.J., Pelosi, R.R., et al., 2003. Response of Florida grapefruit to short-duration heat treatments using vapor heat or hot water dips. Proc. Florida State Hortic. Soc. 116, 405–409.
Ritenour, M.A., Pelosi, R.R., Burton, M.S., Stover, E.W., Dou, H., McCollum, T.G., 2004. Assessing the efficacy of preharvest fungicide applications to control postharvest diseases of Florida citrus. HortTechnology 14, 58–62.
Ritenour, M.A., Miller, W.M., Wardowski, W.W., 2012a. Recommendations for degreening Florida fresh citrus fruits. University of Florida, Gainesville. Electronic Data Information Source (EDIS) Cir1170. http://edis.ifas.ufl.edu/hs195.
Ritenour, M.A., Zhang, J., Dewdney, M.M., Narciso, J., 2012b. Decay control of fresh citrus. Citrus Ind. 93, 6–10.
Salvatore, J.J., Ritenour, M.A., 2007. Effectiveness of different fungicides applied preharvest at reducing postharvest decay of fresh Florida citrus. Proc. Florida State Hortic. Soc. 120, 281–284.
Saunt, J., 1990. Citrus Varieties of the World. Sinclair International Limited, Norwich.
Schirra, M., Cabras, P.M., Angioni, A., Melis, M., 1996. Residue level of imazalil fungicide in lemons following prestorage dip treatment at 20 and 50°C. J. Agric. Food Chem. 44, 2865–2869.
Smilanick, J.L., Brown, G.E., Eckert, J.W., 2006a. The biology and control of postharvest diseases. In: Wardowski, W.F., Miller, W.M., Hall, D.J., Grierson, W. (Eds.), Fresh Citrus Fruits, second ed. Florida Science source, Inc., Longboat Key, pp. 339–396.
Smilanick, J.L., Mansour, M.F., Sorenson, D., 2006b. Pre- and postharvest treatments to control green mold of citrus fruit during ethylene degreening. Plant Dis. 90, 89–96.
Smoot, J.J., 1963. Hot water as a control for decay of oranges. Proc. Florida State Hortic. Soc. 76, 322–327.
Smoot, J.J., Houck, L.G., Johnson, H.B., 1971. Market disease of citrus and other subtropical fruits. Agricultural Research Service, United States Department of Agriculture, Washington, DC.
Stewart, W.S., 1949. Effects of 2,4-dichlorophenoxyacetic acid and 2,4,5-trichlorophenoxyacetic acid on citrus fruit storage. Proc. Am. Soc. Hortic. Sci. 54, 109–117.
Úrbez-Torres, J.R., Leavitt, G.M., Guerrero, J.C., Guevara, J., Gubler, W.D., 2008. Identification and pathogenicity of *Lasiodiplodia theobromae* and *Diplodia seriata*, the causal agents of bot canker disease of grapevines in Mexico. Plant Dis. 92, 519–529.
USDA-NASS, 2011. Florida citrus statistics 2009–2010. www.nass.usda.gov/Statistics by State/Florida/Publications/Citrus/fcs/2009-10/fcs0910.pdf.
US EPA, 2002. Reregistration Eligibility Decision (RED) Benomyl. www.epa.gov/oppsrrd1/REDs/benomyl_red.pdf.
US EPA, 2004a. Fludioxonil; Pesticide tolerance. Fed. Reg. 69, 58084–58091.
US EPA, 2004b. Pyrimethanil; Pesticide tolerance. Fed. Reg. 69, 52434–52444.
US EPA, 2008. Emergency exemption for Florida – thiophanate methyl on citrus. www.epa.gov/opprd001/section18/08FL01.pdf.
US EPA, 2009. Notice of pesticide: Registration/reregistration. Graduate A^+. EPA Registration No. 100–1308.
Voorhees, R.K., 1942. Life history and taxonomy of the fungus *Physalospora rhodina*. Florida Agric. Exp. Station Tech. Bull. 371.

Voorhees, R.K., 1944. Investigations of melanose and stem-end rot of citrus fruit. Florida Agric. Exp. Station Ann. Rep. 1943–44, 188–191.

Wardowski, W.F., Brown, G.E., 1993. Postharvest decay control recommendations for Florida citrus fruit. Florida Cooperative Extension Service, University of Florida. Circular 359-A, pp. 1–5.

Wardowski, W.F., Miller, W.M., Grierson, W., 2006. Degreening. In: Wardowski, W.F., Miller, W.M., Hall, D.J., Grierson, W. (Eds.), Fresh Citrus Fruits, second ed. Florida Science Source, Inc., Longboat Key, pp. 277–298.

Winston, J.R., 1950. Harvesting and handling citrus fruits in the Gulf States. Farmers' Bulletin No. 1763. US Department of Agriculture.

Zhang, J., 2004. Effect of ethylene on natural resistance of citrus fruit to stem-end rot caused by *Diplodia natalensis* and its relation to postharvest control of this decay. Proc. Florida State Hortic. Soc. 117, 364–367.

Zhang, J., 2007. The potential of a new fungicide fludioxonil for stem-end rot and green mold control on Florida citrus fruit. Postharvest Biol. Technol. 46, 262–270.

Zhang, J., 2009. Evaluation for the potential of a new fungicide pyrimethanil for postharvest Diplodia stem-end rot control on Florida citrus fruit. Proc. Florida State Hortic. Soc. 122, 350–353.

Zhang, J., 2012. Comparative effects of postharvest fungicides for Diplodia-stem-end rot control of Florida citrus. Proc. Florida State Hortic. Soc. 125, 243–247.

Zhang, J., Swingle, P., 2005. Effects of curing on green mold and stem-end rot of citrus fruit and its potential application under Florida packing system. Plan Dis. 89, 834–840.

Zhang, J., Timmer, L.W., 2007. Preharvest application of fungicides for postharvest disease control on early season tangerine by hybrids in Florida. Crop Prot. 26, 886–893.

CHAPTER 11

Colletotrichum gloeosporioides (Anthracnose)

Yasmeen Siddiqui[1] and Asgar Ali[2]

[1]Laboratory of Food Crops, Institute of Tropical Agriculture, Universiti Putra Malaysia, Serdang Selangor D. E., Malaysia; [2]Centre of Excellence for Postharvest Biotechnology (CEPB), School of Biosciences, The University of Nottingham Malaysia Campus, Semenyih, Selangor D. E., Malaysia

Contents

OVERALL IMPORTANCE OF THE FUNGUS

Fruits and vegetables are inseparable ingredients of a balanced human diet. Nutritionists and other health care providers are unequivocal in advocating the virtues of including more fresh fruits and vegetables in the diet as they provide nutritionally important, complex chemical compounds called vitamins which cannot be synthesized in our body. The importance of tropical fruits is being recognized internationally, as evident from their nutritional and therapeutic values (Rufino et al., 2010). Unfortunately, a majority of these fruits are perishable and have a short shelf-life, since they continue with physiological functions (e.g. respiration and transpiration) postharvest which leads to their deterioration.

Postharvest Decay
http://dx.doi.org/10.1016/B978-0-12-411552-1.00011-9

The tropical fruits have a global market, and the general practice is to transport the majority of these fruits by sea because of the cheaper cost. However, the *quid pro quo* for this is the longer time entailed, during which the fruits are subjected to handling and storage losses and which ultimately results in a shorter shelf-life. Apart from direct economic considerations, contaminated or diseased produce poses a potential health risk. Virtually all postharvest diseases of fruits and vegetables are caused by fungi and bacteria. In some crops, virus infections may be also present preharvest, however, they are generally not found to be important causes of postharvest diseases.

Colletotrichum is a large genus comprising a number of important species that are among the most prevalent fungal pathogens causing diseases in diverse tropical and subtropical fruits and vegetables. Almost every crop grown throughout the world is susceptible to one or more species of *Colletotrichum*. The genus is associated with economically significant diseases of cereals and grasses, legumes, vegetables and perennial crops, including tree fruits. Recently, the genus *Colletotrichum* was nominated as the eighth most important group of phytopathogenic fungi in the world, based on perceived scientific and economic significance (Dean et al., 2012).

The prominence of *Colletotrichum gloeosporioides* (Penz.) Penz. & Sacc and the anthracnose disease it causes to horticultural produce present a mounting threat to global agriculture. More than 50% of losses of fresh fruits and vegetables are caused by *Colletotrichum* species (Paull et al., 1997; Awang et al., 2011). The delayed onset of disease symptoms caused by the fungus in a crop is often devastating and leads to significant postharvest losses. Important tropical fruit crops in which *C. gloeosporioides* causes severe losses include papaya, mango, and avocado, among others. Recently, it has also been found to affect pitaya, commonly known as dragon fruit. The highly perishable nature of these fruits highlights the urgent need to develop technologies that can reduce their deterioration and also extend their shelf-life to impart value to the buyer or consumer.

TAXONOMY AND MORPHOLOGY

The taxonomic and phylogenetic studies on the genus *Colletotrichum* are in an expansive phase following the epitypification of *C. gloeosporioides* (Penz.) Penz. & Sacc. (Cannon et al., 2008). The species is exceedingly inconstant in its morphology, however, several earlier reports have discussed its differentiation and classification in several groups based on conidial size, setae and sexual state morphology (Simmonds, 1965; Lenné, 1978; Sutton, 1980; Cox and Irwin, 1988). *Colletotrichum gloeosporioides* was first identified in 1884 and its perfect state (telomorph) was given the generic name *Glomerella cingulata* (Stonem) in 1903. More than 600 synonyms and at least seven formae speciales of the fungus were described and are recognized as a heterogeneous group with great variation in morphophysiological characteristics (Von Arx, 1957; Sutton, 1992). The fungus produces hyaline, one-celled, ovoid to oblong, slightly curved or dumbbell shaped conidia with obtuse ends (Fig. 11.1). They are usually borne on distinct, well-developed hyaline conidiophores and measure 12.5–14.8 µm × 4.1–4.7 µm (Xie et al., 2010). The waxy acervuli of *C. gloeosporioides* that develop profusely on diseased parts of plants (petioles, leaves and fruits) are subepidermal. The acervuli have been reported to measure 80–250 µm (Sattar and Malik, 1939) and 115–467 × 95–22 µm (Bose et al., 1973). Under moist conditions, the acervuli, when mature, exude pink or salmon colored masses of conidia.

The similarities of morphological characters and phenotypes among *Colletotrichum* species make classification difficult. Sutton (1992) commented on *C. gloeosporioides* that 'No progress in the systematics and identification of isolates belonging to this complex is likely to be made based on morphology alone'. Attempts were made towards a modern understanding of this name with the designation of an epitype specimen with a culture derived from it to stabilize the application of the name (Cannon et al., 2008). Molecular techniques involving internal transcribed spacer (ITS) sequencing, fragment length polymorphism (RFLP) analysis, polymerase chain reaction (PCR)

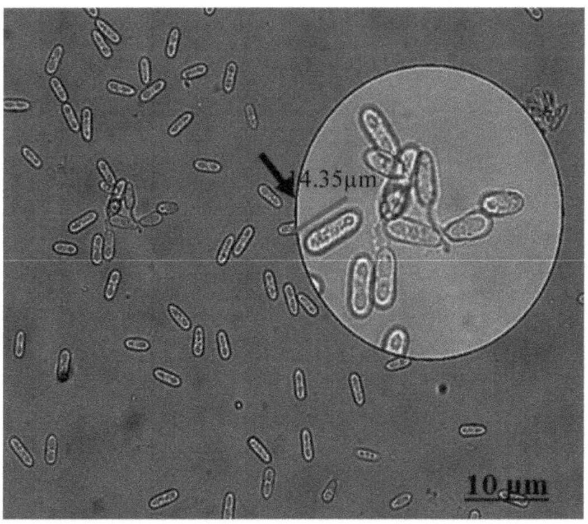

Figure 11.1 Conidial morphology of *Colletotrichum gloeosporioides* isolated from dragon fruit under light microscope, arrow indicates the length of the conidia.

based methods, chromosomal karyotyping and sequencing of the ribosomal RNA genes have been employed to differentiate populations of *C. gloeosporioides* and other species (Johnston and Jones, 1997; Freeman et al., 2000; Photita et al., 2005). These studies have focused on phylogenetic reassessments of species complexes and have determined that what were previously thought to be a single species, comprise multiple distinct lineages. For example, the ex-epitype isolate belongs in a strongly supported clade, distinct from other taxa that have been confused with *C. gloeosporioides* in the past, such as *C. acutatum* and *C. boninense* (Damm et al., 2012a, b; Weir et al., 2012). It is now known that the gloeosporioides clade (*C. gloeosporioides* species complex) comprises more than 22 species (Weir et al., 2012). The species numbers in the major clades are likely to rise, unraveling the cryptic taxa based on multigene phylogenetic analyses and incorporating a large number of isolates in worldwide collections (Cannon et al., 2012). However, biological and genetic relationships within the broad *C. gloeosporioides* clade remain confused and ITS sequences alone are insufficient to resolve them.

BIOLOGY OF *C. GLOEOSPORIOIDES* INFECTION PROCESS

Generally, disease symptoms caused by *Colletotrichum* species are commonly known as anthracnose. The anthracnose pathogen invades preharvest in inflorescences, fruits, leaves and stems of a variety of fruit trees. Since the focus of this chapter is on postharvest disease only, symptom appearance after crop harvest is discussed here. The fungus produces enzymes (polygalacturonase and pectatelyase) that degrade plant cell walls and causes significant economic losses (Rodríguez-López et al., 2009). Typically, disease symptoms on fruits are characterized by small, depressed water-soaked lesions which are subcircular or angular with translucent light brown margins (Alvarez and Nishijima, 1987). The lesions are sometimes referred to as chocolate spots. As infection advances in these circular sunken lesions, erumpent orange to pink conidial masses develop and frequently occur in concentric rings. The internal tissue is firm with grayish white discoloration, which later turns to brown (Alvarez and Nishijima, 1987; Sivakumar et al., 2002; Masyahit et al., 2009). Verhoeff (1974) and Swinburne (1983) described these infections as 'quiescent or dormant parasitic relationship which, after certain duration, becomes active'. The known developmental stages where the fungus might become inactive are during penetration, initiation of germination, germ tube elongation, formation of appressorium or subsequent establishment. Based on the definition of Swinburne (1983), it was widely accepted that failure of spores to germinate or develop beyond any subsequent stage is due to temporary adverse physiological conditions imposed by the host directly on the fungus or indirectly by exhibiting temporary resistance against the fungus.

In papaya, anthracnose infections are commonly introduced in the early fruit developmental stages. However, the fungus remains quiescent as an appressorium along with an infection peg or as subcuticular hyphae until the fruit attains climacteric (Dickman and Alvarez, 1983). Overall, the first symptoms in this fruit are characterized as small well-defined dried pink spots on the surfaces later changing into dark brown to black colored sunken lesions (Fig. 11.2A).

In a similar fashion, mango anthracnose is caused by *Glomerella cingulate* (Stoneman) Spauld and H. Schrenk, anamorph: *C. gloeosporioides* (Penz.) var. minor J. H. Simmonds (Fitzel and Peak, 1984). It has also been suggested that the pathogen causes blossom blight, leaf blight and, in some severe cases, tree dieback (Ploetz et al., 1997). The symptoms appear as rounded brown to black lesions with an indefinite border on the fruit surface (see Fig. 11.2B). However, infections in large fruits do not normally develop into lesions. After initial establishment in the fruit, the fungus usually remains dormant until the fruit begins to ripen. As ripening progresses, dark depressed circular lesions develop and increase rapidly in size to cover almost the entire fruit in extremely severe cases. Arauz (2000) indicated that lesions of different sizes can integrate and cover extensive areas of the fruits, characteristically in a tear-stain pattern, extending from the basal towards the distal end of the fruit. These abrasions are usually limited to the peel but, in advanced stages of infection, the fungus can penetrate the pulp producing acervuli and abundant orange to pink masses of conidia.

Similarly, in avocado (*Persea americana*), anthracnose lesions initially develop in immature fruits following wounding by insects. It has been

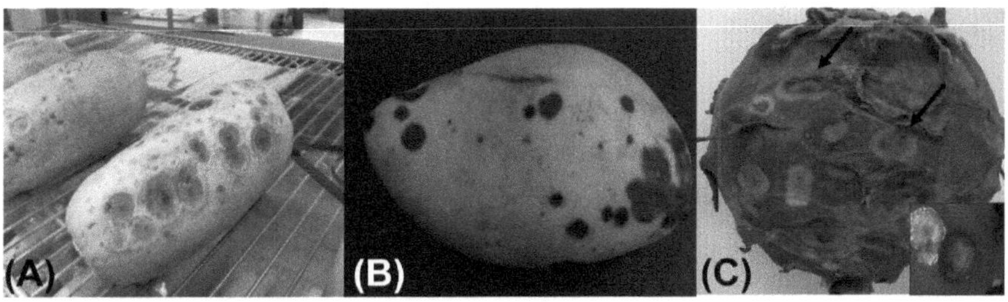

Figure 11.2 Typical anthracnose symptoms on (A) papaya, (B) mango and (C) dragon fruit. Inset (C) showing deep water soaked sunken lesion.

reported that dead leaves entangled in the tree are a main source of inoculum, consequently, high fruit losses can occur in trees with large canopies (Fitzell, 1987). Germinated spores produce appressoria that penetrate the cuticle, which causes the resultant subcuticular hyphae to become quiescent and do not flourish further until the fruit is harvested and ripens. Consequently, round dark sunken lesions are formed that expand rapidly on the fruit skin followed by infection into the pulp causing rot (Nelson, 2008). Dragon fruit, a newly introduced fruit crop, is a tropical climbing cactus producing red- or white-fleshed fruits. It is also called pitaya, strawberry pear and night blooming cereus (Crane and Balerdi, 2005) and, in Malaysia, buah naga. Being a non-climacteric fruit, it has no peak production of ethylene or carbon dioxide (Le Bellec et al., 2006). However, it is highly susceptible to many diseases resulting in extensive postharvest losses during transportation and storage (Masyahit et al., 2009). At ambient temperature, the fruit stays fresh for a few days after which its physical appearance deteriorates and disease symptoms appear resulting in spoilage within 6–7 days (Lau et al., 2001). In Malaysia, red-fleshed dragon fruit seem to be more frequently affected by anthracnose than white-fleshed fruit. Infected plants and fruit have small, water-soaked, sunken, round spots that become black, emit spore masses and cause rotting (see Fig. 11.2C) (Masyahit et al., 2009).

Generally, most fruits can be infected with conidia from isolates of *Colletotrichum* spp. regardless of the original host plant. For example, isolates from avocado, papaya and citrus can cause infection in mango fruits (Freeman and Shabi, 1996). The epidemiological importance of these potential inoculum sources on disease development still needs to be evaluated. Genetic and geographical data seem to suggest that *C. gloeosporioides* was disseminated throughout the world from a single source as an endophyte (Akem, 2006). An increased understanding of the origins and diversity of *C. gloeosporioides* in tropical fruits would have significance to future research on host and control strategies across disease prevalent areas and locations.

AGRONOMIC AND ENVIRONMENTAL FACTORS FAVORING *C. GLOEOSPORIOIDES* INFECTION

The disease development depends not only on the presence of the pathogen but also on agronomic management practices and environmental factors, such as temperature, rainfall and humidity, which have an important bearing on any disease epidemic.

The occurrence of anthracnose in fruit orchards has increased in recent years mainly due to the increasing level of inoculum left in orchards. The development is also influenced by host genotype, duration of wetness period, growth stage and temperature. The infection process of *C. gloeosporioides* in fruits involves field inoculum which can be conidia from infected seed, infested debris, volunteers from previous seasons and petioles of dying infected leaves. The conidia are dispersed locally from infected plants by rain splashes before being disseminated further by wind. The pathogen initially infects intact, non-wounded, immature green fruits in the field. This is followed by rapid spore germination and appressorium formation under conducive environmental conditions resulting in quiescent infection. The fungus, using its appressorium, enzymatically penetrates the cuticle and then remains as subcuticular hyphae until the post-climacteric stage of fruit growth is attained (Chau and Alvarez, 1983). Subsequent incipient fungal infection becomes apparent only as the fruit ripens. Thus, knowledge of the effect of environmental and agronomic conditions on germination, appressorium formation and growth from latent infections are critical in the development of pre- and postharvest disease management strategies of tropical fruits.

It is surprising that very few studies have been carried out to support the relationship between environmental factors on germination and growth of *C. gloeosporioides*. Usually, the environmental conditions favoring the pathogen are high temperature (25–30°C being optimal) and high humidity. It has been established that *C. gloeosporioides* spores must have free water or relative humidity (RH) above 95% for germination and appressorium formation. Spores are only released from acervuli when there is an abundance of moisture. The effect of environmental factors on the growth and development of infection structures of *C. gloeosporioides* isolated from diseased mangoes, both on plant surfaces and *in vitro*, was carried out in the Philippines. It was indicated that the most stimulatory temperature for production of appressoria ranged from 20 to 25°C. However, more appressoria became melanized at 30°C but the frequency of formation of penetration pegs was highest at 25°C. Spores germinated at RH as low as 95%, but the percentage of spores germinating and forming appressoria increased as the RH approached 100%. Studies also showed that approximately 18% of the spores of *C. gloeosporioides* when incubated in lower RH (62 and 86%) for 4 weeks retained viability, and were also capable of forming appressoria when RH increased to 100% (Estrada et al., 2000; Akem, 2006). These findings are in corroboration with those who stated that about 12 h of continual leaf wetness due to rain at temperatures of 15–27°C are necessary for establishing a *C. gloeosporioides* conidial infection in blueberry (Hartung et al., 1981). However, isolates of *C. gloeosporioides* from different locations require considerable variation in the optimal temperature for germination and appressorium formation. Severity of disease depends on the weather, and the fungus is relatively inactive in dry weather. Sunlight, low humidity and temperature extremes (below 18°C or greater than 25°C) rapidly inactivate spores.

POSTHARVEST FACTORS FAVORING *C. GLOEOSPORIOIDES* INFECTION

Stakman and Harrar (1957) emphasized that 'the responsibilities of the plant pathologists do not end with the harvest of satisfactory yields of plant products and that harvesting marks the termination of one phase of plant protection and the beginning of another'. This testimonial indicates that the second phase of plant protection of fruits, vegetables and other economic plant parts, from the time of harvest until they reach the consumer, is also very important. Temperature is perhaps the single most important factor influencing disease development after harvest. Additionally, warm and relative humid conditions largely contribute towards anthracnose development. Optimum production of conidia occurs between 25 and 30°C when free moisture is available. It plays an important role in the initiation of infection and is usually essential for the successful development of *C. gloeosporioides*. Rathod (2011) elucidated that during the storage of mango fruits at 25°C and 100% RH anthracnose severity was highest whereas the fungus failed to establish at 10°C and 30% RH. The physical environment during storage had a significant role since the percentage of anthracnose severity was increased with an increase in relative humidity. Several reports indicated that papayas are high in water content and susceptible to postharvest shriveling which will negatively impact the appearance and market potential of the fruit (Paull et al., 1997; Wills et al., 1998). In order to minimize weight loss and desiccation of the skin, the fruits are generally treated with edible coatings/wraps or stored under a high RH of 90–95%. However, a high RH can induce a greater incidence of decay, particularly if stored for longer durations.

Quite a few authors have reviewed the proper maintenance of hygiene during cultivation, production and postharvest handling of perishable fruits to prevent postharvest decays caused by variety of fungi

(*C. gloeosporioides, Fusarium oxysporum, Rhizopus stolonifer, Alternaria alternata, Penicillium* and *Aspergillus* spp.) which is critical in reducing sources of inoculum (Johnson and Hoffman, 2009, Chowdhury and Rahim, 2009; Sivakumar et al., 2011; Bautista-Baños et al., 2013). Anthracnose initiates as a preharvest infection, hence removing old or diseased fruit, twigs and branches can reduce the source of inoculum in the orchards. However, during postharvest storage, symptoms develop as a result of quiescent infection. During postharvest handling of papaya, the fruits are susceptible to scratches and punctures when in contact with rough or sharp surfaces. These wounds can serve as an infection site for numerous pathogens causing severe decays. In the cleaning process of papaya, sodium hypochlorite is effective when maintained at a concentration of 150 ppm and a pH of 6.5–6.7 per 100 liters of water. The fungicide thiabendazole (TBZ) is also added to the wash water at a concentration of 500 ppm to prevent postharvest decays, specifically anthracnose. Water used for washing the fruits usually becomes soiled with pathogen propagules, latex and dirt particles leaving the diminished activity of sanitizing agents and fungicides. Therefore, the water tank should be changed when necessary and refilled with clean water containing the required concentration of sanitizers and fungicide. Any rejected produce, if not disposed of properly, can also provide an ideal substrate for pathogen development including *C. gloeospoiroides*.

CONTROL

Conventional Strategies

Successful control of postharvest disease in fruits is achieved by adopting good pre- and postharvest management practices, such as running strict orchard hygiene, application of fungicides and temperature management during storage and handling.

Fungicides

Traditionally, synthetic fungicides have been used to protect fruits and vegetables against fungal pathogens causing spoilage of the produce at different stages of harvest. Occurrence of anthracnose due to quiescent infection in fresh produce makes it difficult to manage during the postharvest period. Effectiveness of timing of application and type of fungicide is dependent on the stage at which infection occurs. For pathogens like *C. gloeosporioides* which cause infection before harvest, field application of fungicides is frequently required. Usually, the postharvest rot complex is controlled in the field by application of copper fungicides to reduce inoculum levels (Peterson and Inch, 1980; Darvas et al., 1987; Hartill, 1992, Everett et al., 2007). For example, in New Zealand, benomyl and copper hydroxide were the only registered chemicals for use in avocado orchards. However, only copper fungicides can be used since benomyl has been withdrawn from sale. Copper hydroxide (Blue shield®DF a.i. 200 g 100 L^{-1}) can be applied up to 18 times in a season. Similarly, prevention of mango anthracnose in Sri Lanka is carried out by spraying the trees regularly with a protectant fungicide, such as benlate (60 g a.i. 100 L^{-1}), daconil (200 g a.i. 100 L^{-1}) or Maneb (200 g a.i. 100 L^{-1}) during flowering and development of fruits. However, if it rains during flowering, a systemic fungicide prochloraz (Octave® 462 g kg^{-1}) is applied to deactivate already established infections and to inhibit new infections.

Postharvest, fungicide treatment is used to control established infections in epidermal tissues of the produce or to shield it against infections which may occur during handling and storage. Currently, synthetic fungicides such as prochloraz, benomyl, carbendazim, propineb and difenoconazole have been recommended to control *Colletotrichum* spp. on various fruits and vegetables (Hoa, 2008). These fungicides can be applied as dips, sprays, fumigants, treated wraps and box liners or in waxes and coatings. With washing, waxing and fungicide treatment in the packing house, natural infections were shown to be reduced but not completely eliminated. Some other findings on the use of fungicides for controlling postharvest anthracnose in some fruits are shown in Table 11.1.

Hot Water Treatment (HWT)

Management of the temperature of harvested produce is crucial for the control of postharvest diseases since most of the pathogens flourish at ambient temperatures or above. Generally, heat treatments are utilized to satisfy insect quarantines, reduction of anthracnose and other postharvest decays is an added benefit (McGuire, 1991). HWT as a decay control measure and quality management technique has been practiced widely and has been reflected in a range of the literature (Jaccobi et al., 1995; Anwar and Malik, 2007; Djioua et al., 2009). It is an appropriate method for some fruits where the temperature of the water does not cause decolorization and scalding of fruit skin, change in flavor or softening of fruit pulp. The temperature selected for HWT is slightly higher than the thermal death point of the target pathogen, providing excellent potential for minimizing postharvest losses of various fruits. At high temperature, the pathogen is

Table 11.1 Pre- and Postharvest use of some Fungicides and their Level of Control of Anthracnose Disease Caused by *Colletotrichum Gloeosporioides* in Different Fruits

Fungicide	Application Method	Crop	Causal Agent	Concentration	Level of Control (Disease reduction)	Reference
Thiabendazole	Postharvest	Papaya	*C. gloeosporiodes*	4–8 g L^{-1}(carnabua wax + water in 1:3)	90%	Couey and Farias, 1979
Prochloraz	Postharvest	Avocado	*Glomerella cingulata* var. minor	1 g L^{-1}	1–90%	Muirhead et al., 1982
Maneb Captan Chlorotalonil	Preharvest		*C. gloeosporioides*	4.0 g L^{-1} 1.2 g L^{-1} 2.6 g L^{-1}	70–90% at harvest 40–65% (4 DAH) 20–35% (6 DAH)	Solano and Arauz, 1995
Benomyl Maneb Captan	Preharvest		*C. gloeosporioides*	225 g ha^{-1} 900 g ha^{-1} 750 g ha^{-1}	-	De los Santos et al., 1997
Prochloraz	Postharvest		*C. gloeosporioides*	810 ppm	40–80%	Swart et al., 2002
Prochloraz	Postharvest		*C. gloeosporioides*	450 g L^{-1}	>90 %	Nel et al., 2003
Benomyl Prochloraz	Postharvest	Mango	Multiple including *C. gloeosporioides*	1 g L^{-1} (hot dip for 5 min) for local market 180 mL L^{-1} (20 s at ambient temperature)		
Chlorothalonil Mancozeb Thiophanate-methyl	Preharvest		*C. gloeosporioides*	0.20% p.c. 0.20% p.c. 0.70% p.c.	8–73% (according to season)	Ventura et al., 2004

Continued

Table 11.1 Pre- and Postharvest use of some Fungicides and their Level of Control of Anthracnose Disease Caused by *Colletotrichum Gloeosporioides* in Different Fruits—cont'd

Fungicide	Application Method	Crop	Causal Agent	Concentration	Level of Control (Disease reduction)	Reference
Prochloraz	Postharvest		*C. gloeosporioides*	0.2–1.0g L^{-1}	88–99%	Ventura et al., 2004;
Imazalil	Postharvest			0.05 %	44–85%	Zavala-León et al., 2005
Azoxystrobin	Postharvest				47–74%	
Kresoxym	Postharvest				49–70 %	
methyl	Postharvest				15–62%	
Benomil	Preharvest				58%	
Difenoconazole						
Azoxystrobin	Preharvest		*C. gloeosporioides*	0.25–4 mL L^{-1}		Sundravadana et al., 2006
Benomyl	Preharvest	Tanger-ines	*C. gloeosporioides*	2.2 kg ha^{-1}	37.4–62.6%	Zhang and Timmer, 2007
Benzothiadia-zole	Postharvest		*C. gloeosporioides*	50 mg L^{-1}	46–70%	Lin et al., 2011

DAH: days after harvest

usually killed and most of its enzymes and proteins are denatured. However, the efficacy of HWT depends not only on temperature and the time for which the produce is exposed but also on stage of maturity, size and weight of fruit, cultivar and growing conditions and should be therefore considered when determining the temperature and duration of the treatment (Jaccobi et al., 2001).

This technology has been proven to increase profit, rendering high market value without compromising the quality of the produce (Barrera-Necha et al., 2008). In the Philippines, HWT is accepted as an effective mango disease control strategy, particularly against anthracnose. Findings from four cases (exporters/processors) revealed that they all practice HWT at 52–55°C for 2–7 minutes depending on the growing conditions and variety. In addition, bagging of mango fruits preharvest followed by HWT (52–55°C) for 10 min was reported to reduce *C. gloeosporioides* decay by 83%. With HWT, the companies are assured of lowering anthracnose disease incidence by 10% compared to those with as much as 30% incidence without using the technique. This resulted in higher volume of fruits procured and sold and consequently increased income and profit (Aveno and Orden, 2004).

Similar findings were reported in papaya decay caused by *C. gloeosporioides*. Hot water treatment at 48°C for 20 min has been the principal postharvest treatment for decay control since as early as 1964, when this method was adopted on an industry wide scale (Akamine, 1967; Nishijima et al., 1990). In later studies, Li et al (2013) established that the incidence of anthracnose in hot water (54°C, 4 min) treated papaya (*Carica papaya* cv. 'Sunrise') was significantly reduced by 37.42% when compared to that of untreated controls. Nevertheless, HWT was also effective in reducing the carrier rate of *C. gloeosporioides* in peel by 58.3%. The reduction in papaya anthracnose by 44% was also reported when fruit were subjected to HWT at 54°C for 4 min after 9 days of storage at 25°C (Zhao et al., 2013). In another study for papaya cultivars 'Sunrise Solo' and 'Tainung 1' inoculated with two isolates of *C. gloeosporioides*, the diameter of the lesion was reduced after a hot water treatment of 20, 25 and 30 min at 48°C, and total control of the fungi was achieved by fruit immersion for 20 min at 50°C (Martins et al., 2010). A delay in the appearance of disease symptoms was also reported by Chávez-Sánchez et al. (2013) for up to 11 days in papayas treated at 55°C for 3, 6 and 9 min.

The basic problem with short-term heat treatment is that it involves use of high temperatures, sometimes near the level injurious to the commodity, which necessitates careful management and high accuracy of temperature control. Too high a temperature for a given host may increase the chances of susceptibility to pathogens even though no visible damage is evident and may also affect the ripening process in some fruits (Phillips and Harris, 1979; Paull, 1990). In a heat sensitive mango variety such as 'Tommy Atkins', the addition of 0.1% a.i. imazil to hot water (53°C, 3 min) controlled *C. gloeosporioides* decay and also enabled the temperature to be reduced to below the heat injury level (Spalding and Reeder, 1986). In a separate finding, HWT at 50°C for 5 min in combination with prochloraz applied as a cold dip (20°C) at 810 ppm for the management of mango anthracnose has been found to reduce disease incidence by 66% more than other treatments (thiabenzadole applied either at 1000 or 1500 ppm either as cold or hot dip) with 92–100% fruits affected during overseas shipment (Swart et al., 2002). This improvement might possibly be due to the more effective infiltration of fungicides in the wound site on fruits (Brown, 1984). However, the extended exposure to fungicides such as benomyl (0.1%), a benzimidazole, systemic fungicide which was effective against pre- and postharvest development of mango anthracnose may lead to the emergence of resistant strains of fungi and therefore must be used sparingly (Dodd et al., 1991; Akhtar et al., 1998; Johnson and Hoffman, 2009).

Emerging Technologies for *C. Gloeosporioides* Control

In recent years, some novel or emerging technologies are being explored with the goal of achieving effective control of anthracnose disease of perishable fruits. Among these, the use of generally recognized as safe (GRAS) compounds, sanitizers, biological control, natural inducers, edible coatings and essential oils are discussed in this section.

Generally Recognized as Safe Compounds

The mounting awareness of health hazards and environmental pollution due to indiscriminate use of pesticides has compelled environmental protection agencies of several developed and developing countries to ban the use of numerous chemicals that were frequently applied to fruits and vegetables (Rendall-Dunn, 1991). Furthermore, the possibility of the development of resistance in pathogens to fungicides was another critical factor in the restriction of available chemicals.

Inorganic Salts

Several inorganic salts have shown antifungal activity against fungal pathogens. Some of these compounds, such as sodium bicarbonate, have been broadly used in the food industry and are considered as GRAS compounds. While studying the effects of ammonium carbonate and sodium bicarbonate on the anthracnose disease of papaya cv. 'Ratna' in Sri Lanka, Sivakumar et al. (2002) found that carbonic acid salt caused significant inhibition in radial mycelium growth of *C. gloeosporioides*. Ammonium carbonate at 2% gave complete inhibition, whereas sodium bicarbonate at 2% caused 60% inhibition of radial mycelial growth. Similarly, Gamagae et al. (2003, 2004) used sodium bicarbonate incorporated into wax to control anthracnose of papaya and managed to achieve a significant reduction of disease. It was concluded that a wax formulation containing ammonium carbonate should be developed commercially to extend the storage of papaya for 23 days and allowing 70% of fruits to become more marketable by increasing their shelf-life. However, Deliopoulos et al. (2010) stated that inorganic salts are generally less effective than synthetic fungicides and hence could not replace them.

Compounds that stimulate plant resistance may offer coherent choices to complement postharvest treatments of inorganic salts and improve effectiveness. Preharvest dips of mango fruit in plant defense-inducing chemicals integrated with postharvest treatments with inorganic salts and hot water were evaluated for the management of anthracnose on artificially inoculated mango fruit. Both salicylic acid and potassium phosphonate at 1000 mg L^{-1} combined with a fruit dip for 3 min in 3% aqueous sodium bicarbonate at 51.5°C significantly reduced disease development (Dessalegn et al., 2013). The integrated management of mango anthracnose could be achieved when applied in a two-step treatment. Nevertheless, preharvest fruit dips in plant defense inducing-chemicals are difficult to apply commercially, simpler methods of treatment such as spraying warrants testing. Additional assessment of integrated methods using different commodities under different storage conditions and economic feasibility is also of deemed importance for its practical application to manage fruit rots caused by anthracnose.

Sanitizers

One interesting method to improve safety, quality and shelf-life of postharvest produce is chemical treatment with ozone gas. Research and commercial applications indicate that ozone has the potential to replace chlorine and fungicides. It is a strong oxidant that is effective in controlling bacteria, molds, protozoa and viruses (Kim et al., 2006). A novel approach utilizing gaseous ozone as a potential antifungal preservation technique to overcome anthracnose disease of papaya during cold storage has been investigated by our team. *In vitro* mycelia growth and conidial germination of *C. gloeospoiroides* were greatly affected at various concentrations and periods of ozone exposure. Maximum inhibition of 91.6% in mycelial growth was observed at 4 ppm exposure for 144 h while all ozone treatments (0.04, 1.6 and 4 ppm) totally inhibited conidial germination by 100%. A more detailed picture of inactivation of fungal conidia subjected to ozone exposure is as illustrated in scanning electron micrographs (SEM). There was regularity, smoothness, and integrity in untreated conidia (Fig. 11.3A). No apparent erosion of the

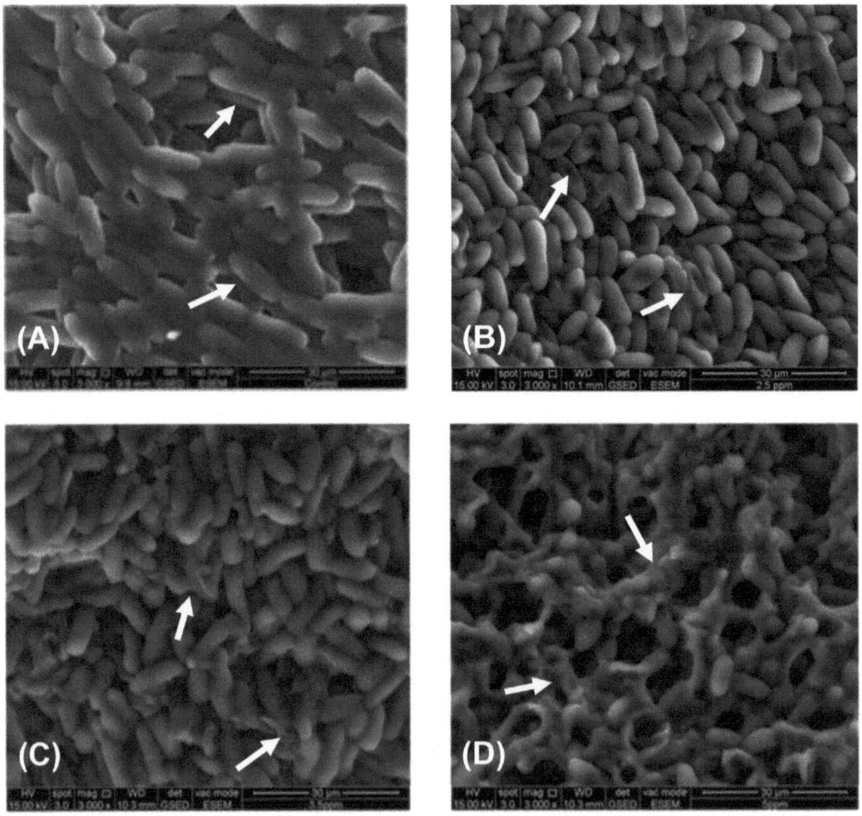

Figure 11.3 SEM micrographs (30 μm) of (A) untreated *C. gloeosporioides* conidia, arrows indicate smooth and regular conidial coat; (B) conidia treated with 0.04 ppm, arrows indicate shrinkage of conidia; (C) conidia treated with 1.6 ppm, arrows indicate rupturing and irregularity in conidial coat; and (D) conidia treated with 4 ppm ozone, arrows indicate conidial deformation and extrusion of inner material after 24 h exposure at 25°C, 70±5% RH.

conidial coat was experienced with 0.04 ppm exposure although shrinkage and shriveling was obvious (Fig. 11.3B), whereas at 1.6 ppm ozone treated conidia appeared to lose the coat regularity (Fig. 11.3C). It was revealed that exposure to 4 ppm ozone fumigation greatly altered the conidial structure. Some conidia were seriously deformed or had lost their original shape and extrusion of their inner material was also detected (Fig. 11.3D). However, in further *in vivo* studies the authors found that exposing the papaya fruit to 1.6 ppm ozone for 96 h delayed the disease incidence and simultaneously decreased it to 40% following 28 days of storage at 12±1°C and 80% RH (Ong et al., 2013). It was suggested that fruit treated with the higher concentration of ozone (4 ppm) had increased fruit decay indicating reduced disease resistance to anthracnose. Longer exposure at higher concentration had resulted in surface browning due to surface oxidation from excessive use of ozone.

Biological Control

With the increasing health concerns expressed by consumers and the intention of governments to regulate pesticide use and their residues in fresh produce, emphasis is being given to innovative approaches to control pathogens and postharvest diseases of papaya and other tropical fruit utilizing biocontrol agents and natural products under controllable postharvest conditions (Pramila and Dubey, 2004). There is a global effort to identify suitable microorganisms and to improve their effectiveness in controlling postharvest anthracnose disease of fruits, and it is anticipated that this disease will be tackled effectively by the use of biocontrol agents (BCAs).

The antagonistic activity of BCAs against pathogens tends to be greater when they are applied prior to pathogen establishment. This makes the control of quiescent field infections caused by *Colletotrichum* sp. using postharvest application of an antagonist more challenging than controlling infections occurring after harvest. Since an antagonist may have some effect on the induction of a host response or an exterminating effect, field applications are often mandatory for successive control of quiescent infections. Shi et al. (2011) demonstrated that papaya fruit treated with the endophytic bacterium *Pseudomonas putida* (MGY2), originally isolated from fruit, showed a significantly lower disease index, disease incidence and lesion diameter than the control fruit. Additionally, this strain maintained the firmness and delayed the peak of ethylene production of harvested papaya fruit stored at 25°C. It also significantly enhanced the activities of phenylalanine ammonia-lyase (PAL), catalase (CAT) and peroxidase (POD) and increased the phenolic content. The expression levels of PAL1, CAT1 and POD genes were elevated in *P. putida* (MGY2) treated fruit. They suggested that papaya fruit is capable of responding to the endophyte *P. putida* (MGY2) which could activate defensive enzymes and genes and thereby induce host disease resistance. Under *in vitro* conditions, the antagonism of *Trichoderma* species against *C. gloeosporioides* isolated from different fruit crops has been well documented (Raheja and Thakore, 2002; Tasiwal et al., 2009). The reduction in mycelial growth of *C. gloeosporioides* from papaya fruits was more than 60% in the presence of *Trichoderma virens* which was superior to other bioagents tested. Other findings dealing with different BCAs on inhibition of mycelial growth or conidial germination gave promising results with different efficacies. For example, *Streptomyces hygroscopicus* suppressed the mycelial growth by 72.8%, 67% and 45 %, while 90% and 87% decrease in spore germination was evident in the presence of *S. noursei* and *S. natalensis*, respectively (Prapagdee et al., 2008; Svetlana et al., 2010).

Although there is no doubt that BCAs are effective, the control levels may differ in *in vitro* and *in vivo* conditions. Based on this, relative research on the efficacy of two bacterial antagonist isolates of *Bacillus firmus* (B10 and B3) was tested under laboratory as well as their direct application on papaya fruits. A significant reduction in the growth of *C. gloeosporioides* by 75.3 and 69.1%, respectively was recorded in the *in vitro* test but there was no effect in controlling anthracnose disease in fruits (Baños-Guevara et al., 2004). However, in a separate study, out of the 30 isolates of various epiphytic yeasts selected from leaves and fruit of the same papaya, one isolate (CEN63) *Criptococcus magnus* used at a concentration of 10^7 to 10^8 cells mL^{-1} did show significant control in anthracnose disease development as indicated by disease progress curve after 7 days of storage (de Capdevile et al., 2007). This could be because biocontrol efficacy is directly affected by the amount of pathogen inoculum present (Roberts, 1994). Preferably, the efficacy of an antagonist should be against a broad range of pathogens of fruits and vegetables, but the effectiveness is pathogen specific. Additionally, formulations of such antagonists should have a long shelf-life with low manufacturing costs. Since some successful biocontrol agents are opportunistic human pathogens too, very strict food safety considerations demand careful examination of products for the possible presence of toxic or unacceptable substances produced by BCAs. Due to all these factors, biological control agents should be combined with other disease control strategies if they are to provide acceptable control. For example Osman et al. (2011) reported that

Bacillus amyloliquefaciens applied after papaya fruits were treated with a 100 μL 1-methylcyclopropene concentration, controlled anthracnose by up to 50% stored at 10°C, 85% RH for 14 days and ripening at 25°C. The added advantage of this combination was visualized as delay in ripening, maintenance of firmness and color retention. A few other examples of BCAs alone or in combination with other control measures to suppress *C. gloeosporioides* and anthracnose disease are listed in Table 11.2.

Natural Inducers

Disease control by inducing host resistance through the activation of defense mechanisms in harvested fresh produce has also been explored (Johnson and Hofman, 2009). In a study in Sri Lanka, the resistance of unripe avocado fruits attacked by *C. gloeosporioides* was attributed to the presence of the preformed antifungal diene compound (1-acetoxy-2-hydroxy-4-oxo-heneicosa-12, 15- diene). *In vitro* inhibition of spores by 15% was observed when treated with 450 μg mL^{-1} of antifungal diene compound whereas 100 % inhibition was achieved at exposure to 790 μg mL^{-1}. In the peel of unripe fruits, concentration of this diene was 1200 μg g^{-1} fresh weight of peel which was fungitoxic in nature (Prusky et al., 1990; Prusky and Keen, 1993). However, there was a decline in these compounds to sublethal concentration once the fruit ripening commenced leading to pathogen development. In a further study, it was also noted that the level of these diene compounds can be enhanced, inducing resistance to *C. gloeosporioides* in avocado, by challenge inoculation of the fruits with the non-pathogenic mutant strain of *Colletotrichum magna* isolated from watermelon (Prusky et al., 1994).

Similarly, β-aminobutyric acid treatment (100 mM) effectively suppressed the development of anthracnose lesion diameter on mango fruits during storage at 25°C. Furthermore, β-aminobutyric acid treatment enhanced the activities of β-1, 3-glucanase (GLU), chitinase (CHT) and phenylalanine ammonia lyase (PAL). It also contributed to the accumulation of hydrogen peroxide (H_2O_2), simultaneously decreasing the rate of superoxide radical (O_2^-) production. Additionally, an increase in superoxide dismutase (SOD) activity, and inhibition in catalase (CAT) and ascorbate peroxidase (APX) activities were observed. The authors concluded that increased disease resistance of mango fruit after β-aminobutyric acid treatment during storage could be due to an elicitation of defense responses involved in the enhancement of defense-related enzyme and variation of antioxidant system activities (Zhang, et al., 2013).

Table 11.2 Effect of Biocontrol Agents Alone or in Combination with other Control Measures on Suppression of *Colletotrichum Gloeosporioides* and Anthracnose Disease

Antagonist/ Concentration	Crop/Fungus	Level of Suppression (%)	Reference
Bacillus subtilis	Avocado	26–50	Korsten et al., 1995
Candida oleophila (10^3 cell mL^{-1}) + sodium carbonate 2%	Papaya	60	Gamagae et al., 2003
Pseudomonas fluorescens (10^8 cfu mL^{-1}) + chitin (10 mg mL^{-1})	Mango (preharvest and postharvest during 15 days of storage at 28±2°C)	60	Vivekanathan et al., 2004
Burkholderia cepacia	*C. gloeosporioides*	74.1 (mycelial growth) 100 (spore germination)	Rahman et al., 2007
Pseudomonas aeruginosa	*C. gloeosporioides*	68.4 (mycelial growth)	Rahman et al., 2007
Burkholderia cepacia (10^9 CFU mL^{-1}) + 0.75% chitosan+ 3% calcium chloride	Papaya	100	Rahman et al., 2009
Streptomyces sp. (MJM5763)	*C. gloeosporioides*	64 (mycelial growth)	Palaniyandi et al., 2011
	C. gloeosporioides	90 (spore germination)	
Streptomyces sp. (A1022)	Pepper	90	Lee et al., 2012
Meyerozyma caribbica	Yam (in field)	86	Bautista-Rosales et al., 2013
Wickerhamomyces anomalus (10^8 cells mL^{-1})	Papaya	31	Rabelo de Lima et al., 2013
Meyerozyma guilliermondii (10^8 cells mL^{-1})		41	
Bacillus pumilus	*C. gloeosporioides*	89	Zheng et al., 2013
Bacillus thuringiensis	Mango	94	
	C. gloeosporioides	80	
	Mango	87	

Many authors have reported on the use of salicylic acid (SA), a well-known natural inducer of disease resistance in plants, and effective against mango anthracnose. Mango cultivar 'Kensington Pride', when treated with SA at a concentration of 2000 mg L^{-1}, showed significantly reduced anthracnose severity and also delay in senescence (Zainuri et al., 2001). In another study, Zeng et al. (2006) demonstrated that SA enhanced the resistance in mango fruits when treated with 1 mmol L^{-1} SA solution under vacuum infiltration for 2 min under low pressure (−80 kPa) and an additional 10 min at air pressure. There was reduction in disease incidence and lesion diameter by 37.5 and 21%, respectively after 4 days of incubation at 13°C, 85–95% RH. They further showed that there was 6- or 0.9-fold increase in PAL and GLU activities, respectively. After assessing the level of hydrogen peroxide (H_2O_2) or superoxide radical (O_2^-) generation in SA-treated fruit, a similar conclusion was drawn suggesting their role in enhancement of disease resistance in mango fruits. Similar findings were in corroboration with those who stated that the induction of local acquired resistance (LAR) by SA, 2, 6-dichloroisonicotinic acid, or benzo-(1,2,3)-thiadiazole-7-carbothioic acid S-methyl ester pretreatment prevents anthracnose infections in unripe pepper fruits. Salicylic acid induced LAR, and suppressed appressoria formation resulting in protection against *C. gloeosporioides* (Lee et al., 2009).

Edible Coatings

Recently, much attention has been paid to edible surface coatings for fruit storage in order to reduce the volume of packaging materials and risks compared to those with plastic film wraps (Han et al., 2004; Tanada-Palmu et al., 2005; Valverde et al., 2005; Durango et al., 2006). In addition to a reduction in respiration, water loss and oxidation (Park, 1999), the changes are likely to maintain the quality and extend the shelf-life of the produce (Li and Barth, 1998; Bai et al., 2003; Pen and Jiang, 2003; Dong et al., 2004; Valverde et al., 2005; Ali et al., 2011). Therefore, an ideal coating is defined as one that can extend the storage life and maintain the quality of fresh produce by not causing anaerobiosis and by reducing decay (El-Ghaouth et al., 1997).

Antimicrobial coatings are a rapidly emerging technology that can be employed for controlling the microbiological decay of perishable food products (Del-Nobile et al., 2008). These coatings contribute towards environmentally friendly technology that may enhance the food quality, safety and mechanical handling properties by retarding moisture loss, and enhancing oxygen and solute transport.

Chitosan is a promising antimicrobial coating. It is a biodegradable polymer derived from chitin found in the exoskeleton of crustaceans such as shrimps and crabs (Devlieghere et al., 2004; Kim et al., 2006). It has been widely used in agriculture and the food industry mainly because of its antimicrobial and film-forming properties that allow its use as an edible coating (El-Ghaouth et al., 1991; Romanazzi et al., 2003; Ben-Shalom et al., 2003). Chitosan has direct antifungal properties, ranging from fungistatic to fungicidal, against a wide range of fungi at various stages of their growth (El-Ghaouth et al., 1997, 2000; Bautista-Baños et al., 2003, 2004; Molloy et al., 2004; Ali and Mahmud, 2008). A very interesting study was carried out by Bautista-Baños et al. (2003) on the effect of chitosan on the growth of *C. gloeosporioides* and quality of papaya fruits in Mexico. It was elucidated that *in vitro* inhibition of mycelial growth of *C. gloeosporioides* was 100% after 7 days incubation at 2.5 and 3% chitosan concentration while a fungistatic effect on the fruits during storage was observed with 60% control of the disease by application of 1.5% chitosan. These findings were supported by the reports of Ali et al. (2010b, 2011) which showed that chitosan used at 1.5% has the potential to delay postharvest anthracnose and extend shelf-life of 'Exotica II' papaya for up to 5 weeks. However, at concentrations more than 1.5% fruits failed to ripen at all and subsequently rotted. Coating may improve retention of some aspects of quality in papaya fruits, but it appears to predispose to ripening problems when treated with higher concentrations. It was suggested that coatings may be so restrictive to gas exchange that it might have induced physiological disorders on the skin, possibly by inducing fermentation, and accumulation of toxic metabolites, cellular death of tissues and increase the incidence of decay in papaya fruits during storage (Ali et al., 2010b).

Several other studies involving the use of chitosan on mycelial growth and spore germination inhibition have been well documented. Zhong et al. (2009) demonstrated the efficacy of carboxymethyl chitosan and chitosan sulfate derivatives on suppression of *C. gloeosporioides* with an inhibitory index of 43%. Moreover, the inhibitory index of all of the compounds was enhanced with increase in concentration, and the highest antifungal activity was observed at 500 μg mL^{-1}. Others reported the inhibition of radial mycelial growth when concentration of chitosan increased from 0.75 to 6.0 mg mL^{-1} (El Ghaouth et al., 1997; Ali and Mahmud, 2008). Traditionally, chitosan coatings have been used as a barrier to minimize water loss and delay the decaying process. However, the new generation of chitosan and other edible coatings is being specifically designed to include and/or control the release of antioxidants, antimicrobials, colorants, flavors, vitamins, nutrients and spices (Pranoto et al., 2005; Puoci et al., 2008) by means of the application of promising technology such as nanoencapsulation (Vargas et al., 2008).

In this sense, in more recent studies, Zahid et al. (2012) and Ali et al. (2013) proposed that instead of applying chitosan in a conventional form, it could be more effective as a biofungicide for controlling anthracnose of fresh fruits when applied in the form of nanoemulsions. An *in vitro* evaluation of a nanoemulsion form of chitosan (droplet size of 600 nm) at concentrations of 1% and above significantly reduced the mycelial growth and spore germination of *C. gloeosporioides* isolated from papaya and dragon fruits. Moreover, it has a great impact on the morphology of the fungus as revealed as hyphal agglomeration (Fig. 11.4B) and abnormal spores (Fig. 11.4D) under scanning electron microscopic studies when compared to untreated controls (Fig. 11.4A and C). In further studies, a chitosan-loaded nanoemulsion at 1% concentration and droplet size of 600 nm when used as a coating material on papaya and dragon fruits showed the best results in delaying the onset of anthracnose and maintaining quality of all the fruits for up to 28 days of cold storage. In another study, the combination of chitosan at 1% combined with ammonium carbonate (3%) and sodium carbonate (2%) reduced *C. gloeosporioides* infection in papaya and maintained the quality during storage at 13.5°C and 95% RH for 14 days, and for 2 days under simulated market conditions (25°C, 75% RH) (Sivakumar et al., 2005).

Gum arabic (GA), another polysaccharide, is of interest as a potential film or coating component because of its unique emulsifying properties (Ali et al., 2010a). In a recent study, we found that GA alone had no antifungal effects against *C. gloeosporioides* infection on papaya. However, a composite coating developed using 10% GA formulated with lemon grass (0.05%) and cinnamon oil (0.4%) resulted in controlling anthracnose decay by 71%. Moreover, this coating had great potential to preserve papaya quality for up to 28 days at 12±1°C at 80–90% RH with an additional 5 days at 25±2°C, 60% RH (for ripening). It was suggested that the coating with GA and 0.4% cinnamon oil might be particularly suitable for papaya growers and exporters as a postharvest biopesticide (Maqbool et al., 2011).

Interestingly, propolis, the natural glue produced by honey-bees comprised of resins (flavonoids, phenolics and their esters), waxes, vitamins and essential oils (Juliano et al., 2007), is also being tested as a coating material. Due to its chemical composition and antimicrobial properties, it is widely

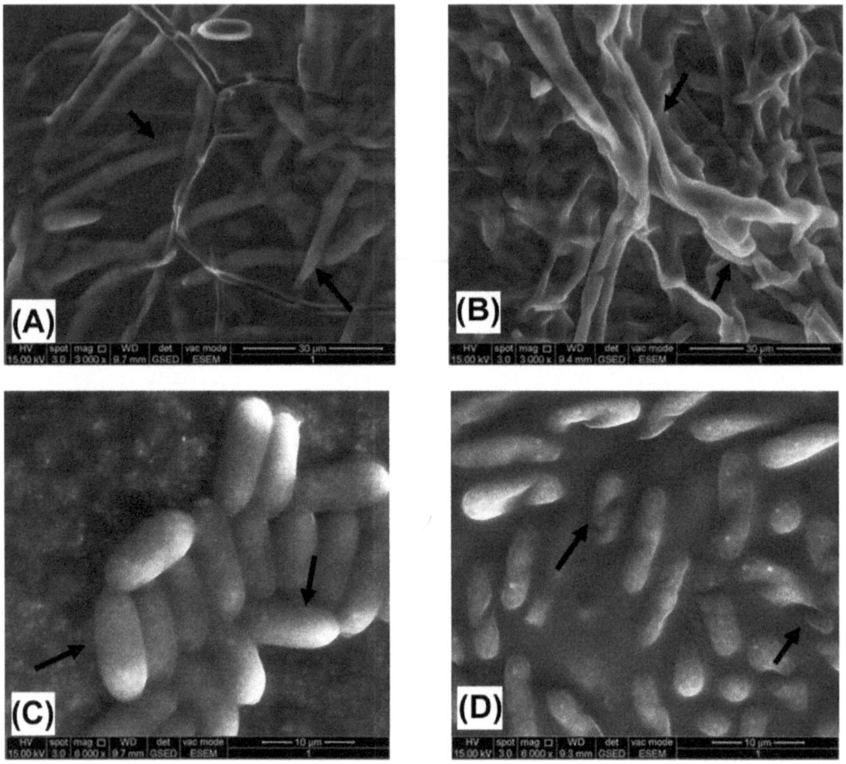

Figure 11.4 Scanning electron micrographs of hyphal and spore morphology of *C. gloeosporioides* isolated from dragon fruits. Arrows indicate (A) normal hyphae (untreated control); (B) hyphal agglomeration (2% chitosan nanoemulsion at 600 nm droplet size); (C) healthy spores (untreated control); and (D) abnormal and shriveled conidia (2% chitosan nanoemulsion at 600 nm droplet size).

used in the pharmaceutical industry. The effects of different concentrations of ethanolic extracts of propolis (EEP) have been tested on maintaining the shelf-life of dragon fruits. It was elucidated that fruits treated with 0.50% EEP (Fig. 11.5C) and stored at 20±2°C and 80±5% relative humidity had a greater shelf-life than the untreated fruits as well as other concentrations tested (Fig. 11.5A, B, D and E). Furthermore, EEP at 0.50% slowed down the ripening process in the fruits as evident from a decreased level of ethylene production and increased the biosynthesis of nutritional components such as total flavonoids and total antioxidants. However, higher concentrations of EEP (0.75 and 1.0%) showed some phytotoxic effects (Fig. 11.5 D and E) on the fruit surface as early as 8 days of storage, which may reduce their sensory as well as aesthetic quality (Zahid et al., 2013). It was suggested

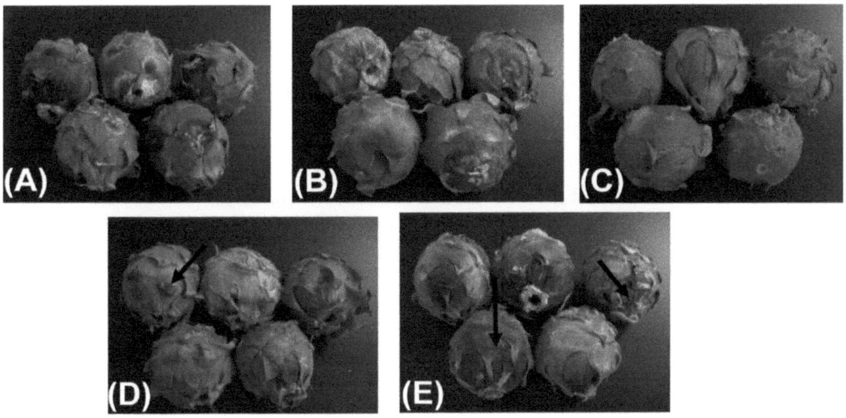

Figure 11.5 Effect of propolis on dragon fruits after 20 days of storage at 13±2°C and 80±5% RH. (A) Untreated control; fruits treated with (B) 0.25%, (C) 0.50%, (D) 0.75%, and (E) 1.0% propolis. Arrows show the phytotoxic effects of propolis on dragon fruits at higher concentration.

that the decrease in total phenolic compounds in control fruit and fruit treated with higher concentrations of EEP could be due to the higher rate of respiration, which resulted in the degradation of certain phenolic compounds (El Ghaouth et al., 1991), while the retention of total flavonoids in 0.50% EEP-treated fruit could be due to the slower respiration rates as compared to the fruit treated with higher concentrations.

Essential Oils

Plant essential oils contain a variety of chemicals, including terpenoids, which are known to be involved in plant defense mechanisms against pathogens (Lee et al., 2003). The essential oil obtained from castor beans had a significant suppressive effect on *in vitro* sporangial germination and mycelial growth of *C. gloeosporioides* and on anthracnose development in papaya fruits during storage. It has been suggested that the activity of the castor oil is due to a mixture of fatty acids as well as their corresponding salts (Skrivanova et al., 2005; Siqueira et al., 2012). A comparative study on the efficacy of essential oil extracted from lemon grass and fungicidal properties on postharvest control of mango fruits carried out at Rajamangala University of Technology in Thailand showed that 4000 ppm lemon grass oil in hot water treatment resulted in minimal anthracnose disease score when compared to 100 ppm carbendazim in hot water treatment (Raweewon, 2008). Interestingly, a combination of essential oils (*Lippia scaberrima)* rich in R-(−)-carvone, (d)-limonene and 1, 8-cineole with a commercial coating (Avoshine®) was recommended by Regnier et al. (2010) as an acceptable postharvest treatment for the organic market to control anthracnose in avocados. However, application of essential oils in vapor phase is preferred over liquid phase due to its volatility, which leads to higher efficacy, use of lower concentration and lack of unwanted sensory properties of treated produce (Laird and Phillips, 2012). Similar results by Sellamuthu et al. (2013) concluded that postharvest application of volatile thyme oil significantly enhanced the activity of antioxidants and induced defense-related enzymes, including chitinase, 1,3-β-glucanase, phenylalanine ammonia-lyase and peroxidase. It also maintained a higher phenolic content in fruit tissues and thereby enhanced resistance against invasion and reduced the spoilage due to anthracnose in avocados. Cinnamon oil is another essential oil obtained from *Cinnamomum zeylanicum* with cinnamaldehyde as its major active component. Barrera-Necha et al. (2008) reported the *in vitro* antifungal efficacy of this compound against *C. gloeosporioides* isolated from papaya fruits. From the above, it is evident that incorporation of essential oils in edible coatings could provide a safe inexpensive method to control postharvest diseases. Several other essential oils tested on different commodities have shown promise for the control of anthracnose diseases (Table 11.3).

CONCLUDING REMARKS

This chapter has reviewed that combating the devastating effects of anthracnose disease requires additional effort in terms of developing more specific treatments that take into account environmental and health factors. Conventional methods, such as synthetic chemical fungicides, are dependable for treating anthracnose, but the possibility of fungal resistance and health problems still poses a problem. In addition, not all fruits with symptoms of anthracnose are receptive to fungicides. Current approaches in treating the postharvest disease have included less environmentally destructive methods to prolong shelf-life, such as hot water treatment, having less destructive effects to the surrounding environment and consumers. Yet timely applications and high accuracy are crucial in both fungicide and hot water treatments and, consequently, require more careful protocols and management. This chapter has highlighted recent studies that have developed safer alternatives utilizing biological control agents and naturally occurring products. Essential oils and biodegradable chitosan in the form of nanoemulsions have

Table 11.3 Effect of Essential Oils on Suppression of *Colletotrichum Gloeosporioides* and Anthracnose Disease in Papaya and Mango Fruits

Essential Oil	Concentration	*C. gloeosporioides*/ Crop	Disease/Mycelial Growth Reduction (%)	Storage Time (days)	Reference
Scaligeria tripartita	20 mg mL^{-1}	*C. gloeosporioides*	6.5	–	Tabanca et al., 2007
Cinnamomum zeylanicum	250 µg mL^{-1}	Papaya	17.3	5	Barrera-Necha et al., 2008
Syzgium aromaticum	50 µg mL^{-1}	Papaya	13.2	5	Barrera-Necha et al., 2008
Cinnamomum zeylanicum	250 µg mL^{-1}	Papaya	12.4	8	Barrera-Necha et al., 2008
Syzgium aromaticum	50 µg mL^{-1}	Papaya	3.3	8	Barrera-Necha et al., 2008
Carnauba+*Lippia scaberrima* oil	2000 µl L^{-1}	Mango	5.0	6	Regnier et al., 2008
Lippia scaberrima	1000 µl L^{-1}	*C. gloeosporioides*	100	6	Regnier et al., 2010
Salvia rosifolia	20 mg mL^{-1}	*C. gloeosporioides*	8.5	–	Ozek et al., 2010
Thyme oil	0.12%	Papaya	50	9	Bosquez-Molina et al., 2010
	0.06%	*C. gloeosporioides*	90	7	
Thyme oil	500 µl L^{-1}	*C. gloeosporioides*	100	7	Combrinck et al., 2011
Cinnamon oil	0.4%	Papaya	71		Maqbool et al., 2011
		C. gloeosporioides	85 (spore inhibition) 70 (mycelial growth)	7 h 7	

been identified as effective in inhibiting the spread of anthracnose and reducing the losses in stored fruit. Alternative treatments, such as the use of bacteria, inducing host resistance, edible and antimicrobial coatings, as well as ozone gas, provide a greener substitute in postharvest disease management. Employing these strategies synergistically is sometimes needed to achieve reliable results. Some of these technologies are still in the stages of infancy, and future work will most likely visualize the need to bridge the gap between employing effective and specific and, at the same time, environmentally friendly technologies.

REFERENCES

Akamine, E.K., 1967. History of the hot water treatment of papaya. Hawai Farm Sci. 16, 4–6.

Akem, C.N., 2006. Mango anthracnose disease: Present status and future research priorities. Plant Pathol. J. 5, 266–273.

Akhthar, K.P., Khan, I., Khar, I.A., Khan, S.M., 1998. Studies on the incident and pathogenesis of *Colletotrichum gloeosporioides* Penz.causing anthracnose of mango and chemical control. Pak. J. Phytopathol. 10, 42–44.

Ali, A., Mahmud, T.M.M., 2008. The potential use of locally prepared chitosan to control in vitro growth of *Colletotrichum gloeosporioides* isolated from papaya fruits. Acta Horticulturae 804, 177–182.

Ali, A., Maqbool, M., Ramachandran, S., Alderson, P.G., 2010a. Gum arabic as a novel edible coating for enhancing shelf-life and improving postharvest quality of tomato *Solanum lycopersicum* L. fruit. Postharvest Biol. Technol. 58, 42–47.

Ali, A., Muhammad, M.T.M., Sijam, K., Siddiqui, Y., 2010b. Potential of chitosan coating in delaying the postharvest anthracnose (*Colletotrichum gloeosporioides* Penz.) of Eksotika II papaya. Int. J. Food Sci. Technol. 45, 2134–2140.

Ali, A., Muhammad, M.T.M., Sijam, K., Siddiqui, Y., 2011. Effect of chitosan coatings on the physicochemical characteristics of Eksotika II papaya (*Carica papaya* L.) fruit during cold storage. Food Chem. 124, 620–626.

Ali, A., Zahid, N., Sivakumar, M., Siddiqui, Y., Alderson, P.G., Maqbool, M., 2013. Effectiveness of submicron chitosan dispersions in controlling anthracnose and maintaining quality of dragon fruit. Postharvest Biol. Technol. 86, 147–153.

Alvarez, A.M., Nishijima, W.T., 1987. Postharvest diseases of papaya. Plant Dis. 71, 681–686.

Anwar, R., Malik, A.U., 2007. Hot water treatment affects ripening quality and storage life of mango (*Mangifera indica* L. Pak. J. Agric. Sci. 44, 304–311.

Arauz, L.F., 2000. Mango anthracnose: Economic impact and current options for integrated management. Plant Dis. 84, 600–611.

Aveno, J.L., Orden, M.E.M., 2004. Hot water treatment of mango: A study of four export corporations in the Philippines. KMITL Sci. Technol. J. 4, 1685–2044.

Awang, Y., Ghani, M.A.A., Sijam, K., Mohamad, R.B., 2011. Effect of calcium chloride on anthracnose disease and postharvest quality of red-flesh dragon fruit (*Hylocereus polyrhizus*). Afri. J. Microbiol. Res. 5, 5250–5259.

Bai, J., Alleyne, V., Hagenmaier, R.D., Mattheis, J.P., Baldwin, E.A., 2003. Formulation of zein coatings for apples *Malus domestica* Borkh. Postharvest Biol. Technol. 28, 259–268.

Baños-Guevara, P.E., Zavaleta-Mejía, E., Colinas-León, M.T., Luna-Romero, I., Gutiérrez-Alonso, J.G., 2004. Control biológico de *Colletotrichum gloeosporioides* [(Penz.) Penz.y Sacc.] en papaya Maradol roja (*Carica papaya* L.) y fisiología postcosecha de frutos infectados. Mex. J. Phytopathol. 22, 198–205.

Barrera-Necha, L.L., Bautista-Baños, S., Flores-Moctezuma, H.E., Rojas-Estudillo, A., 2008. Efficacy of essential oils on the conidial germination, growth of *Colletotrichum gloeosporioides* (penz.) Penz. and Sacc. and control of postharvest diseases in papaya (*Carica papaya* L. Plant Pathol. J. 7, 174–178.
Bautista-Baños, S., Hernández-López, M., Bosquez-Molina, E., Wilson, C.L., 2003. Effect of chitosan and plant extracts on growth of *Colletotrichum gloeosporioides*, anthracnose levels and quality of papaya fruit. Crop protect. 22, 1087–1092.
Bautista-Baños, S., Hernández-López, M., Bosquez-Molina, E., 2004. Growth inhibition of selected fungi by chitosan and plant extracts. Mex. J. Phytopathol. 22, 178–186.
Bautista-Baños, S., Sivakumar, D., Bello-Pérez, A., Villanueva-Arce, R., Hernández-López, M., 2013. A review of the management alternatives for controlling fungi on papaya fruit during the postharvest supply chain. Crop Protect. 49, 8–20.
Bautista-Rosales, P.U., Calderon-Santoyo, M., Servín-Villegas, R., Ochoa-Álvarez, N.A., Ragazzo-Sánchez, J.A., 2013. Action mechanisms of the yeast *Meyerozyma caribbica* for the control of the phytopathogen *Colletotrichum gloeosporioides* in mangoes. Biol. Control 65, 293–301.
Ben-Shalom, N., Ardi, R., Pinto, R., Aki, C., Fallik, E., 2003. Controlling grey mould caused by *Botrytis cinerea* in cucumber plants by means of chitosan. Crop Prot. 22, 285–290.
Bose, S.K., Sindhan, G.S., Pandey, B.H., 1973. Studies on dieback disease of mango in Tarai region of Kumaon. Progressive Horticulture 70, 557–584.
Brown, G.E., 1984. Efficacy of citrus postharvest fungicides applied in water or resin solution water wax. Plant Dis. 68, 415–418.
Cannon, P.F., Buddie, A.G., Bridge, P.D., 2008. The typification of *Colletotrichum gloeosporioides*. Mycotaxon. 104, 189–204.
Cannon, P.F., Damm, U., Johnston, P.R., Weir, B.S., 2012. *Colletotrichum* – current status and future directions. Stud. Mycol. 73, 181–213.
Chau, C.F., Alvarez, A.M., 1983. A histological study of anthracnose on *Carica papaya*. Phytopathol. 73, 1113–1116.
Chávez-Sánchez, I., Carrillo-López, A., Vega-García, M., Yahia, E.M., 2013. The effect of antifungal hot-water treatments on papaya postharvest quality and activity of pectinmethylesterase and polygalacturonase. J. Food Sci. Technol. 50, 101–107.
Chowdhury, M.N.A., Rahim, M.A., 2009. Integrated crop management to control anthracnose *(Colletotrichum gloeosporioides)* of mango. J. Agric. Rural Dev. 7, 115–120.
Combrinck, S., Regnier, T., Kamatou, G.P.P., 2011. In vitro activity of eighteen essential oils and some major components against common postharvest fungal pathogens of fruit. Ind. Crops Prod. 33, 344–349.
Couey, H.M., Farias, G., 1979. Control of postharvest decay of papaya. Hortscience 14, 719–721.
Cox, M.L., Irwin, J.A.G., 1988. Conidium and appressorium variation in Australian isolates of the *Colletotrichum gloeosporioides* group and closely related species. Aust. Syst. Bot. 1, 139–149.
Crane, J.H., Balerdi, C.F., 2005. The pitaya (*Hylocereus undatus* and other spp.) in Florida. University of Florida, IFAS Extension, Miami Dade County.
Damm, U., Cannon, P.F., Woudenberg, J.H.C., Crous, P.W., 2012a. The *Colletotrichum acutatum* species complex. Stud. Mycol. 73, 37–113.
Damm, U., Cannon, P.F., Woudenberg, J.H.C., et al., 2012b. The *Colletotrichum boninense* species complex. Stud. Mycol. 73, 1–36.
Darvas, J.M., Kotze, J.M., Wehner, F.C., 1987. Pathogenicity of fungi causing pre- and postharvest diseases of avocado fruit. Phytophylactica 19, 489–493.
Dean, R., Van Kan, J.A., Pretorius, Z.A., et al., 2012. The top 10 fungal pathogens in molecular plant pathology. Mol. Plant Pathol. 13, 414–430.
de Capdevile, G., Texeira Jr, S.M., Pereira, S.J.R., de Paula, M.S., Rodrigues, C.A., Goncalves, T.F.A., 2007. Selection and testing of epiphityc yeasts to control anthracnose in postharvest of papaya fruit. Sci. Horticulturae 111, 179–185.

Del Nobile, M.A., Conte, A., Incoronato, A.L., Panza, O., 2008. Antimicrobial efficacy and release kinetics of thymol from zein films. J. Food Eng. 89, 57–63.

De los Santos, de la, R.F., Becerra, L.E.N., Mosqueda, V.A., Vargas, G.A.B. 1997. Manual de Producción de papaya en el estado de Veracruz. Inst. Nac. de Investigaciones Forestales, Agrícolas y Pecuarias, Centro de Investigación Regional Golfo Centro, Campo Experimental, Cotaxtla. Veracruz, Ver. México, p. 87.

Deliopoulos, T., Kettlewell, P.S., Hare, M.C., 2010. Fungal disease suppression by inorganic salts: a review. Crop Protect. 29, 1059–1075.

Dessalegn, Y., Ayalew, A., Woldetsadik, K., 2013. Integrating plant defense inducing chemical, inorganic salt and hot water treatments for the management of postharvest mango anthracnose. Postharvest Biol. Technol. 85, 83–88.

Devlieghere, F., Vermeiren, L., Debevere, J., 2004. New preservation technologies: Possibilities and limitations. Int. Dairy J. 14, 273–285.

Dickman, M.B., Alvarez, A.M., 1983. Latent infection of papaya caused by *Colletotrichum gloeosporioides*. Plant Dis. 67, 748–750.

Djioua, T., Charles, F., Lopez-Lauri, F., et al., 2009. Improving the storage of minimally processed mangoes (*Mangifera indica* L.) by hot water treatments. Postharvest Biol. Technol. 52, 221–226.

Dodd, J.C., Bugante, R., Koomen, I., Jeffries, P., 1991. Pre-and post-harvest control of mango anthracnose in the Philippines. Plant Pathol. 40, 576–583.

Dong, H., Cheng, L., Tan, J., Zheng, K., Jiang, Y., 2004. Effects of chitosan coating on quality and shelf life of peeled litchi fruit. J. Food Eng. 64, 355–358.

Durango, A.M., Soares, N.F.F., Benevides, S., et al., 2006. Development and evaluation of an edible antimicrobial film based on yam starch and chitosan. Packaging Technol. Sci. 19, 55–59.

El Ghaouth, A., Arul, J., Ponnampalam, R., Boulet, M., 1991. Chitosan coating effect on storability and quality of fresh strawberries. J. Food Sci. 56, 1618–1620.

El Ghaouth, A., Arul, J., Wilson, C., Benhamou, N., 1997. Biochemical and cytochemical aspects of the interactions of chitosan and *Botrytis cinerea* in bell pepper fruit. Postharvest Biol. Technol. 12, 183–194.

El-Ghaouth, A., Smilanick, J.L., Wilson, C.L., 2000. Enhancement of the performance of *Candida saitoana* by the addition of glycolchitosan for the control of postharvest decay of apple and citrus fruit. Postharvest Biol. Technol. 19, 103–110.

Estrada, A.B., Dodd, J.C., Jeffries, P., 2000. Effect of humidity and temperature on conidial germination and appressorium development of two Philippine isolates of the mango anthracnose pathogen *Colletotrichum gloeosporioides*. Plant Pathol. 49, 608–618.

Everett, K.R., Boyd, L.M., Pak, H.A., Cutting, J.G.M., 2007. Calcium, fungicide sprays and canopy density influence postharvest rots of avocado. Australas. Plant Pathol. 36, 22–31.

Fitzell, R.D., 1987. Epidemiology of anthracnose disease of avocados. South Afri. Avocado Growers' Assoc. Yearbook 10, 113–116.

Freeman, S., Shabi, E., 1996. Cross-infection of subtropical and temperate fruits by *Colletotrichum* species from various hosts. Physiol. Mol. Plant Pathol. 49, 395–404.

Freeman, S., Minz, D., Jurkevitch, E., Maymon, M., Shabi, E., 2000. Molecular analyses of Colletotrichum species from almond and other fruits. Phytopathology 90, 608–614.

Gamagae, S.U., Sivakumar, D., Wijeratnam, R.S., Wijesundera, R.L.C., 2003. Use of sodium bicarbonate and *Candida oleophila* to control anthracnose in papaya during storage. Crop Prot. 22, 775–779.

Gamagae, S.U., Sivakumar, D., Wijesundera, R.L.C., 2004. Evaluation of post-harvest application of sodium bicarbonate-incorporated wax formulation and *Candida oleophila* for the control of anthracnose of papaya. Crop Protect. 23, 575–579.

Han, C., Zhao, Y., Leonard, S.W., Traber, M.G., 2004. Edible coatings to improve storability and enhance nutritional value of fresh and frozen strawberries *Fragaria ananassa* and raspberries *Rubus ideaus*. Postharvest Biol. Technol. 33, 67–78.

Hartill, W.F.T., 1992. Post-harvest rots of avocado in New Zealand and their control. In Brighton Crop Protection Conference. Pests Dis. 3, 1157–1162. The British Crop Protection Council, Farnham, pp. 1157–1162.

Hartung, J.S., Burton, C.L., Ramsdell, D.C., 1981. Epidemiological studies of blueberry anthracnose disease caused by *Colletotrichum gloeosporioides*. Phytopathology 71, 449–453.

Hoa, N.V., 2008. Current research activities and the development of good agricultural practice (GAP) for pitaya in Vietnam. Pitaya Production. Market Export Challenges Prospects 20. October 2008, Putrajaya, Malaysia.

Jaccobi, K.K., Wong, L.S., Giles, J.E., 1995. Effects of fruit maturity on quality and physiology of high humidity hot air treated 'Kensington' mango (*Mangifera indica* Linn.). Postharvest Biol. Technol. 5, 149–159.

Jaccobi, K.K., MacRae, E.A., Hetherington, S.E., 2001. Postharvest heat disinfestation treatment of mango fruit. Sci. Horticulturae 89, 171–193.

Johnson, G.I., Hofman, P.J., 2009. Postharvest Technology and Quarantine Treatments. In: Litz, R.E. (Ed.), The Mango, Botany, Production and Uses, second ed. CAB International, Wallingford, pp. 529–605.

Johnston, P.R., Jones, D., 1997. Relationships among Colletotrichum isolates from fruit-rots assessed using rDNA sequences. Mycologia 89, 420–430.

Juliano, C., Pala, C.L., Cossu, M., 2007. Preparation and characterisation of polymeric films containing propolis. J. Drug Deliv. Sci. Technol. 17, 177–181.

Kim, D.G., Jeong, Y.I., Choi, C., et al., 2006. Retinol-encapsulated low molecular water-soluble chitosan nanoparticles. Int. J. Pharmaceutics 319, 130–138.

Korsten, L., De Jager, E.S., De Villers, E.E., Lourens, A., Kotze, J.M., Wehner, F.C., 1995. Evaluation of bacterial epiphytes isolated from avocado leaf and fruit surfaces for biocontrol of avocado postharvest diseases. Plant Dis. 79, 1149–1156.

Laird, K., Phillips, C., 2012. Vapour phase: a potential future use for essential oils as antimicrobials. Lett. Appl. Microbiol. 54, 169–174.

Lau, C.Y., Othman, F., Eng, L., 2001. The effect of heat treatment, different packaging methods and storage temperatures on the shelf life of dragon fruit (Hylocereus spp.). http://www.doa.sarawak.gov.my/modules/web/download_show.php?id=548.

Le Bellec, F., Vaillant, F., Imbert, E., 2006. Pitaya (*Hylocereus* spp): A new fruit crop, a market with a future. Fruits 61, 237–250.

Lee, S., Hong, J.C., Jeon, W.B., Chung, Y.S., Sung, S., Choi, D., Joung, Y.H., Oh, B.H. 2009. The salicylic acid -induced protection of non-climacteric unripe pepper fruit against *Colletotrichum gloeosporioides* is similar to the resistance of ripe fruit. Plant Cell Rep. 28 (10), 1573–1580.

Lee, E., Ahn, Y.J., Lee, H.S., Chung, N., 2012. Biocontrol of pepper anthracnose by a new *Streptomyces* sp. A1022 under greenhouse conditions. J. Korean Soc. Appl. Biol. Chem. 55, 447–449.

Lee, J.Y., Park, H.J., Lee, C.Y., Choi, W.Y., 2003. Extending shelf-life of minimally processed apples with edible coatings and anti-browning agents. LWT Food Sci. Technol. 36, 323–329.

Lenné, J.M., 1978. Studies on the biology and taxonomy of Colletotrichum species. Doctoral dissertation. University of Melbourne.

Li, P., Barth, M.M., 1998. Impact of edible coatings on nutritional and physiological changes in lightly-processed carrots. Postharvest Biol. Technol. 14, 51–60.

Lin, J., Gong, D., Zhu, S., Zhang, L., Zhang, L., 2011. Expression of PPO and POD genes and contents of polyphenolic compounds in harvested mango fruits in relation to Benzothiadiazole-induced defense against anthracnose. Sci. Horticulturae 130, 85–89.

Li, X., Zhu, X., Zhao, N., et al., 2013. Effects of hot water treatment on anthracnose disease in papaya fruit and its possible mechanism. Postharvest Biol. Technol. 86, 437–446.

Maqbool, M., Ali, A., Alderson, P.G., Mahmud, T.M.M., Siddiqui, Y., Zahid, N., 2011. Postharvest application of gum arabic and essential oils for controlling anthracnose and quality of banana and papaya during cold storage. Postharvest Biol. Technol. 62, 71–76.

Martins, D.M.S., Blum, L.E.B., Sena, M.C., et al., 2010. Effect of hot water treatment on the control of papaya (*Carica papaya* L.) postharvest diseases. Acta Horticulturae 864, 181–185.

Masyahit, M., Sijam, K., Awang, Y., Satar, G.M., 2009. First report on bacterial soft rot disease on dragon fruit (*Hylocereus* spp.) caused by *Enterobacter cloacae* in peninsular Malaysia. Int. J. Agric. Biol. 11, 659–666.

McGuire, R.G., 1991. Concomitant decay reductions when mangoes are treated with heat to control infestations of Caribbean fruit flies. Plant Dis. 75, 946–949.

Molloy, C., Cheah, L.H., Koolard, J.P., 2004. Induced resistant against *Sclerotinia sclerotiorum* in carrots treated with enzymatically hydrolysed chitosan. Postharvest Biol. Technol. 33, 61–65.

Muirhead, I.F., Fitzell, R.D., Davis, R.D., Peterson, R.A., 1982. Post-harvest control of anthracnose and stem-end rots of Fuerte avocados with prochloraz and other fungicides. Aust. J. Exp. Agric. Animal Husbandry 22, 441–446.

Nel, A., Krause, M., Khelawanlall, N., 2003. A Guide for the Control of Plant Diseases, second ed. Directorate: Food Safety and Quality Assurance, Pretoria.

Nelson, S.C., 2008. Anthracnose of avocado. Plant Dis. 58, 1–6.

Nishijima, W.T., Ebersole, S., Fernandez, J.A., 1990. Factors influencing development of postharvest incidence of Rhizopus soft rot of papaya. Acta Horticulture 269, 495–502.

Ong Mei, K., Feroz, K.K., Forney, C.F., Ali, A., 2013. Effect of gaseous ozone on papaya anthracnose. Food Bioprocess Technol. 6, 2996–3005.

Osman, S.M., Sivakumar, D., Korsten, L., 2011. Effect of biocontrol agent *Bacillus amyloliquefaciens* and 1-methyl cyclopropene on the control of postharvest diseases and maintenance of fruit quality. Crop Prot. 30, 173–178.

Ozek, G., Demirci, F., Ozek, T., et al., 2010. Gas chromatographic-mass spectrometric analysis of volatiles obtained by four different techniques from *Salvia rosifolia* Sm., and evaluation for biological activity. J. Chromatogr. A 1217, 741–748.

Palaniyandi, S.A., Yang, S.H., Cheng, J.H., Meng, L., Suh, J.W., 2011. Biological control of anthracnose (*Colletotrichum gloeosporioides*) in yam by *Streptomyces* sp. MJM5763. J. Appl. Microbiol. 111, 443–455.

Park, H.J., 1999. Development of advanced edible coatings for fruits. Trends Food Sci. Technol. 10, 254–260.

Pen, L.T., Jiang, Y.M., 2003. Effects of chitosan coating on shelf life and quality of fresh-cut Chinese water chestnut. Food Sci. Technol. 36, 359–364.

Paull, R.E. 1990. Postharvest heat treatments and fruit ripening. Postharvest News and Info., 1,355–353.

Paull, R.E., Nishijima, W., Reyes, M., Cavaletto, C., 1997. Postharvest handling and losses during marketing of papaya *(Carica papaya)*. Postharvest Biol. Technol. 11, 165–179.

Peterson, R.A., Inch, A.J., 1980. Control of anthracnose on avocados in Queensland. Queensland. J. Agric. Animal Sci. 37, 79–83.

Phillips, D.J., Harris, C.M., 1979. Post-harvest brown rot of peaches and inoculum density of *Monilinia fructicola*. USDA AgriRes. Serv. Publication ARR-W-9.

Photita, W., Taylor, P.W., Ford, R., Hyde, K.D., Lumyong, S., 2005. Morphological and molecular characterization of *Colletotrichum* species from herbaceous plants in Thailand. Fungal Diversity 18, 117–133.

Ploetz, R.C., Benscher, D., Vazquez, A., Colls, A., Nagel, J., Schaffer, B., 1997. Mango decline: Research in Florida on an apparently wide-spread disease complex. Acta Horticulturae 455, 547–557.

Pramila, T., Dubey, N.K., 2004. Exploitation of natural products as an alternative strategy to control postharvest fungal rotting of fruit and vegetables. Postharvest Biol. Technol. 32, 235–245.

Pranoto, Y., Salokhe, V.M., Rakshit, S.K., 2005. Physical and antibacterial properties of alginate-based edible film incorporated with garlic oil. Food Res. Int. 38, 267–272.

Prapagdee, B., Kuekulvong, C., Mongkolsuk, S., 2008. Antifungal potential of extracellular metabolites produced by Streptomyces hygroscopicus against phytopathogenic fungi. Int. J. Biol. Sci. 4, 330–337.

Prusky, D., Keen, N.T., 1993. Involvement of preformed antifungal compounds in the resistance of subtropical fruits to fungal decay. Plant Dis. 77, 114–119.

Prusky, D., Karni, L., Kobiler, I., Plumbley, R.A., 1990. Induction of the antifungal diene in unripe avocado fruits: effect of inoculation with *Colletotrichum gloeosporioides*. Physiol. Mol. Plant Pathol. 37, 425–435.

Prusky, D., Freeman, S., Rodriguez, R., Keen, N.T., 1994. A non-pathogenic mutant strain of *Colletotrichum magna* induces resistance to *Colletotrichum gloeosporioides* in avocado fruits. Mol. Plant Microbe Int. 7, 326–333.

Puoci, F., Lemma, F., Spizzirri, U.G., Cirillo, G., Curcio, M., Picci, N., 2008. Polymer in agriculture: A review. Am. J. Agric. Biol. Sci. 3, 299–314.

Rabelo de Lima, J., Barros Gonçalves, L.R., Rocha Brandão, L., Rosa, C.A., Marto Pinto, V.F., 2013. Isolation, identification, and activity in vitro of killer yeasts against *Colletotrichum gloeosporioides* isolated from tropical fruits. J. Basic Microbiol. 53, 590–599.

Raheja, S., Thakore, B.B.L., 2002. Effect of physical factor, plant extracts and bio-agent on *Colletotrichum gloeosporioides* Penz., the causal organism of anthracnose of yam. J. Mycol. Plant Pathol. 32, 293–294.

Rahman, M.A., Kadir, J., Mahmud, T.M.M., Abdul Rahman, R., Begum, M.M., 2007. Screening of antagonistic bacteria for biocontrol activities on *Colletotrichum gloeosporioides* in papaya. Asian J. Plant Sci. 6, 12–20.

Rahman, M.A., Mahmud, T.M.M., Kadir, J., Abdul Rahman, R., Begum, M.M., 2009. Enhancing the efficacy of *Burkholderia cepacia* B23 with calcium chloride and chitosan to control anthracnose of papaya during storage. Plant Pathol. J. 25, 361–368.

Rathod, G.M., 2011. Effect of physical factors on development of anthracnose of mango fruits. Curr. Bot. 2, 15–16.

Raweewon, D., 2008. Natural essential oils from lemon grass (*Cymbopogon citratus*) to control postharvest anthracnose of mango fruit. Int. J. Biotechnol. 10, 104–108.

Regnier, T., Plooy, W., Combrinck, S., Botha, B., 2008. Fungitoxicity of *Lippia scaberrima* essential oil and selected terpenoids components on two mango postharvest spoilage pathogens. Postharvest Biol. Technol. 48, 254–258.

Regnier, T., Combrinck, S., Plooy, W.D., Botha, B., 2010. Evaluation of *Lippia scaberrima* essential oil and some pure terpenoid constituents as postharvest mycobiocides for avocado fruit. Postharvest Biol. Technol. 57, 176–182.

Rendall-Dunn, A.J., 1991. General news. Postharvest News Info 2, 3.

Roberts, R.G., 1994. Integrating biological control into postharvest disease management strategies. HortScience 29, 758–762.

Rodríguez-López, E.S., González-Prieto, J.M., Mayek-Pérez, N., 2009. La infección de *Colletotrichum gloeosporioides* Penz. Penz. and Sacc. en aguacatero (*Persea americana* Mill.): Aspectos bioquímicos y genéticos. Mex. J. Phytopathol. 27, 53–63.

Romanazzi, G., Nigro, F., Ippolito, A., 2003. Short hypobaric treatments potentiate the effect of chitosan in reducing storage decay of sweet cherries. Postharvest Biol. Technol. 29, 73–80.

Rufino, M.S.M., Alves, R.E., Brito, E.S., Pérez-Jiménez, J., Saura-Calixto, F., 2010. Bioactive compounds and antioxidant capacities of 18 non-traditional tropical fruits from Brazil. Food Chem. 121, 996–1002.

Sattar, A., Malik, S.A., 1939. Some studies on anthracnose of mango caused by *Glomerella cingulata* (Stonem.) Spaulding and von Schrenk, (*Colletotrichum gloeosporioides* Penz.) in Punjab. Indian J. Agric. Sci. 9, 511–521.

Sellamuthu, P.S., Mafune, M., Sivakumar, D., Soundy, P., 2013. Thyme oil vapour and modified atmosphere packaging reduce anthracnose incidence and maintain fruit quality in avocado. J. Sci. Food Agric. 93, 3024–3031.

Shi, J., Liu, A., Li, X., Feng, S., Chen, W., 2011. Inhibitory mechanisms induced by the endophytic bacterium MGY2 in controlling anthracnose of papaya. Biol. Control 56, 2–8.
Simmonds, J.H., 1965. A study of the species of Colletotrichum causing ripe fruit rots in Queensland. Queensland. J. Agric. Animal Sci. 22, 437–459.
Sivakumar, D., Hewarathgamagae, N.K., Wijeratnam, R.W., Wijesundera, R.L.C., 2002. Effect of ammonium carbonate and sodium bicarbonate on anthracnose of papaya. Phytoparasitica 30, 486–492.
Sivakumar, D., Sultanbawa, Y., Ranasingh, N., Kumara, P., Wijesundera, R.L.C., 2005. Effect of the combined application of chitosan and carbonate salts on the incidence of anthracnose and on the quality of papaya during storage. J. Horticultural Sci. Biotechnol. 80, 447–452.
Sivakumar, D., Jiang, Y., Yahia, E.M., 2011. Maintaining mango (*Mangifera indica* L.) fruit quality during the export chain. Food Res. Int. 44, 1254–1263.
Skrivanova, E., Marounek, M., Dlouha, G., Kaňka, J., 2005. Susceptibility of *Clostridium perfringens* to $C_2 - C_{18}$ fatty acids. Lett. Appl. Microbiol. 41, 77–81.
Solano, V., Arauz, L.F., 1995. Combate de antracnosis en frutos de papaya mediante aplicaciones de fungicidas en el campo en la zona atlántica de Costa Rica. Agronomia Costarricense 19, 25–30.
Spalding, D.H., Reeder, W.F., 1986. Decay and acceptability of mangos treated with combinations of hot water, imazalil, and γ-radiation. Plant Dis. 70, 1149–1151.
Stakman, E.C., Harrar, J.G., 1957. Principles of Plant Pathology. Ronald Press Co., New York.
Sundravadana, S., Alice, D., Kuttalam, S., Samiyappan, R., 2006. Control of mango anthracnose by azoxystrobin. Tunisian J. Plant Prot. 1, 109–114.
Sutton, B.C., 1980. The Coelomycetes. Fungi imperfecti with pycnidia, acervuli and stromata. Commonwealth Mycological Institute, Kew, Surrey.
Sutton, B.C., 1992. The genus *Glomerella* and its anamorph *Colletotrichum*. In: Bailey, J.A., Jeger, M.J. (Eds.), *Colletotrichum* Biology, Pathology and Control. CAB International, Wallingford, pp. 1–26.
Svetlana, Z., Stojanović, S., Ivanović, Ž., Gavrilović, V., Tatjana, P., Jelica, B., 2010. Screening of antagonistic activity of microorganisms against *Colletotrichum acutatum* and *Colletotrichum gloeosporioides*. Arch. Biol. Sci. 62, 611–623.
Swart, S.H., Serfontein, J.J., Kalinnowski, J., 2002. Chemical control of postharvest diseases of mango – the effect of prochloraz, thiobendazole and fludioxonil on soft brown rot, stem-end rot and anthracnose. S.A. Mango Growers' Association Year Book 22, 55–62.
Swinburne, T.R., 1983. Quiescent infections in post-harvest diseases. In: Dennis, C. (Ed.), Post-harvest pathology of fruits and vegetables. Academic Press, London, pp. 1–21.
Tabanca, N., Demirci, B., Baser, K.H., et al., 2007. Characterization of volatile constituents of *Scaligeria tripartita* and studies on the antifungal activity against phytopathogenic fungi. J. Chromatogr. B Analyt. Technol. Biomed. Life Sci. 850, 221–229.
Tanada-Palmu, P.S., Carlos, R., Grosso, F., 2005. Effect of edible wheat gluten-based films and coatings on refrigerated strawberry *Fragaria ananassa* quality. Postharvest Biol. Technol. 36, 199–208.
Tasiwal, V., Benagi, V.I., Hegde, Y.R., Kamanna, B.C., Naik, K.R., 2009. *In vitro* evaluation of botanicals, bioagents and fungicides against anthracnose of papaya caused by *Colletotrichum gloeosporioides* (Penz.) Penz. Sacc. Karnataka J. Agric. Sci. 22, 803–806.
Valverde, J.M., Valero, D., Martínez-Romero, D., Guillén, F., Castillo, S., Serrano, M., 2005. Novel edible coating based on Aloe vera gel to maintain table grape quality and safety. J. Agric. Food Chem. 53, 7807–7813.
Vargas, M., Pastor, C., Chiralt, A., McClements, J., Gonzalez-Martinez, C., 2008. Recent advances in edible coatings for fresh and minimally processed fruits. Crit. Rev. Food Sci. Nutr. 48, 496–511.

Ventura, A.J., Costa, H., da Silva, T.J., 2004. Papaya diseases and integrated control. In: Naqvi, S.A.M.H. (Ed.). Diseases of Fruits and Vegetables, vol. II. Kluwer Academic Publisher, The Netherlands, pp. 201–268.
Verhoeff, K., 1974. Latent infections by fungi. Ann. Rev. Phytopathol. 12, 99–110.
Vivekanathan, R., Ravi, M., Ramanathan, A., Samiyappan, R., 2004. Lytic enzymes induced by *Pseudomonas fluorescens* and other biocontrol organisms mediate defense against the anthracnose pathogen in mango. World J. Microbiol. Biotechnol. 20, 235–244.
Von Arx, J.V., 1957. Die Arten der Gattung Colletotrichum Corda. Phytopathol. Zeitschrift 29, 413–468.
Weir, B., Johnston, P.R., Damm, U., 2012. The *Colletotrichum gloeosporioides* species complex. Stud. Mycol. 73, 115–180.
Wills, R., Barry, M.G., Doug, G., Daryl, J., 1998. Postharvest: An Introduction to the Handling of Fruits, Vegetable and Ornamentals, fourth ed. University of New South Wales, Australia.
Xie, L., Zhang, J.Z., Wan, Y., Hu, D.W., 2010. Identification of *Colletotrichum* spp. isolated from strawberry in Zhejiang Province and Shanghai City, China. J. Zhejiang Univ. Sci. B Biomed. Biotechnol. 11, 61–70.
Zahid, N., Ali, A., Manickam, S., Siddiqui, Y., Maqbool, M., 2012. Potential of chitosan–loaded nanoemulsions to control different *Colletotrichum* spp. and maintain quality of tropical fruits during cold storage. J. Appl. Microbiol. 113, 925–939.
Zahid, N., Ali, A., Siddiqui, Y., Maqbool, M., 2013. Efficacy of ethanolic extract of propolis in maintaining postharvest quality of dragon fruit during storage. Postharvest Biol. Technol. 79, 69–72.
Zainuri, D.C.J., Wearing, H., Coates, L., Terry, L., 2001. Effects of phosphonate and salicylic acid treatments on anthracnose disease development and ripening of 'Kensington Pride' mango fruit. Aust. J. Exp. Agric. 41, 805–813.
Zavala-León, M.J., Tun-Suárez, J.M., Cristobal-Alejo, J., et al., 2005. Control postcosecha de la antracnosis en papaya y sensibilidad de *Colletotrichum gloeosporioides* (Penz.) Sacc. a fungicidas organosintéticos. Rev. Chapingo Serie Horticultura 11, 251–255.
Zeng, K., Cao, J., Jiang, W., 2006. Enhancing disease resistance in harvested mango (*Mangifera indica* L. cv.'Matisu') fruit by salicylic acid. J. Sci. Food Agric. 86, 694–698.
Zhang, J., Timmer, L.W., 2007. Preharvest application of fungicides for postharvest disease control on early season tangerine hybrids in Florida. Crop Prot. 26, 886–893.
Zhang, Z., Yang, D., Yang, B., et al., 2013. β-Aminobutyric acid induces resistance of mango fruit to postharvest anthracnose caused by *Colletotrichum gloeosporioides* and enhances activity of fruit defense mechanisms. Sci. Horticulturae 160, 78–84.
Zhao, N., Li, X., Chen, W., Shi, J., 2013. Effect of hot water treatment on the inhibition of anthracnose, PG, PME activity and PGIP gene expression in harvested papaya fruits. Acta Horticulturae 975, 487–493.
Zheng, M., Shi, J., Shi, J., Wang, Q., Li, Y., 2013. Antimicrobial effects of volatiles produced by two antagonistic Bacillus strains on the anthracnose pathogen in postharvest mangos. Biol. Control 65, 200–206.
Zhong, Z., Li, P., Xing, R., Liu, S., 2009. Antimicrobial activity of hydroxylbenzenesulfonailides derivatives of chitosan, chitosan sulfates and carboxymethyl chitosan. Int. J. Biol. Macromol. 45, 163–168.

INDEX

Note: Page numbers with "*f*" denote figures; "*t*" tables.

N

O

P

R

S

T

U

V

W

Y

FSC
www.fsc.org
MIX
Papier | Fördert
gute Waldnutzung
FSC® C083411

Kolibri 360 GmbH
Ferdinand-Jühlke-Straße 7
99095 Erfurt, Deutschland
produktsicherheit@kolibri360.de